Computational Approaches in Molecular Radiation Biology

Monte Carlo Methods

BASIC LIFE SCIENCES

Ernest H. Y. Chu, Series Editor
The University of Michigan Medical School
Ann Arbor, Michigan

Alexander Hollaender, Founding Editor

Recent Volumes in the series:

Volume 49	OXYGEN RADICALS IN BIOLOGY AND MEDICINE Edited by Michael G. Simic, Karen A. Taylor, John F. Ward, and Clemens von Sonntag
Volume 50	CLINICAL ASPECTS OF NEUTRON CAPTURE THERAPY Edited by Ralph G. Fairchild, Victor P. Bond, and Avril D. Woodhead
Volume 51	SYNCHROTRON RADIATION IN STRUCTURAL BIOLOGY Edited by Robert M. Sweet and Avril D. Woodhead
Volume 52	ANTIMUTAGENESIS AND ANTICARCINOGENESIS MECHANISMS II Edited by Yukiaki Kuroda, Delbert M. Shankel, and Michael D. Waters
Volume 53	DNA DAMAGE AND REPAIR IN HUMAN TISSUES Edited by Betsy M. Sutherland and Avril D. Woodhead
Volume 54	NEUTRON BEAM DESIGN, DEVELOPMENT, AND PERFORMANCE FOR NEUTRON CAPTURE THERAPY Edited by Otto K. Harling, John A. Bernard, and Robert G. Zamenhof
Volume 55	*IN VIVO* BODY COMPOSITION STUDIES: Recent Advances Edited by Seiichi Yasumura, Joan E. Harrison, Kenneth G. McNeill, Avril D. Woodhead, and F. Avraham Dilmanian
Volume 56	NMR APPLICATIONS IN BIOPOLYMERS Edited by John W. Finley, S. J. Schmidt, and A. S. Serianni
Volume 57	BOUNDARIES BETWEEN PROMOTION AND PROGRESSION DURING CARCINOGENESIS Edited by Oscar Sudilovsky, Henry C. Pitot, and Lance A. Liotta
Volume 58	PHYSICAL AND CHEMICAL MECHANISMS IN MOLECULAR RADIATION BIOLOGY Edited by William A. Glass and Matesh N. Varma
Volume 59	PLANT POLYPHENOLS: Synthesis, Properties, Significance Edited by Richard W. Hemingway and Peter E. Laks
Volume 60	HUMAN BODY COMPOSITION: *In Vivo* Methods, Models, and Assessment Edited by Kenneth J. Ellis and Jerry D. Eastman
Volume 61	ANTIMUTAGENESIS AND ANTICARCINOGENESIS MECHANISMS III Edited by G. Bronzetti, H. Hayatsu, S. DeFlora, M. D. Waters, and D. M. Shankel
Volume 62	BIOLOGY OF ADVENTITIOUS ROOT FORMATION Edited by Tim D. Davis and Bruce E. Haissig
Volume 63	COMPUTATIONAL APPROACHES IN MOLECULAR RADIATION BIOLOGY: Monte Carlo Methods Edited by Matesh N. Varma and Aloke Chatterjee

---

Computational Approaches in Molecular Radiation Biology

Monte Carlo Methods

Edited by

Matesh N. Varma
U.S. Department of Energy
Gaithersburg, Maryland

and

Aloke Chatterjee
Lawrence Berkeley Laboratory
Berkeley, California

Plenum Press • New York and London

Library of Congress Cataloging-in-Publication Data

On file

Proceedings of a DOE Workshop on Computational Approaches in Molecular Radiation Biology: Monte Carlo Methods, held April 26–29, 1993, in Irvine, California

ISBN 0-306-44995-1

A Division of Plenum Publishing Corporation
233 Spring Street, New York, N.Y. 10013

Printed in the United States of America

PREFACE

The Office of Health and Environmental Research (OHER) has supported and continues to support development of computational approaches in biology and medicine. OHER's Radiological and Chemical Physics Program initiated development of computational approaches to determine the effects produced by radiation of different quality (such as high energy electrons, protons, helium and other heavy ions, etc.) in a variety of materials of biological interest—such as water, polymers and DNA; these include molecular excitations and sub-excitations and the production of ionization and their spatial and temporal distribution. In the past several years, significant advances have been made in computational methods for this purpose. In particular, codes based on Monte Carlo techniques have been developed that provide a realistic description of track-structure produced by charged particles. In addition, the codes have become sufficiently sophisticated so that it is now possible to calculate the spatial and temporal distribution of energy deposition patterns in small volumes of subnanometer and nanometer dimensions. These dimensions or resolution levels are relevant for our understanding of mechanisms at the molecular level by which radiations affect biological systems.

Since the Monte Carlo track structure codes for use in radiation chemistry and radiation biology are still in the developmental stage, a number of investigators have been exploring different strategies for improving these codes. Nevertheless, the scientific community here in the United States and the European communities felt that all the international experts working in this field ought to be brought together to discuss the computational advances, evaluate the state of the art, and to provide directions for future research in this field. In addition, the scientific community and OHER recognized that discussions of the ongoing computational approaches in radiological and chemical physics should also consider other major advances in the field of computational biology, evaluate the impact of the increased technology and computational methods that have taken place in the last few years, and suggest new approaches to enhance this unique capability for further advancing the goals of track structure methods.

Considering the importance of this research and the unprecedented opportunities that are available, OHER organized and sponsored a workshop on computational approach Monte Carlo methods in molecular radiation biology at the National Academy of Sciences facilities in Irvine, California. In order to have sufficient information and input from outside experts, the organizers made special efforts to bring internationally recognized experts in other fields, such as structural biology, computational biology, statistics and mathematics to the workshop. By all accounts, interaction and exchange of information and ideas between the participants was outstanding. Eighteen invited scientific papers were presented and ample time was provided for discussions. These proceedings provide a complete record of the reports and discussions at the workshop, documents the progress in this field and points to future research directions. We are extremely thankful to Lawrence Berkeley Laboratory and the staff of the Life Sciences Division for untiring help in the local arrangements of the workshop, and for editing and proofreading the proceedings. Special thanks are due Ms. Cathy Sage and Ms. Martha Franklin for coordinating the prepublication of these proceedings.

Matesh N. Varma
Aloke Chatterjee

CONTENTS

SIGNIFICANCE OF COMPUTATIONAL BIOLOGY

INITIAL PHYSICAL AND CHEMICAL STUDIES

TRACK STRUCTURE CODE DEVELOPMENT

COMPARISON OF TRACK STRUCTURE CODES

MODELING OF BIOLOGICAL EFFECTS

COMPUTATIONAL BIOLOGY OPPORTUNITY AND CHALLENGES FOR THE FUTURE

John C. Wooley and Matesh N. Varma

US Department of Energy
Washington, D.C. 20585

ABSTRACT

Recent developments in high performance computers and computing methods have opened new avenues for tackling serious, important and challenging problems in biology and medicine. Only a few years back these problems were considered too complex and difficult, if not impossible to solve. An understanding of cross-disciplinary knowledge will be a prerequisite for applications of this enormous computing capability to enhance our understanding of governing principals in biology and medicine. We will show some specific research areas where computational biology can be applied effectively and then provide some ideas on future applications.

INTRODUCTION

The goal of computational biology, which crosses all biological and biomedical research activities in the Office of Health and Environmental Research, is to link the ongoing revolution in the biological sciences with that in computer and information science. Analogous to older fields like computational chemistry and computational physics, computation biology is undergoing rapid growth. Exploiting the tools of mathematics, computer technology, scientific databases, and network communication, computational biology offers insights into research problems too complex for traditional analysis. Almost all research fields within the biological sciences are ripe for extensive application of the tools of computational biology.

Monte Carlo methods are currently being applied in many areas of biology, as well as in the specific work on molecular radiation biology described in this workshop. We recognize the importance of this work, while noting the applications of simulation approaches and modeling are very familiar to the workshop participants. As a consequence, we wish to describe related areas of computational biology and bioinformatics (use of information theory and scientific databases for studying biological systems). In particular, the value of using scientific databases within biology is rapidly changing. The general problem of how science deals with information overload,as our knowledge of the natural

Computational Approaches in Molecular Radiation Biology
Edited by M.N. Varma and A. Chatterjee, Plenum Press, New York, 1994

world continues to increase dramatically, has been termed, "learning to drink from the fire hose."[1]

COMPUTING, INFORMATION TECHNOLOGY AND LIVING SYSTEMS

Biology is inherently information-rich because of the complexity and variety of living systems. Understanding these systems requires information about their organization, structure and function at a multitude of levels from the macroscopic to the molecular. Examples of the complexity of living systems:the human genome contains 3×10^9 bases; a single Purkinje cell is thought to have in excess of 1,000,000 independent inputs; the human central nervous system is thought to contain 10^{10} cells, and 10^{18} synapses. Computer technology provides the ideal tool for organizing the knowledge of living systems, so that experimental scientists and theoreticians can pose and answer critical biological questions. For some time, mathematical and computational approaches to biology were not widely accepted; however, the maturation of experimental tools, and the development of powerful computational approaches has led to significant changes in the role of computational in support of experimental approaches to biology, especially in the new found role for information technology or informatics.

In fact, the complexity of living systems at any level has long been recognized, but only recently has been the level of collection, integration, and understanding of the relevant data been adequate to justify this perception. Indeed, this hallmark of hierarchical, interrelated information, rather than the sheer volume of the data, is what distinguishes biology from other disciplines. Gilbert[2] has termed this sea-change in how biologists use computers "a paradigm shift in [molecular] biology." Increasingly, scientists need to access major community databases in order to plan an experimental approach or understand the implications of their own recent findings.

BIOINFORMATICS: BUILDING AND EXPLOITING SCIENTIFIC DATABASES IN BIOLOGY

In sum, emerging scientific paradigms will require substantial bodies of data organized into large and dynamic databases to support ongoing biological research. Indeed, some aspects of modern biology (e.g., the Human Genome Project, and protein structure-function studies with Structural Biology) are now utterly dependent upon database and computer technology. To this end, the Department of Energy (DOE) has supported the development and enhancement of a number of databases, the largest of which, often termed community databases, are described in Table 1. Notably, DOE has actively supported the Genome DataBase (GDB) for storage and access to chromosomal mapping information, the Protein DataBase (PDB) for storage and access to macromolecular structural information, and the Genome Structure DataBase (GSDB) for DNA sequence information. One goal is to link these distributed databases via the internet into a federation of relational databases, which would allow the use of a standard language for discovering information through the use of the database. This language, Structured Query Language (SQL), coupled with internet access, allows one to ask questions requiring access to more than one database; our goal is to make that type of approach for fundamental research, databases as discovery tools, readily available to scientists around the country. (A meeting report, a preliminary step toward a white paper on Federated Databases for Biology, termed the DOE Informatics Summit, describes recommendations made to DOE for building robust links between major biological knowledge resources for genome and structural biology.) Continued enhancement of these

Table I. Some of the Community Databases for Biology.

Protein Identification Resource	10,000 entries; all complete sequenced proteins, including modifications in known
Geonome Sequences DataBase	15,000 entries; 400 million nucleotide sequences (DNA)
Protein DataBank	Over 1,000 structural entries; 930 protein structures by x-ray; 20 by protein structures by NMR; 60 DNA or RNA structures; also few carbohydrate structures
On-Line Mendelian Inheritance in Man	Over 5,000 traits and genetic disorders, links to literature
Genome DataBase	13,113 mapped segments; over 3,647 genes (1/12/93)

*Entries as of Mid-1993.

three international resources remain pressing needs, including support for collaborations on new developments to increase their efficiency, timeliness, and general utility.

The hierarchical complexity of biological phenomenon also means that biological knowledge is described and understood in graphical and pictorial terms. Visualization tools for the understanding of complex data, representation of closely related information in databases, and the exploration of experimental simulations are essential complements to advancing scientific databases for the management of large, complex biological datasets. A close coupling and interaction of experimental observation, theory, and simulation/ computational tools, as shown in Figure 1, will be needed to truly exploit all the advances in the field of computational biology.

The growing complexity of the community databases for biology are given in Figure 2. Figure 3 shows the growth rate of the number of structures in the Protein DataBank, which can be fit by an exponential curve. The GSDB at Los Alamos National Laboratory maintains an internet accessible relational database version of all known DNA sequences, sharing its information with DDBJ (Japan), EBI (Europe) and GenBank (NCBI), the other

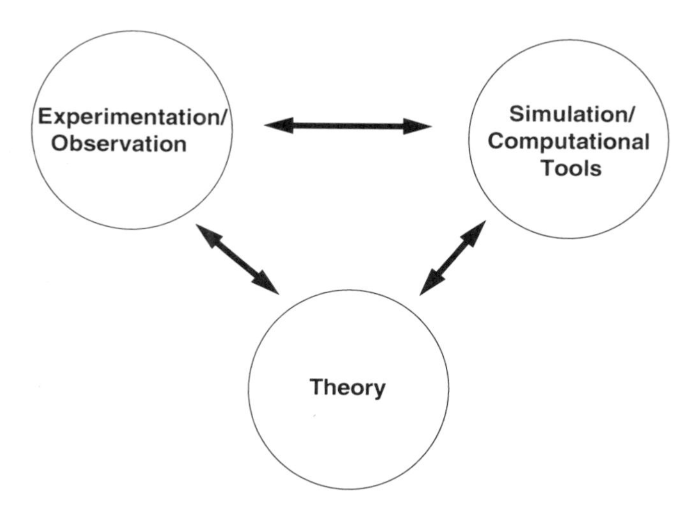

Figure 1. Interaction and complementation of knowledge obtained from theory, experimental observation, and simulation/computational tools.

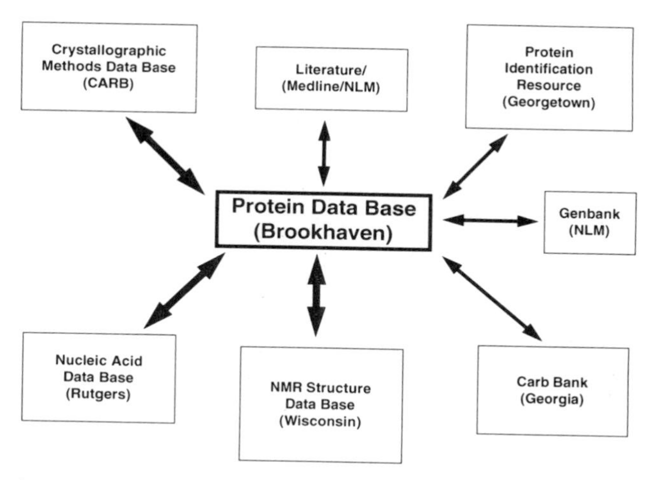

Figure 2. Shows an example of complexity of the community databases for biology.

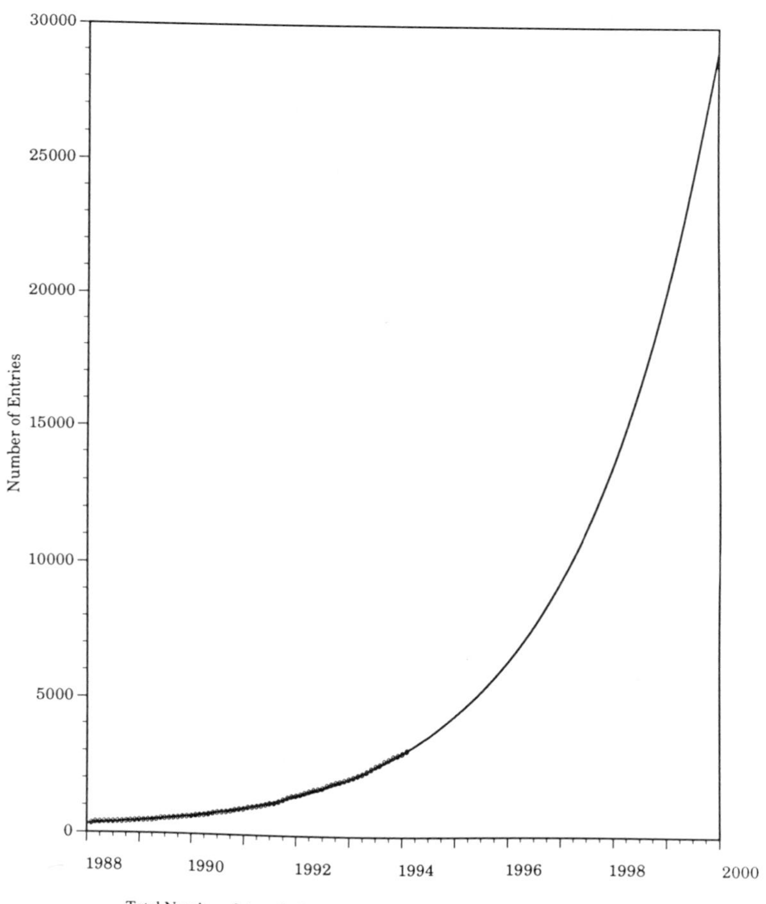

Figure 3. Shows the growth rate of the structures in the Protein Data Bank. It is clearly seen that an exponential curve can be fitted to the data.

three sites receiving DNA sequence information. GDB is a central tool for physical mapping, providing a discovery tool, with reagents and other information available to guide researchers in obtaining higher resolution maps of chromosomes. The physical and genetic map data, when added to the morbid anatomy of genetic diseases, provides the clinical information that will lead to a new generation of medical approaches. This information resides in On-Line Mendelian Inheritance in Man (OMIM) linked intellectually and by geographic proximity to GDB. Ultimately, the clinically useful fruits of the Genome Project will wind up in OMIM, for physicians all over the world to access to facilitate diagnoses. The Protein Identification Resource (PIR), at Georgetown University National Biomedical Research Foundation contains sequences of proteins derived from gene sequences, proteins directly sequenced, and modifications of proteins when known.

GSDB, GDB, and OMIM will all be federated together with other databases, such as PIR, PDB, and EGAD. The Expressed Gene Anatomy Database, a collection of cDNA sequences referenced to the tissue or cell type of origin, developed by The Institute for Genome Research (TIGR).

GOAL: A NATIONAL INFORMATION INFRASTRUCTURE FOR THE BIOLOGICAL SCIENCES

The Genome Information System of tomorrow, illustrated in a cartoon in Figure 4, is expected to encompass databases of the genome of human and model organisms, macromolecular structure databases, medical databases, and many other specialized biological knowledge resources; these should be connected in a seamless way, transparent to the user and maintained in a distributed fashion around the world. User-friendly interfaces must provide computer-novices with the ability to exploit internet connections among these

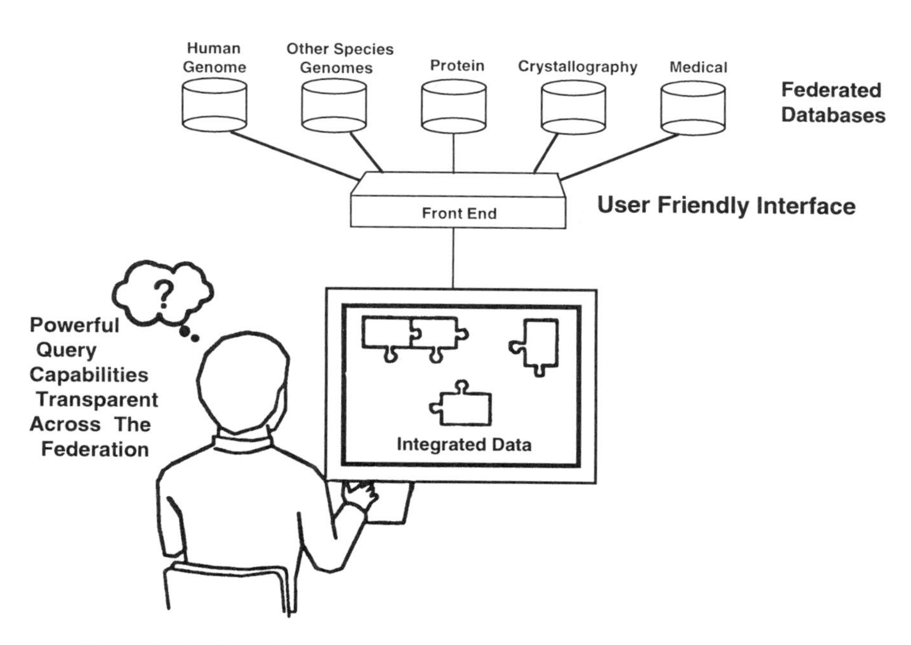

Figure 4. An idea on how the genome information system of tomorrow might appear.

database or biological knowledge resources. Those without access to such internet databases will no longer be competitive in research as Gilbert has pointed out (loc. cit.), or able to provide up-to-date medical care. At the same time, the growth of our core understanding of living systems will provide new opportunities for theoretical biology, including computational approaches.

PROTEIN FOLDING: AN EXAMPLE OF A GRAND CHALLENGE IN COMPUTATIONAL BIOLOGY

Protein folding can be examined as an illustrative example of an area within computational biology, in terms of the potential importance for fundamental biology, the impact on a broad range of areas of interest to DOE and the nation, and the multidisciplinary nature of the research. Generally speaking, beginning to develop a basis for deeper understanding of structure-function relationships in proteins is not only the central research problem for structural biology, but also represents a grand challenge in advanced scientific computing, a fundamental requirement for significant application of commercial biotechnology and for the development of new biological materials for industrial use.

One of the fundamental goals of biotechnology involves obtaining proteins targeted for specific tasks, or "designer proteins." Along with medical applications, proteins hold special promise as new materials for engineering and technology applications. The tools of molecular biology allow the ready design of a protein of any sequence, but many steps remain before targeted development can be a commercial reality. The potential reward for understanding how to optimize proteins for specific tasks is very high. For example, enzymes act billions of times faster than chemical catalysts. Spider silk protein has higher tensile strength and yet also greater elasticity than steel. The properties of proteins, just like their biological function, depend on their three dimensional architecture; that is, on their unique shape and the chemical map of their amino acids. Proteins lose activity when they lose their organized shape. Yet 80 percent of proteins produced currently by genetic engineering do not fold properly, leading to severe product loss and unsuitable products. A better understanding of protein structure-function relationships is key to progress, along with new approaches for scaled up production and for purification.

Protein architecture is irregular and complex. With time, however, many key elements and patterns of protein structure have been uncovered. Protein folding converts a linear sequence of amino acids into an organized, stable and functional three dimensional structure. All proteins are constructed in accordance with instructions coded by the sequence of DNA in genes. Understanding the nature of this genetic information, the genetic code, was the grand challenge that led to the molecular biology revolution. The next step, understanding why and how proteins fold, is also a grand challenge, the final unsolved part of the genetic code. We are a long way from being able to tackle the ultimate folding problem , predicting three dimensional architecture of a large protein or complex accurately, but much progress is being made on understanding pieces of the puzzle. The most immediate problem is often considered to be the inverse protein folding problem, that is, predicting the amino acid sequence necessary to provide the chemical properties and architectural arrangement needed for a given functional application (e.g., see Cohen[3] and Eisenberg[4]). The inverse problem is at the core for the rational design of new proteins for commercial biotechnology, for new biomaterials, and other industrial applications. The complexity of the chemistry confronts any predictive effort on the stable active structure and on the folding pathways it takes to achieve this structure. However, the use of databases of experimental and theoretical information provides a basis for structure aided drug design and reflects an application of computational biology of particular commercial relevance. Computers and databases can speed the way to designing new drugs and peptides for biotechnology.

Advances in X-ray and NMR data analysis will lead directly to rapid developments in the field of protein folding that will be synergistic with developments in other areas of biology itself, and especially computational biology. Common problems of data representation, search strategy, pattern recognition and data visualization appear in many fields. There is a particularly exciting synergistic relationship between the simulations and theoretical work being carried out in the protein folding field and those of structure determination by X-ray crystallography and 2-D NMR. Each field will benefit from rapid advances in the other disciplines. Improved folding algorithms provide a new way to attack the phase problem in crystallography, and new, more carefully refined protein structures provide rich new insights into protein folding.

GENETICS AND GENOME TOOLS

A wide range of tools are being developed or extended to look for informational patterns in DNA sequences, to facilitate chromosomal mapping, and genetic marker analysis. This work is already supported through the genome effort. Informatics tools are needed just to organize the work in major genome efforts, as well as to discover the information inherent in genome sequences and exploit evolutionary relationships to drive medical advances. Reviews of recent advances in algorithms for studying genetic information, and analogous approaches to looking at protein structure, are found in Doolittle[5,6] and Waterman.[7]

To be fully useful in a meaningful quantitative sense, the molecular studies now being carried out require powerful computer simulation, modeling, and graphical representation. Common to all of biological research is the need for good visualization of data. Visualization is necessary because the sequence analysis phase for a molecular biologist is equivalent to exploratory analysis for a statistician. It is at this point that the experimentalist gains the feeling for, and understanding of, a sequence which may then guide many months of experimental work. The complexity inherent in biological systems is so great that very sophisticated methods of analysis are required. These are the tools which must be readily accessible to molecular and cellular biologists untrained in computer technology.

COMPUTATIONAL NEUROSCIENCE AND MAPPING FORM AND FUNCTION OF THE BRAIN

Recent advances in experimental and computational neurobiology give us the hope of integrating the large, complex collection of anatomical and physiological information about the nervous system, which is now available in functional or structural terms in many disparate formats. The development of software to provide standards for brain images by any one technique, such as PET, as well as methods for related images from different modalities, is essential. Integration of the experimental data, coupled with the new computational simulations, should facilitate a deeper understanding of brain function; this can be accomplished through the development of interrelated brain maps or brain databases, which include the anatomical structure, physiological functions, neuronal connectivity, and molecular biology of human brains. A number of new simulations relating brain activity to events, which use massively parallel computers, are also promising leads to be developed.

The Department already has a strong base in the instrumentation and in the computational tools for studies on brain mapping and modeling. This work will make specific, testable predictions in the areas of sensory perception (visual, olfactory, and auditory), memory, learning, and motor control. Above all, it will lead to the integration of all these aspects, along with pharmacology and molecular biology, to provide an eventual understanding of the total functioning of the nervous system. Such integration can be expected to provide new insights that will lead to improvements in the treatment of diseases of the nervous system at all levels, from neuropharmacology to psychotherapy. In addition, studies of this kind may be expected to contribute to major advances in artificial intelligence and practical robotics. Several U.S. agencies are working together, on an initiative called the Human Brain Project, with parallels to the Genome project, to develop methods for mapping brains of human and model organisms.

The Human Brain Project grew out of the recommendations of the National Academy of Sciences Mapping the Brain. The NAS recommended that a highly complex linked set of databases, capturing the information necessary to map brain anatomy and functional correlates be established with the long-term objective of integrating information on three-dimensional computerized maps and models of structure, function, neuronal connectivity, pharmacology, and molecular biology of human, rat, and monkey brains across developmental stages and reflecting both normal and diseased states. The information could then be used in an effective manner by neurosurgeons and scientists involved in basic research, for the understanding and treatment of mental disorders and diseases.

STRUCTURAL PROPERTIES AND DRUG DESIGN

It has long been recognized that conformation, configuration and organization of biomolecules is relevant to the regulation of replication and gene expression and other biological activities of interest. Modern computational methods and techniques provide an efficient and productive means of simulating structures of biomolecules and studying changes in these structures as a function of environmental insults, such as radiation or chemicals. This change in structure can be correlated with biological activity of the organism. Understanding of this structure-activity of biomolecules would be very helpful in new drug designs. For this purpose the computational biology tool is extremely productive and useful.

OFFICE OF HEALTH AND ENVIRONMENTAL RESEARCH EXPECTATIONS

The interdisciplinary and multidisciplinary research intrinsic to computational biology will require that we encourage research teams to tackle challenging problems in biology and medicine. Figure 5 shows one view of the interlinking of various programs supported by OHER; in reality, the potential interconnections among programs are even more extensive. In addition, new approaches for representing and studying biological knowledge through the use of internet-connected, scientific databases will be developed and maintained. The computational biology effort will build on the core expertise and strengths at our national laboratories and universities. To this end, the advances in application of Monte Carlo methods that were presented during the workshop will be very useful. The research teams in institutional settings will provide valuable training for students and opportunities for young investigators. Computational biology and methodology provides an unprecedented opportunity with numerous challenges for application in the field of biology and medicine. Advancement in the computer hardware and software make it possible to tackle important scientific problems for wide areas of biotechnology, drug design, treatment of mental disorders, and other human diseases.

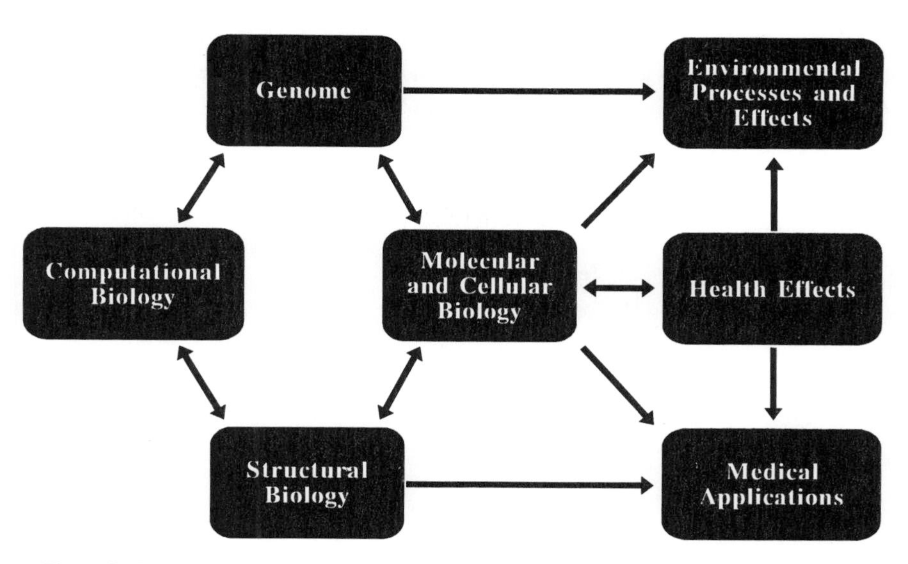

Figure 5. How various programs supported by OHER interlink and couple with each other.

REFERENCES

1. "Learning to Drink from the Fire Hose," (1990) *Science Magazine 248*, pp. 674–5.
2. Gilbert, W. (1991), "Towards a Paradigm Shift in Biology," *Nature 349*, pp. 99.
3. Ring, C.S., and Cohen F.E., (1993), "Modeling Protein Structures: Construction and Their Application," *FASEB Journal 7*, pp. 783–90.
4. Eisenberg, D., Bowie J.U., and Choe, S. (1992), "Three dimensional Profiles for Analyzing Protein Sequence-Structure Relationships," *Faraday Discussions 92*, pp. 25–34.
5. Doolittle, R.F. (1989), "Similar Amino Acid Sequences Revisited," *Theory in Biochemical Sciences 14*, pp. 244–245.
6. Doolittle, R.F. (1990), "The Structure and Evaluation of Vertebrate Fibrinogen: A Comparison of the Lamprey and Mammalian Proteins," *Advances in Experimental Medicine and Biology 281*, pp. 25–37.
7. Waterman, M.S., and Jones, R. (1990), "Consensus Methods for DNA and Protein Sequence Alignment," *Methods in Enzymology 183*, pp. 221–237.

OVERVIEW OF SIGNIFICANT CHALLENGES IN MOLECULAR BIOLOGY AMENABLE TO COMPUTATIONAL METHODS

Robert M. Glaeser

Lawrence Berkeley Laboratory
University of California
Berkeley, CA 94720

ABSTRACT

Many challenging but significant opportunities exist for the development of theoretical approaches in modern Cell and Molecular Biology. The creation of data bases which contain extremely large amounts of information has proven to be an unexpectedly important factor in gaining acceptance and respectability for theoretical work that builds on nothing more than what is in the data base itself, such as theoretical work involving the analysis of known protein structures, or the development of more powerful homology searches. Other opportunities, not yet accepted by a broad community, involve work on complex networks (metabolic, genetic, immunologic, and neural networks) and work on the "physics of how things work." The DOE National Laboratory System represents the ideal institution that would be well suited to the role of being an "incubator" for the creation of a theoretical and computational discipline within modern biology.

INTRODUCTION

Databases are a very highly respected part of the institutionally supported infrastructure within modern cellular and molecular biology. I think everybody recognizes that databases are very important. They accept them.

In the case of the Brookhaven Protein Structure database, however, that acceptance is not much more than about five years old. Previous to the point when the concept of a database became really accepted, the idea in structural biology was that you solved the structure and kept that information very close to your vest, so that you could protect the opportunity to look at it and say something profound about it. Nobody else got to see it.

This older practice is the classic paradigm, according to which the individual investigator does his or her own theory, designs their own experiments, carries out the work at the bench themselves, analyzes the results, and fits the results into some larger context. In structural biology people were possibly more jealous than in any other field about their

Computational Approaches in Molecular Radiation Biology
Edited by M.N. Varma and A. Chatterjee, Plenum Press, New York, 1994

primary data. They literally would not let it go. Suddenly, about five years ago, there was a major paradigm shift. Journals began to require authors to submit their solved structures to the database. Now it is difficult to publish in a journal unless you put your data into a database.

What I want to focus on first in my talk is something that Walter Gilbert has observed recently about the impact of databases. The advent of databases is bound to drive a wedge into the wall that has separated the way in which theoretical biology used to be done, i.e., as an integral part of individual-investigator science, and the way in which it is done routinely in the physical sciences. The traditional metaphor in biology is going to have to be rewritten. There is going to be a paradigm shift.

I'm delighted to see interest within the Department of Energy in creating support for a larger field of computational biology, which I'm going to call theoretical biology. Obviously, you can't do much theory without computations, although it would be possible to do a lot of computations without theory. Therefore, I would prefer to speak in terms of theoretical biology. The phrase "theoretical biology" is very badly tarnished however, and that's why people are not using it anymore. I think that we should forget about the fact that it has been tarnished.

In January 1991, Gilbert published a one-page opinion piece in "Nature". Gilbert makes the point that, according to the current paradigm of biology, the attack on problems is viewed as being solely experimental. The "correct" approach is to identify a gene through some direct, experimental procedure. A new paradigm, Gilbert says, will have to emerge once all the genes are known. We do indeed hope and believe that all genes will be known in as little as, some people say, a year and a half. Maybe it will be five years, perhaps it will be ten years. It doesn't really matter; they will be known. And they'll be known in the sense that the genes will be resident in a database that will be available electronically.

Quoting directly, Gilbert says "The starting point of biological investigation will then be theoretical. An individual scientist will begin with a theoretical conjecture, only then turning to experiment to test that hypothesis". Referring generally to the information that will be available in databases, Gilbert also says that to use "this flood of knowledge that will pour across computer networks of the world, biologists will not only have to become computer literate, but also change their approach to the problem of understanding life".

I want to say a little bit more about the so-called "correct approach" in the current paradigm of molecular biology. I heard somebody say recently during a symposium at the American Society of Cell Biology meeting that what is going on in modern cell biology is nothing different, intellectually, from bird watching. Exciting though it is, modern cell biology is doing little more than to describe what is there, and which one of those molecules interacts with another molecule.

Now, the most pejorative thing that a reviewer can say on a study section about somebody else's grant that is being reviewed, is that it is merely descriptive. But, in fact, most of what is happening in molecular biology is truly descriptive. There is, of course, hypothesis-testing that drives what is being done. But in the end, people are just describing what is there. They like nothing better, when they have sequenced a gene, than to search through a database and find that this gene, whose biochemical function they did not know, is homologous to some other gene whose function is already known.

The illusion of learning that comes with such identifications is really just an illusion, in my opinion. Because of the excitement that exists over how much is being learned, in this descriptive fashion, it seems as though deep understanding will soon be gained. But in the explosion of discovery and description, there is no truly quantitative theory, or even the framework of a theory, to test whether we understand what is going on.

To elaborate on this, we could use one of the most thoroughly well-described examples of metabolic regulation that is known in biochemistry, the regulation of glycogen storage and glycogen utilization. You can ask a biochemist, "Do you understand how this metabolic regulation occurs?" With great enthusiasm, a biochemist could say, "We sure do", and point to pages in Stryer's textbook in biochemistry that describe how this works.

Hormones bind to a receptor. The receptors signal across the membrane and activate adenylate cyclase, which makes cyclic AMP out of ATP. Cyclic AMP, in conjunction with other things such as calcium, activates a protein kinase. The protein kinase phosphorylates another kinase, which is phosphorylase kinase. This kinase subsequently phosphorylates glycogen phosphorylase, the enzyme that uses phosphate to cleave glucose from the glycogen, giving glucose phosphate.

The same cascade leads to phosphorylation of glycogen synthase. In the phosphorylated state, one of these two enzymes is shut off, and the other one is turned on. In addition, there is a second half of this story which involves another network which cleaves off those phosphates resulting in a "futile cycle." Therefore, these reactions have to be regulated so they just don't spin around constantly, phosphorylating and dephosphorylating. On top of this, a number of small molecules such as ATP, ADP and citrate act as allosteric effectors to regulate this whole process. It turns out to be an enormously complicated network, but the biochemists can describe all of it in exciting detail, and then at the end say, "I understand: I've now given you this half-hour lecture, and that is how it works."

If somebody in the audience then says, "What quantitatively verifiable predictions have you made, based on your theoretical understanding of how this system works?" You would not get an answer, because there aren't any answers. People do not even think in those terms. They don't believe that it is expected, in modern cellular and molecular biology, to be able to make quantitatively verifiable predictions, or falsifiable predictions, based on what we understand about how a system works.

At the moment, the question, "What verifiable or falsifiable predictions have been made, based on your theoretical understanding," is not even a part of the culture of modern molecular and cellular biology. You don't see it raised in Stryer's Biochemistry. You don't see it raised in Alberts' Molecular Biology of the Cell.

In the new paradigm that I believe Gilbert is talking about, we will need to have a quantitative understanding of metabolic networks. We will also need to know how the performance of a network is affected when one element of a cascade can no longer be turned off, such as occurs with the *ras* oncogene. In some cases, of course, we know that mutations which eliminate the ability of a protein to be switched off can turn that cell into a cancer cell. But we don't understand theoretically why that is true. In order to understand, you first need to know how to describe the metabolic network, obviously, but you also need to have constant dialogue between experiment and theory. If an experimentalist believes that they're finished, and if we really had a theory that worked, then you could apply the theoretical analysis to see whether we could properly account for the experimental facts. If the facts cannot be accounted for, there are two possibilities. One is that the theorist does not know how to do things right, and the other is that the experimental description is not yet finished. Hopefully, the role of theory could be to say, 'I've tried every which way to describe what happens experimentally and it just won't work. Evidently something is still missing.' A really powerful theory would even point to the direction where one should look for the missing experimental facts that are needed to complete the theoretical description. There are a lot of cases where a tremendous amount is known about the metabolic networks, and where the textbooks say 'we understand.' But we really do not know if those 'understandings' are complete, because nobody has a proper theoretical description.

I would like to generalize from what I have discussed so far, to say that there is a need for a mathematics, or what I would call a "testable logic," of networks. This need clearly applies in the realm of metabolic networks which I have used as an example, but it also exists in the realm of networks of gene expression. Genes turn on and turn off other genes, and different gene products interact with each other to synergistically alter the expression of a third gene. In the end, the logic is like that of a metabolic network. There are also networks in the immune system. I won't go into the descriptions of different aspects of the immune system, where there are complex networks of self-reference and cross-reference, both at the cellular level and in terms of the different classes of antibodies, antibodies to antibodies, and that whole sort of thing. In the nervous system there are already well-accepted ideas of the existence of neural networks. There has even been a lot of excitement created by Hopfield's application of the physical theory of spin glasses to nerve-cell networks, which can lead to the establishment of memory. The whole field of neural network theory goes well beyond Hopfield's contributions, of course.

While there is enthusiasm among scientists about the opportunities afforded by modern computing power, there is skepticism that a sound theory has yet been advanced of how the brain's neurophysiology and neurochemistry produces its complex performance. While we understand a great deal about how our neurochemistry functions, we have no idea how to explain the difference between the mind and brain. What is encouraging however, is that theoretical work in this direction is now being supported institutionally. People are being asked by NIH to apply for grants to tackle that very, very difficult problem. A recent multi-institute call for proposals, supported by mental health, drug abuse, the National Science Foundation, aging, child health, deafness, etc., asks that a database of investigator initiated research on informatics resources should be established. The goal is that informatics resources could be used to facilitate research on computer storage and manipulation of neuroscience research information.

So the whole paradigm of there being databases of gene sequences and protein structures is now being adopted in the context of neuroscience. There are many other examples of the same paradigm shift emerging in other areas in the past few years. NIH had a big push a couple of years ago to try to get a 'database of everything.' The effort eventually collapsed because it just was not workable. What I'm building here however, is the idea that—as Gilbert pointed out—we should be expecting a change in the paradigm of modern cell and molecular biology.

Now I'm going to shift a little bit from the business about networks and talk about some other opportunities for theoretical biology. The topic I'm going to pick up on is one that I characterize as being "the theoretical biology of how things work." In particular, one needs to ask what physical principles are involved in the way in which things work, and what are the physical limitations? What limitations do common-sense physics and chemistry impose on how things might work?

One example within this category of theoretical biology would be to ask the question "How are visual transduction and acoustic sensory transduction systems able to work down to such low signal levels that they even function effectively when they are below the level of thermal noise and, in fact, push down to the level of quantum fluctuation?" Bill Bialek and others are very active in addressing these puzzles. In the case of vision, it is well-known that the human eye can see a single photon. In acoustic transduction the mean position of the sensory hairs can be deflected by as little as a tenth of an angstrom and this is detected as auditory sensation. That is way below the amount by which thermal motion causes those hairs to knock around. How can that be? It seems contradictory to physics, so how does it work?

The answer has to do with active filters which are built into the transduction system, and individual cells being tuned so as to pick up a very narrow band. Thermal noise is spread over many frequencies, of course, instead of a single frequency. Although the total power from thermal noise, spread over all frequencies, is much greater than the power needed to sense a tone, the thermal noise at the single frequency of a tuned transducer remains lower than the signal.

Another theoretical puzzle occurs when we ask how is length and size determined in biological structures such as viruses? We actually understand why the tobacco mosaic virus is always 3,000 angstroms long. The reason is that it contains an RNA molecule that serves as a ruler on which the proteins assemble. But there are lots of other examples of virus structures where we can't find any such "ruler molecule" that determines the size or shape. In these cases we don't understand why a complex assembly has a certain length, and why, when we make a mutant, it suddenly gets much longer than it was before.

We would like to know how macromolecular assemblies are self-limited to have specific shapes. For example, the centriole, the organizing center for the mitotic apparatus, always looks the same. How is this? What is making it so? Microvilli in the intestines are almost always the same length. In this case the organizing structure is made up of actin bundles on the inside. In our skeletal muscles, the "Z" bands are always the same distance apart. What limits the size of the myofibrils so that in one organism they're longer and in another, they're shorter? What are the physical mechanisms that determine the self-assembly of macromolecular structures, and what principles are involved in defining their assembly so that they become completely functional?

The replication of chromosomes, and even the replication of some bacterial plasmids, is so tightly regulated that there is *always* precisely one copy in a cell, never more and never less. Why isn't it sometimes two, except in the trivial case just prior to cell division? What prevents the replication apparatus from running away and making lots and lots of copies? It's pretty astonishing that a chromosome or a plasmid can be replicated once, and then not a second time. What can we think about, then, in terms of physical limitations that can circumscribe the possibilities and thus explain these astonishing phenomena?

A totally different class of physical limits arises when we look at the motility of ATPase motors, like myosin and the microtubule motors. How do they work? We always think of them as being proteins that undergo a conformational change when ATP is hydrolyzed. The energy of hydrolysis of the ATP causes this change in protein conformation, and therefore causes the cross-bridge to swing. There has been a recent revolution in thinking, however, which says 'Maybe that's not how it goes.' Maybe what's happening instead is that the cross-bridge type of motors are, in fact, thermal ratchets. They could be devices just like the thing you use to jack up your car. The idea is that there is one direction in which such devices can move easily, and in the opposite direction they cannot move at all. Given this physical boundary condition, based upon a ratchet-like structure, Brownian motion would become rectified.

According to this new idea, an ATPase motor is nothing but a Maxwellian demon that knows right from left. When "kT" knocks it to the right, it goes. When "kT" tries to knock it to the left, it doesn't go. As stated, of course, this idea would violate the conservation of energy and result in a perpetual motion machine. More precisely, if you had a Brownian ratchet with no energy supply other than thermal motion driving it as it moved unidirectionally, it would cool down the bath in which it was sitting and eventually get so cold that it couldn't move anymore. In the thermal ratchet model, hydrolysis of ATP does not provide the energy for the power stroke; that comes from thermal fluctuations. Hydrolysis of ATP can play the role, instead, of serving as the source of entropy needed to rectify the fluctuation.

Another major area of computational biology involves the prediction of the structure of a new protein that has just been sequenced. One would like to be able to predict the structure just from the sequence information.

One approach that has been rather productive, as reflected in the work of Ken Dill at San Francisco, Skolnick at San Diego, and others, is the use of lattice models for small peptides. These simplified models deal with the mathematics of a bunch of beads on a string, where the beads have to sit at one position on a lattice. If two beads are adjacent, they can be allowed to interact with either an attractive or a repulsive energy. What lesson will we learn from studying this mathematical problem? Even though this type of simplified modeling doesn't claim to solve the folding problem, it is trying to open up some little facet of the folding problem.

Molecular dynamics is a somewhat related field in computational molecular biology. Massive computations are used to ask questions such as 'How does oxygen get to the binding site in hemoglobin?' The puzzle here is to take the solved structure of hemoglobin and try to see if you can get an oxygen molecule into it. If you do not allow the molecule to twist and vibrate, you find that you have to pay a hundred kilocalories, and that just isn't going to happen. But if you allow the molecule to vibrate, as we know from physics it must do, one can actually open up transient windows through which the oxygen molecule can zip in perfectly fine. Then, once the oxygen satisfies the ligand field requirement of the iron in the hemoglobin, the heme iron moves into the plane of the porphyrin ring, and that tugs on a histidine residue that is lying on the other side. The final result is a conformational change which propagates to the other subunits and leads to cooperative binding at neighboring sites.

There is a question whether the use of molecular dynamics in structural biology is taking on a life of it's own, unconnected to reality. There are things being said and conclusions being drawn that one is never going to be able to test experimentally. How are we going to find out if they are real? Physics has had a similar crisis; is computational physics actually physics? Since nobody is there to countervene and say, "It ain't true," you take it as being the truth. We're beginning to cross a very, very dangerous line here. An October 1990 paper by Karplus and Petsko expresses, with more enthusiasm that I would be prepared to allow, the view that a combination of an analytical and computational approach has given rise to "theoretical biophysics, a rapidly growing field that is having a significant impact on many areas".

Nevertheless, the demonstration that calculations based on simple physical models can give new and testable insights into internal motions of proteins, has convinced many biochemists that biophysical theory has indeed become relevant. Structural biology is an area in which the professional theorist is now, in fact, an accepted part of the paradigm.

Creating a similar paradigm shift in other areas of cell and molecular biology is not going to be easy. The reason is that the present culture regarding theory follows the paradigm of what I call a "cottage industry." One is supposed to take the work home with you, after you've done your eight-to-five work at the factory, and work on it at night and the weekends. This is a paradigm in which a good scientist is expected to be self-sufficient in every aspect. You conceive the experiment, do the work at the bench, analyze the results, and synthesize those results with previous understanding. Every person who is working on a problem is supposed to be wholly self-sufficient. One does not split some of the work off to theoretical people; that's not the culture. As a result, people are going to resist a paradigm shift that changes their culture.

Furthermore, the attitude within the current paradigm is that professional theorists don't understand biology at all. Theorists are seen as people who make irrelevant models, either based on ignorance or based on necessary but untenable simplifications. Whatever the theorists produce, then, isn't any good. Furthermore, the image is that theorists try to tackle problems before the problems are ready for theory. That is the ruling view.

The summary is that the world view of science, as it is perceived from biology, does not fit the world view as it is perceived from physics, math, and chemistry. Biologists will argue, "But biology is not physics. It is not like physics and chemistry. It can't be done the same way." In the end, that is actually a vitalists' sentiment. Whether people who oppose theoretical biology realize it or not, they're denying that biology can, in fact, be reduced to a physical and mathematical understanding, because biology is somehow too complicated. Of course, I'm not really saying that people intend to be vitalists when they dismiss theory as a serious and worthy enterprise, but logically, if one really thinks about it, I see no other alternative.

Structural biology is, however, an example in which the prevailing paradigm has already been broken. As I said at the beginning, the old paradigm in structural biology used to be that one did a structure and kept that information to oneself. Eventually, however, the peer pressure built up and said that you've got to deposit your coordinates in a database. Suddenly, once that happened, it was like a binary flip. Structural biologists now rejoice in getting their coordinates into the data base within two weeks of the paper being accepted. As a result, there is a huge mass of information that is too much for any experimentalist to cope with, and the experimentalists are glad to let professional, full-time theorists take over. That's the point that Gilbert was making, of course, that the flood of information which forced the creation of the database in the beginning has suddenly caused a paradigm shift, as well. People who had one view about theory three years ago, now have a totally different view.

Two last opinions of mine. First of all, what do I think will be required of a successful theoretical biologist? There's nothing terribly profound that I have to say about it. Clearly, successful theoretical biology requires lateral thinking; that is, you have to understand the knowledge base across many disciplines. I regard myself as an experimentalist. I think I know my own discipline pretty well, but I also know that there are lots of other disciplines that I don't know anything about at all. I often go to the seminars in modern molecular genetics and I don't understand a word that they are saying. The only thing that saves my ego and self-respect is the fact that I could give a seminar to those same scientists and they wouldn't understand my world any better than I understand theirs.

The theorist has to understand all of this information, however, in order to see connections and come to a synthesis. That will only be possible if one has the time available to do it. By making it your professional work to take the time, theorists can start to do something that most experimentalists cannot do. It becomes a full-time occupation to be able to gather all the related facts, and pull things from different fields together in a way that somebody who is just working to advance the frontier of their own field would never be able to do.

The second point is that the theorist must participate in experimental work themselves, at least at a level such that the graduate students they are supervising have to work in the lab of a collaborating experimentalist. Theorists still must have their hands on the reality of the experiments. Without a direct anchor to experimental work, a theorist is working with a tremendous handicap in the effort to describe experimental work.

How will the development of theoretical biology get started; how will we begin to break the dominant paradigm in cell and molecular biology? I think the National Laboratories are precisely the appropriate institution where we should begin to establish centers of computational biology. Two points. I feel first of all that there needs to be a critical mass and a supportive environment. People can't just work as one person in an office, surrounded by experimentalists doing all kinds of weird and incomprehensible things. Theorists need other people to interact with. Universities still do not appreciate theory at this point in biology. If they appoint one theorist on their campus, that's going to be a major breakthrough for theory. Universities are not going to appoint a critical mass of theorists. My second point is that theoretical, computational biology is going to need to have long-term support without the expectation of immediate success. It is going to be a tough job to get theory to the stage where it can really contribute more than what experiments alone can contribute. The "RO1" peer review system doesn't accept that, and as a result theoretical biology is not going to survive in the RO1 peer review process. But the National Laboratories do have the possibility of providing long-term, nurturing support, without expecting immediate results and therefore they are the only place where theory can begin to germinate.

For both of these reasons, I think the timing in the development of theoretical biology is right for the National Labs. Ten years from now, or maybe as long as 30 years from now, theoretical biology may move into the mainstream of academic work, but for the time being, the ball is in the court of the National Labs.

DISCUSSION

Chatterjee: Well, you have given a lot of examples. It appears that the theorists are going to utilize the experimental data banks. I think you seemed to imply that the theorists can come up with novel ideas, but I didn't see many examples of that. Is it a reflection of the fact that the theorists are interested mostly in using the experimental data bank and not paying enough attention to developing novel ideas?

Glaeser: I didn't have time to show one of the viewgraphs that listed some of the classic triumphs of theoretical biology like the Hodgkin Huxley and kinetic equations for the nerve action potential. There were five examples that are by now so integrated into modern biology that people don't even think of it as theory. But there is an important point as well: those five examples have come from particularly successful individuals who had strong mathematical insight and who were experimentalists, not theorists. These examples of theory were developed in the old paradigm where each scientist was a complete scientist, who did theory as well as experiments. I think what you are suggesting, if I can restate it, is that there are not very many important examples outside of structural biology in which theory or computation has made a significant contribution. To me, this simply reflects the fact that having professional theorists try to advance science is just not an accepted and supported activity in biology. Of course, it is well supported in chemistry and physics. If it is not supported, then there is not going to be any accomplishment.

Chatterjee: How can we contribute into the world of experimental biology when most of the biologists typically ask the question in terms of yes or no answers. Theoreticians would like to make a generalized sense from all the data that exists.

Glaeser: Well, theoretical biologists are going to have to earn the respect of experimentalists before theory will be supported. That's only fair. Clearly, a way to be successful and to earn respect is to tackle a problem and come up with an effect that can be experimentally tested. Initially, these experiments have to be things that people can almost do on a Saturday afternoon. Experimentalists are not going to set aside half a year to test some crazy idea that the theorist has published. But if the prediction of the theory, in fact, is borne out by the experiment in contrast to the intuition of the experimentalist, that will make waves. People will say, "All right. I'll let you try something a little more challenging next time.'

Miller: If you're going to harness the resources that are necessary to establish groups of scientists working at national labs, you need to identify what it is that the public want to know enough so that they are willing to spend the money. What do you see as the driving forces behind this kind of theory? Is it drug design? Is it risk assessment?

Glaeser: To me, the driving force is the scandal of the fact that the current cellular and molecular biology is really just a description of what is there. There are many topics in modern biology where I can describe what is going on, but I really cannot say that I understand what is going on in biology. If you claim that you actually understand something, then you must be able to predict quantitatively verifiable things that you haven't yet measured and described. I talked about several examples where we have a need for a quantitative theory. If we had a predictive theory for the network that regulates cell growth and division in immortalized cells, perhaps we would be able to see where the best point would be to intervene (with drugs and inhibitors) so as to stop unregulated growth. Indeed, it is hard for me to see how we can dream of effective medical control of metastatic cancer without having a proper theoretical understanding of the regulatory network that is involved in cell proliferation.

A very speculative, "visionary" thing to say about the development of theoretical biology is to reflect on the nature of biological evolution. When you really think about it, the paradigm shift that Gilbert is talking about is nothing other than taking the step from random mutations and selections according to relative fitness of reproducing yourself, to a logically conceived engineered evolution.

One can imagine the human species evolving in ways that are designed by genetic engineering, once you understand how biological networks work. Once you understand how a complex network functions you can even manipulate a multigene characteristic. Maybe it's a plant, maybe it's a human. Maybe we're talking about a plant that doesn't need to be fertilized, or maybe we're talking about a way to increase human intelligence. When you really understand, in terms of a verifiable theory, how a complex biological phenomenon works, as an engineer, you can go in and change that. Is this something we want to do or is it something, when we realize the logical consequence of what we are playing with — maybe we want to stop and not do it. Those are pretty powerful consequences. They may not be the original reasons why you want to develop theory, but they will be the consequences.

Varma: But given that you have some modern computational technology and high performance computers, the question we want to ask is what other scientific areas do you feel this type of technology can be used for specifically? I know that one example can be the study of brain function. Are there other areas which may be amenable to using modern computational methods?

Glaeser: I think the central message in my presentation is contained in my viewgraph where I showed all the different kinds of biological networks. I'm looking for a logic of networks. I'd like to understand theoretically the performance of a network, whether it's a metabolic network, an immune system network, a neural network, or a network of regulated expression of genes. We need theoretical biology to develop a mathematical logic of networks that plays the same role that statistical mechanics played for the development of physics.

I don't know what that logic is; I don't know what that mathematics is, but I sure know we need it. If we could start saying profound things and make predictions that are testable about how complicated self referencing and cross referencing networks perform, then suddenly we'd be able to open up a lot of vistas in biochemistry and biology.

Steele: There's a research question of George Polya. I think he says, "What is the simplest problem that you can't solve?" If you look in this area of networks, what would be that nitch where you think enough of the basic sciences has been done, where it would be possible for a relative outsider to participate, and which comes close to the notion of being among the simplest problems you can't solve?

Glaeser: All I can say is that that's the perfect question to ask. I'm not a theoretical biologist. If I were going to change my career and become a theoretical biologist, I could not want a better question as a starting point than the one you just posed.

Holley: Who might know the answer to that?

Glaeser: You may have to find your own answer.

BASIC PHYSICAL AND CHEMICAL INFORMATION NEEDED FOR DEVELOPMENT OF MONTE CARLO CODES

Mitio Inokuti

Argonne National Laboratory
Argonne, Illinois 60439

ABSTRACT

It is important to view track structure analysis as an application of a branch of theoretical physics (i.e., statistical physics and physical kinetics in the language of the Landau school). Monte Carlo methods and transport equation methods represent two major approaches.

In either approach, it is of paramount importance to use as input the cross section data that best represent the elementary microscopic processes. Transport analysis based on unrealistic input data must be viewed with caution, because results can be misleading. Work toward establishing the cross section data, which demands a wide scope of knowledge and expertise, is being carried out through extensive international collaborations. In track structure analysis for radiation biology, the need for cross sections for the interactions of electrons with DNA and neighboring protein molecules seems to be especially urgent.

Finally, it is important to interpret results of Monte Carlo calculations fully and adequately. To this end, workers should document input data as thoroughly as possible and report their results in detail in many ways. Workers in analytic transport theory are then likely to contribute to the interpretation of the results.

INTRODUCTION

In matter subjected to any ionizing radiation, energetic particles occur invariably. These include primary particles, charged or uncharged, and secondary particles such as electrons ejected in ionization processes. In track structure analysis, spatial distributions of collision processes of all the energetic particles and their consequences for molecules in matter are studied. The analysis amounts to an application of a branch of theoretical physics called statistical physics[1] (or physical kinetics[2] by the Landau school). Many of the known principles and results of the general theory of stochastic processes[3] are also valuable in particle transport analysis. The neutron transport theory[4] developed for the design of nuclear reactors also gives us important ideas and methods for treating electron tracks.

Monte Carlo methods[5–7] and methods of analytic transport equations[8] represent two major approaches to electron track analysis. These methods complement each other, rather than competing. To solve a complicated problem, e.g., a problem involving complex geometry, the Monte Carlo methods are more practical. To understand principles of physics, methods of transport equations are often transparent and effective.

In either approach, it is essential to use cross section and other data that correctly represent at least the most frequently occurring elementary processes. The primary purpose of the present discussion concerns the basic data necessary for track structure analysis.

KINDS OF BASIC DATA NEEDED

The basic data needed can be classified into three kinds: (1) data on atomic structure of materials, (2) spectroscopic data of materials, and (3) cross section data.

The data on atomic structure of materials must not be taken for granted. All too often data given in handbooks are incorrect or inappropriate for the problem at hand. An illustrative case was encountered during a comprehensive survey of stopping-power data of materials of radiological and dosimetric interest for the International Commission on Radiation Units and Measurements.[9,10] Data in the literature on carbon and silicon solids showed gross inconsistencies depending on the group of authors. The inconsistencies turned out to be largely attributable to measuring samples of different crystalline structures, and hence to different number densities of atoms.

With spectroscopic data, even advanced practitioners tend to take them for granted. Spectroscopic data, which concern energy levels of excited or ionized states of molecules, are highly nontrivial. The energy levels determine possible values of energy transfer from an energetic particle to the molecule. The character of each excited and ionized state, designated by a set of quantum numbers, governs the pertinent cross section, and also modes of decay such as dissociation into fragments, internal conversion, and fluorescence. Knowledge of electronically excited states of all but the simplest small molecules is far from complete. Spectroscopic information about electronic excitation of DNA, RNA, proteins, and other biomolecules is extremely limited. Extensive research is needed as a prerequisite for mechanistic studies. Moreover, the energy levels depend on the environment of the molecule in question (the gas phase, the liquid phase, or the solid phase and pressure and temperature in the gas phase).[11,12]

CROSS SECTION DATA

At the Woods Hole Conference I gave a summary[13] of current knowledge about cross section data. (See in particular pp. 32–38 of Ref. 13.) To supplement that summary, I report here on three activities in which I am participating.

Work with the International Commission on Radiation Units and Measurements

One of the goals of the International Commission on Radiation Units and Measurements (ICRU) is to survey data on physical quantities basic to radiology and dosimetry, to reach an international consensus on the best recommendable values, and to disseminate results and conclusions in the form of ICRU reports. A subject of continuing study over two decades is the stopping power of materials for various charged particles. The first report[9] (on electrons and positrons) was published in 1984, and the second report[10] (on protons and alpha particles as well as muons and pions) in 1993. Both reports were prepared by a committee chaired by M.J. Berger. A new committee chaired by P. Sigmund has begun to prepare a report on heavier ions.

The stopping power is the best studied of quantities describing the penetration of particles in matter, and knowledge of it is indispensable for quantification of radiation fields in dosimetry and is fundamental for modeling of radiation effects. Stopping power alone does not suffice to characterize track structure, but a result of a track structure analysis must be consistent with stopping-power data; otherwise it is untrustworthy.

The two ICRU reports[9,10] present the best possible values available and thus are standard references. They indicate a range of uncertainties in the recommended values and discuss important topics such as theoretical evaluation, methods of measurement, and the influence of chemical binding in molecules and of atomic aggregation in condensed matter.

A new ICRU program concerning the energy distribution of secondary electrons resulting from ionizing collisions of charged particles is now well underway under the chairmanship of M.E. Rudd. In view of extensive work over nearly three decades, as reviewed by Toburen[14] and by Rudd et al.,[15] it is now appropriate for the ICRU to prepare a comprehensive report, which I expect to see in print in 1995.

Work with the International Atomic Energy Agency

The International Atomic Energy Agency (IAEA) has long been engaged in surveys and data dissemination in selected areas such as dosimetric data for radiation protection, nuclear data for reactor research, and atomic data for fusion research. In 1985, the IAEA began to extend its activities to atomic and molecular cross section data for radiotherapy and related radiobiology and organized a meeting on this topic at Rijswijk.[16] In 1988, a second meeting was held in Vienna.[17]

Following recommendations of the Vienna meeting,[17] the IAEA initiated the Coordinated Research Program on Atomic and Molecular Data for Radiotherapy. The purpose of the program is to conduct an extensive survey of atomic and molecular data pertinent to radiology and dosimetry and eventually to prepare a comprehensive report. The participants are the following:

M.J. Berger, Bethesda, Maryland
Hans Bichsel, Seattle, Washington
Ines Krajcar Bronić, Ruder Bošković Institute
D.T. Goodhead, Medical Research Council, Radiobiology Unit, Chilton
Yoshihiko Hatano, Tokyo Institute of Technology
M. Hayashi, Gaseous Electronics Institute, Nagoya
Zdenek Herman, J. Heyrovsky Institute of Physical Chemistry and Electrochemistry
Mitio Inokuti, Argonne National Laboratory (Chairman of the Program)
I.G. Kaplan, Universidad Nacional Autonoma de Mexico
N.P. Kocherov, International Atomic Energy Agency
(Scientific Secretary of the Program)
Franz Linder, Universität Kaisenslautern
Tilmann Märk, Universität Innsbruck
H.G. Paretzke, GSF-Forschungszentrum für Strahlenschutz und Umweltforschung
Helmut Paul, Johannes-Kepler Universität, Linz
Pascal Pihet, Institut de Protection et Sûreté Nucleaires, Fontenay-aux-Roses
Leon Sanche, Université de Sherbrooke
Dušan Srdoč, Brookhaven National Laboratory
Michel Terrissol, Université Paul Sabatier, Toulouse
L.H. Toburen, National Academy of Sciences
Ernst Waibel, Physikalisch-Technische Bundesanstalt, Braunschweig
André Wambersie, Université Catholique de Louvain, Brussels

The work is close to completion; a final meeting to discuss a draft for the report was held in June 1993 at Vienna. The contents of the report will be as follows:

Chapter 1. Development of charged-particle therapy and requirements for atomic and molecular data (Wambersie, Goodhead, and Pihet)
Chapter 2. Ionization cross sections for charged particles (Toburen)
Chapter 3. Electron collision cross sections. (Märk, Hatano, Linder, and Hayashi)
Chapter 4. Interactions of low-energy electrons with condensed matter (Sanche, Hatano, and Märk)
Chapter 5. Photoabsorption, photoionization, and photodissociation (Hatano and Inokuti)
Chapter 6. Rapid conversion of initial ions and excited species in collision with other molecules (Herman, Hatano, Sanche, and Märk)
Chapter 7. Stopping powers, ranges, and straggling (Paul, Berger, Paretzke, and Bichsel)
Chapter 8. Yields of ionization and excitation in irradiated matter (Srdoč, Inokuti, Bronić, Waibel, Hatano, and Kaplan)
Chapter 9. Track-structure quantities (Paretzke, Goodhead, Terrissol, and Kaplan)
Chapter 10. Concluding remarks.

A Monograph on Cross Section Data

Another activity concerning cross-section data from a broader and more basic point of view is underway. I edited a monograph entitled *Cross-Section Data*, to be published in 1994 as Volume 33 of *Advances in Atomic, Molecular, and Optical Physics* by Academic Press. The volume will contain 11 articles on various efforts to determine cross section data through experimental and theoretical studies, on needs for the data in selected applications, and on efforts toward compilation and dissemination of the data. However, the volume does not include extensive tables, graphs, or other presentations of the data themselves. Because of the great bulk of current data, they can be presented better in the form of computer databases than in a printed book. Similarly, no attempt has been made to cover all areas of atomic collision research in this volume; instead, the selected topics for the articles are illustrative rather than comprehensive.

Nevertheless, the volume should provide a guide to those who need to use cross section data with the best judgment and discretion, as well as to those who wish to produce better data through experimental or theoretical work. The volume will also convey a sense of the charms and challenges of what I would call *data physics*, a field that often fails to receive the appreciation and recognition that its importance warrants. I hope that the volume will serve educational purposes. The contents of the volume will be as follows:

Benchmark measurements of cross sections for electron collisions: Optical methods (Lin, McConkey, Anderson, and Filippelli)
Benchmark measurements of cross sections for electron collisions: Analysis of scattered electrons (Trajmar and McConkey)
Benchmark measurements of cross sections for electron collisions: Electron swarm methods (Crompton)
Benchmark measurements of cross sections for simple heavy-particle collisions (Gilbody)
Benchmark calculations of cross sections (Schneider)
Analytic representation of cross-section data (Inokuti, Dillon, Kimura, and Shimamura)
Electron collisions with N_2, O_2, and O: What we do and do not know (Itikawa)
Needs for cross sections in fusion plasma research (Summers)

Needs for cross sections in plasma chemistry (Capitelli)
Guide for users of data resources (Gallagher)
Guide to bibliographies, books, reviews, and compendia of data on atomic collisions (McDaniel and Mansky)

Recommendations and Outlook

During the work on cross section data via the three above avenues, I found many issues that require study in the future. The following are three major issues.

1. Cross sections for charged-particle collisions with molecules in the gas phase remain far from well established. Even for simpler molecules such as H_2O, data are uncertain. For many of the polyatomic molecules of interest to radiation biology, cross section data are scarce. Cross sections of basic building blocks[18] of the DNA and proteins seem to be appropriate targets for immediate study.

Cross section data for individual molecules are important for interpreting and validating cross section data for condensed phases, which are generally harder to obtain directly from experiments.

2. Methods for determining cross sections, in both theory and experiment, are advancing greatly. Expanded support for cross section determinations will be highly productive for several years in the future, in view of the current status of techniques. One new area of study will be the influence of temperature, (i.e., rotational and vibrational excitations in the initial states of molecules due to thermal agitations) on cross sections for electron interactions with molecules. This topic is crucial to a full understanding of electron thermalization and recombination with ions. Pioneering work[19,20] shows the feasibility of experiments on this topic, and is also stimulating to basic theoreticians.

3. Cross sections of molecules in condensed matter differ appreciably from those of isolated molecules, under circumstances that are generally understood. One circumstance concerns collisions involving slow electrons in either initial or final states, or both. Slow electrons in this context have sufficiently low kinetic energies (tens of eV and below) to be affected by forces due to condensed-matter structure. Knowledge about this topic is only beginning to be developed, as summarized in the forthcoming IAEA report. However, much remains to be done in both experiment and theory before fully convincing and comprehensive cross section data are established for any material.

Finally, I present the following recommendations to the workers in Monte Carlo studies of track structure and related topics:

1. Documentation of cross section data used in the published literature is generally too sketchy to allow reported studies to be reproduced. I certainly recognize that full documentation of all the input data will demand considerable effort. However, this effort is necessary to establish the credibility of the work and to improve its quality.

2. Results of Monte Carlo studies must be documented in the greatest detail possible. All too often we see such results presented only in figures and small tables. We can seldom identify the fraction of the uncertainty in the results that is due to limited sampling or the truly stochastic nature of the track structure. Documentation is desirable not only of final results but also of intermediate results. Analysis of intermediate results from a new point of view will be helpful for credibility and for deeper understanding of the physics involved.

ACKNOWLEDGMENTS

This work was supported in part by the U.S. Department of Energy, Office of Health and Environmental Research, Office of Energy Research, under Contract W-31-109-Eng-38.

REFERENCES

1. L.D. Landau and E.M. Lifshitz. Statistical Physics. Translated by J.B. Sykes and M.J. Kearsley. Pergamon Press, Oxford (1969).
2. E.M. Lifshitz and L.P. Pitaevskii. *Physical Kinetics.* Translated by J.B. Sykes and R.N. Franklin. Pergamon Press, Oxford (1981).
3. N.G. van Kampen. *Stochastic Processes in Physics and Chemistry.* North-Holland, Amsterdam (1981).
4. A.M. Weinberg and E.P. Wigner. *The Physical Theory of Nuclear Chain Reactors.* The University of Chicago Press, Chicago (1958).
5. H.G. Paretzke. Radiation Track Structure Theory. *Kinetics of Nonhomogeneous Processes*, G.R. Freeman (ed.), pp. 89–170. John Wiley & Sons, New York (1987).
6. R.H. Ritchie, R.N. Hamm, J.E. Turner, H.A. Wright, and W.E. Bloch. Radiation Interactions and Energy Transport in the Condensed Phase. *Physical and Chemical Mechanisms in Molecular Radiation Biology*, W.A. Glass and M.N. Varma (eds.), pp. 99–135. Plenum Press, New York (1991).
7. M. Zaider. Charged Particle Transport in the Condensed Phase. *Physical and Chemical Mechanisms in Molecular Radiation Biology*, W. A. Glass and M. N. Varma (eds.), pp. 137–162. Plenum Press, New York (1991).
8. M. Kimura, M. Inokuti, and M.A. Dillon. Electron Degradation in Molecular Substances. *Advances in Chemical Physics, Vol. 84*, I. Prigogine and S. A. Rice (eds.), pp. 193–292.
9. International Commission on Radiation Units and Measurements. *Stopping Powers for Electrons and Positrons*, ICRU Report 37. Bethesda, Maryland (1984).
10. International Commission on Radiation Units and Measurements. *Stopping Powers and Ranges for Protons and Alpha Particles*, ICRU Report 49. Bethesda, Maryland (1993).
11. L.G. Christophorou. Radiation Interactions in High-Pressure Gases. *Physical and Chemical Mechanisms in Molecular Radiation Biology*, W.A. Glass and M.N. Varma (eds.), pp. 183–230. Plenum Press, New York (1991).
12. M. Inokuti. How is Radiation Energy Absorption Different Between the Condensed Phase and the Gas Phase? *Radiat. Effects and Defects in Solids* 117:143–162 (1991).
13. M. Inokuti. Atomic and Molecular Theory. *Physical and Chemical Mechanisms in Molecular Radiation Biology*, W.A. Glass and M.N. Varma (eds.), pp. 29–50. Plenum Press, New York (1991).
14. L.H. Toburen. Atomic and Molecular Physics in the Gas Phase. *Physical and Chemical Mechanisms in Molecular Biology*, W.A. Glass and M.N. Varma (eds.), pp. 51–98. Plenum Press, New York (1991).
15. M.E. Rudd, Y.-K. Kim, D.H. Madison, and T.J. Gay. Electron Production in Proton Collisions with Atoms and Molecules: Energy Distributions. *Rev. Mod. Phys.* 64:441-490 (1992).
16. Nuclear and Atomic Data for Radiotherapy and Related Radiobiology. International Atomic Energy Agency. Vienna, 1987.
17. Atomic and Molecular Data for Radiotherapy. International Atomic Energy Agency, IAEA-TECDOC-506. Vienna, 1989.
18. M.A. Dillon, H. Tanaka, and D. Spence. The Electronic Spectrum of Adenine by Electron Impact Methods. *Radiat. Res.* 117:1-7 (1989).
19. W.M. Johnstone, N.J. Mason, and W.R. Newell. Electron Scattering from Vibrationally Excited Carbon Dioxide. *J. Phys. B* 26:L147-L152 (1993).
20. K.T. Stricklett and P.D. Burrow. Electron Scattering from Laser-Excited SF_6. *J. Phys. B* 24:L149-L154 (1991).

DISCUSSION

Chatterjee: Stopping powers in what materials?

Inokuti: ICRU Report 49 treats 25 elements and 48 compounds or mixtures. Those are the materials of interest to therapy and dosimetry.

Ritchie: What energy range did you cover?

Inokuti: The energy range covered begins at 1 keV and up. We make a short remark about stopping powers below 1 keV.

Steele: Just for my curiosity, what is the motivation for studying methane and propane?

Inokuti: Methane and propane are standard gases used in gas ionization chambers. They are very important for dosimetry.

Holley: Are these tables in a computerized data base?

Inokuti: I'm glad you asked that question. Of course, ICRU Report 49 is in the form of a book, but in addition the ICRU sells a diskette.

Wilson: Why don't you consider making them available via internet?

Caswell: ICRU needs the funds.

Inokuti: That's just a simple next step. Because I think once we have it available in any electronic form, to put it into a network is a very simple thing.

Wilson: The advantage of putting it into a data base on internet is you can keep it updated. If one relys on diskettes, you must always ask if your copy is current.

Inokuti: I might elaborate a little more by saying the following. The stopping power tables come out in disks; the numbers are produced by a computer program written by Martin Berger. This program will also be available. The ICRU refrains from selling it because of some legal problems; however, you can get it from Martin Berger. The program takes input, and it shows how to change input in the light of any new information.

Dubois: You had a slide that has some needs and recommendations. Could you comment on data that exist in those three categories, or can you be a little more specific as to needs? Where would you say the emphasis should be?

Inokuti: Well, you know inference of atomic and molecular aggregation, I deliberately didn't talk very much, because it's been said already to a great extent. That's certainly important. Also, about that problem, we would have to clearly delineate what sort of things depend on aggregation and what do not. All of this, I have already talked at Woods Hole and other places; I tried to be brief, so I just spent more time on Point No. 1. Point No. 1, of course, is a study of the polyatomic molecules from the cross-section point of view. It's always very difficult. I think the basic reason is that we don't know enough about excited states of polyatomic molecules. I might show one or two things that come out from Mike Dillon in our lab. We are beginning to produce electron energy-loss-data for different components of DNA. (See Ref. 18.)

Glaeser: Are these transmission spectra or are they reflection spectra?

Inokuti: These are spectra of scattered electrons analyzed in terms of energy losses.

Glaeser: How thick were the samples?

Inokuti: These are all in vapor. We also study thin films. Some features are very similar, and others aren't quite the same.

Glaeser: At these energies, you will probably have multiple scattering at almost any thickness of condensed film.

Inokuti: Yes, you do get the multiple scattering. We begin with this polyatomic molecule which goes into vapor when we mildly heat it.

Varma: With respect to these measurements, is there correlation being made with respect to the structure of different bases?

Inokuti: Yes. That's what we are really beginning to do. Our understanding is really incomplete. We are trying to do quantum-chemistry calculations on different energy levels.

Varma: The question is always asked how the information on isolated bases is relevant for damage for DNA. Can you say how the cross-section data that we get from the individual bases can be related to the whole DNA?

Inokuti: We have compared these things with data of dry DNA, and we can make correlations. Not on every possible feature, but some can be correlated. Doing all this takes a great deal of research. This is a prerequisite for doing more realistic track analysis.

Varma: If you carry the above reasoning a little further and try to project how this type of cross section data can be used in our Monte Carlo code track development, and if we use these cross-sections, then how much effect does it produce in terms of radical distribution in time and space?

Paretzke: I would think, you should put a little bit more emphasis on our needs for information on two additional points, that is the energy migration and the decay of excited states into new chemical species. These appear for me to be the weakest points in our calculations. The energy loss collision, which happens to the electron, is only the first step. But the migration of the transferred energy within the molecule and then the decay of an excited state into a new chemical species in a condensed phase can be very important and that's where our assumptions are weakest.

Inokuti: I completely agree with Herwig. I deliberately stayed out of that problem, because in that field, I'm really unable to make very good suggestions. I don't have enough ideas to really make a breakthrough in the experimental study. Of course, one can always concoct something out of theory, but results should be compared with experiments.

Glaeser: Just back on the energy loss spectrum, have you analyzed the data from the point of view of seeing whether the measured electron energy loss spectra are in agreement with the same information that you would get from UV absorption spectra?

Inokuti: Yes. That is very much so. We didn't ever find gross inconsistency with optical data. The situation is really the same as with any other molecule. However, often there is some discrepancy between photon data and electron energy loss data. There are several well-known reasons. For example, when the photon measurement doesn't have enough energy resolution, then the intensity comes out too low; whereas, electron energy loss method gives out the total intensity quite all right, even with modest resolutions.

Glaeser: The optical measurements are so easy to make and can be made in condensed phases and mixtures.

Inokuti: The optical measurements up to the near-vacuum UV may be relatively straightforward to do. But once you pass UV and go to far UV and soft X-ray regions, the optical measurements are not always easy or very good.

Glaeser: But the Advanced Light Source (ALS) at LBL will make a big difference.

Inokuti: Yes, very much so. However, the availability of photons is only one of the requirements. Also, we have to generate enough interest so that people do take such measurements.

Miller: I just want to respond to Herwig's comment to say that at PNL, we are addressing this question of electron transfer in DNA, and we're coming up with some extremely interesting results. This is mainly the work of Al Fuciarelli at our Laboratory. We are finding an unusual effect that electron transfer in DNA has a directionality.

Chatterjee: Eventually, the electron ends up only being on one base.

Miller: That is correct. The electron is more likely to be detected if the acceptor, (BudR), in our case, is on the 3'end of the adducted strand.

Ritchie: My impression was that Isaacson has done this sort of thing some years ago and was particularly interested in trying to identify the different base materials and had not really found very much difference. He was trying to fingerprint. It looks as if that's a similar conclusion here.

Inokuti: I think we need to look at this with higher resolution. Also, the same sort of work as Isaacson's should really need to be repeated because one difficulty with electron energy loss data is that it often has a wide-range distortion.

Ritchie: What do you mean by distortion?

Inokuti: For example if we compare well-known cases, the spectra often suffer from distortion, because scattered electrons are analyzed at different energies, and transmission may not be the same. If you do it in a straightforward way and the electron optics depends on the kinetic energy at which the electron is analyzed, that generally causes a general distortion of the data. That's always what we have to watch out for. This is very different from the photon case. In photo-absorption measurements the number one problem is a possibility of saturation due to very narrow lines. There, often we'll miss the intensity. Another difficulty with photo measurement even with light source is the fact that we have to go from one spectrometer to another, depending on the region. Very often, the different spectrometers used in different regions have different characteristics and data don't go smooth. Both electron measurements and photon measurements have their own difficulties, and it's always good to conduct both measurements. They tend to complement each other.

Varma: John, just a question to see if they can be correlated with the prior studies in terms of looking at the electrons. Doesn't that depend upon the initial production of those electrons at what angles or what energies?

Miller: These are electron-adducts to DNA. A hydrated electron interacts with DNA forming an electron adduct to DNA and that electron adduct migrates along the DNA chain to find its most comfortable place to stay.

Varma: Would it be fair to say that in order to determine the probability of production of these electrons, we need to know the cross sections in the DNA so that you have the number?

Miller: Certainly.

Paretzke: We have one different degree of complexity: an ionization leaves behind a positive hole and that as was well as the electron can migrate within a macromolecule and between molecules.

Miller: This is true. Fuciarelli's work does not include the influence of the positive hole on the migration of electrons in DNA because the ionization that ejected the electron that eventually adducted to DNA was in the aqueous medium and usually very far from a DNA molecule.

Varma: I think this point is very important because of what Bob Glaeser mentioned this morning. Biologists measure ultimate effects of electrons with DNA. They cannot tell you how the probability of effect will change, given a set of initial conditions. Monte Carlo programs can help in arriving at these probabilities. The biologist or the biochemist can then measure that probability and provide for confirmation of the modeling.

Chatterjee: I think the reason we are not appreciated is because the biologists feel we are very, very far from the real situations. Our model systems do not come close to biology or biological situations, even if we may study some special DNA cell units or some special molecules or solutions. We can come up with some basic, very important physical properties. The question is how those relate to the DNA in a cell? That's what biologists would like to know.

Varma: That's not what I am saying. What I'm saying is what Bob Glaeser mentioned, which I believe is probably very true, that the biologist doesn't appreciate that the information we are generating could be useful for them. It is an educational process, and we should work in cooperation with biologists so that we can use each others data for advancing scientific knowledge.

Glaeser: What I was saying is that you ultimately have to come up with effects that experimentalists intuitively would not expect, but which relate to their interests and involve experiments that aren't terribly difficult for them to do. Then you'll really start to get their respect.

Varma: I think the point is that you have to integrate your knowledge with the biologists. I think you cannot continue to work in isolation

Zaider: Supporting what you said, I think one of the best things we can do is to actually use physics to put restrictions on what biologists can think as to what produces a certain effect. An example which comes to mind is oncogene activation by radiation. From what we know, this is an impossibility. Now, biologists will speak ad nauseam about radiation-induced oncogenes and look for them. Very simple arguments show that it is highly unlikely any dose of radiation that we normally use in radiobiology can actually localize energy deposition on a particular piece of DNA of several kilobases and activate it to produce a single-base mutation. Herwig's statement as to the possibility of energy migration along with DNA might very well be an explanation of what's happening. But, again, we know close to nothing about that. The idea of soliton movement along the DNA has been proposed previously. There are thousands of papers on solitons, yet there's not one single experiment showing energy transports by solitons.

Varma: I don't think you are saying you want to work in isolation. You have to make the connection and show where this might be useful for biology.

Zaider: Yes.

Varma: If you cannot convince biologists why this is a useful thing for them to do then I think you are missing the point.

Zaider: But we can put important restrictions as to what mechanism they (biologists) should look for, what kind of doses, what kind of radiation.

Caswell: I agree with Marco's general premise. On the other hand, I think the physicists need to be very careful. An example that I have seen is the effects of low frequency electromagnetic fields. There are a number of physicists who feel they can prove that there is no such effect. Many biologists think there is an effect. The question may be that the physicist hasn't though of the right mechanism. I don't know the answer. It's very controversial, but I think the physicists need to be sure that they have looked at all the aspects of the problem before they make pronouncements.

INTERACTIONS OF LOW-ENERGY ELECTRONS WITH CONDENSED MATTER: RELEVANCE FOR TRACK STRUCTURE

R.H. Ritchie,* R.N. Hamm, and J.E. Turner*

Oak Ridge National Laboratory
Oak Ridge, TN 37831-6123

W. E. Bolch

Texas A&M University
College Station, TX 77843

ABSTRACT

Here we review the state of knowledge about the transport of subexcitation electrons in water. Longitudinal optical phonon and acoustical phonon interactions with an electron added to the system can give rise to an increase in the effective mass of the electron and to its damping. We generalize the pioneering treatment of thermalization given by Frölich and Platzman to apply to both long-range, polarization-type interactions, and as well to short-range interactions of low-energy electrons in water. When an electron is ejected from a molecule in condensed matter quantum interference may take place between the various excitation processes occuring in the medium. We have estimated the magnitude of this interference effect, together with the effect of Coulombic interactions on the thermalization of subexcitation electrons generated in the vicinity of a track of positive ions in water.

INTRODUCTION

We review the state of knowledge about the transport of subexcitation electrons in water. Little detailed information on the energy dependence of thermalization lengths, transport and inelastic cross sections, or possible trapping cross sections is available at present. For the present purposes we consider a water medium to consist of a system of extended boson states; longitudinal optical phonon and acoustical phonon interactions with an electron added to the system give rise to an increase in the effective mass of the electron and to its damping. The latter is due to creation and annihilation of bosons in the various

* Also with Department of Physics, University of Tennessee, Knoxville, TN 37996.

Computational Approaches in Molecular Radiation Biology
Edited by M.N. Varma and A. Chatterjee, Plenum Press, New York, 1994

fields present. Our approach generalizes the pioneering treatment of thermalization given by Frölich and Platzman (1953) to apply to both long-range, polarization-type interactions, and to short-range interactions of low-energy electrons in water. Energy losses in the generation of acoustical phonons is found to be negligible compared to losses to longitudinal optical modes, but appreciable contribution to the elastic interaction probability is found at lower energies. We have investigated the self-energy and effective mass of electrons in water. When an electron is ejected from a molecule in condensed matter quantum interference may take place between the various excitation processes occuring in the medium. Intrinsic and extrinsic excitation of the medium is said to happen due to the sudden appearance of the positive ion and of the electron, respectively. We have estimated the magnitude of this interference effect on energy loss by the electron. Preliminary calculations of the effect of Coulombic interactions on the thermalization of subexcitation electrons generated in the vicinity of a track of positive ions in water are discussed.

BACKGROUND

Extensive Monte Carlo calculations of electron transport in biologically significant media have been carried out by several groups. The code OREC, developed at Oak Ridge National Laboratory, was designed from the beginning to apply to water in the liquid form, and utilizes interaction cross sections inferred from experimental data on this material (Ritchie, et al., 1990). It has been used for a great many interesting problems, and results from this code have been compared with experiment (Bolch, et al., 1988) and with results from Monte Carlo calculations by other groups (Paretzke, et al., 1991).

The interaction of subexcitation electrons with water is of great importance to the understanding of basic chemical and physical mechanisms involved in the radiolysis of water. After being liberated in a medium by ionizing radiation, electrons lose energy by creating electron-hole pairs and electronic excitations in that medium. When their kinetic energy has decreased below the lowest excitation energy of the constituent molecules, further significant energy loss can occur only through energy transfer to nuclear motion. Platzman (1955), (Hart and Platzman, 1961) early emphasized the importance of such subexcitation electrons. Because of their small loss rate they can explore relatively large regions of space around their points of origin before thermalizing; thus they can be important in determining geminate ion pair separations and in initiating reactions in the medium (Voltz, 1991). Kaplan and Miterev (1987) have given a thorough discussion of the state of experiment and theory in regard to thermalization length of subexcitation electrons in water. Here we designate by $\ell_{th}(E)$ the mean distance covered by a subexcitation electron with energy E in slowing down to thermal energy.

THEORY

Attention has been given to the theory of subexcitation electron interactions in condensed matter (Voltz, 1991). The interaction of a subexcitation electron with a dipolar medium was studied early on by Frölich and Platzman (1953), who employed a semiclassical model in which the electron was approximated as a point charge in uniform motion. The response of the medium was taken from its measured frequency-dependent, complex dielectric function as known at that time. They emphasized the importance of dielectric relaxation in the loss of energy. Magee (1977) early recognized the importance of electron-phonon coupling in the subexcitation range. He noted that in all molecular media, energy can be lost in direct or short-range collisions and in indirect or distant collisions. He considered carefully the division of energy losses between those due to 'indirect' or distant interactions with the electric field induced in the medium by the electron, and 'direct' interactions via short-range forces. Also, qualitative consideration was given to the question of localization in a single event in the liquid through the transfer of a relatively large amount of energy into vibrational modes. Fano and collaborators (1991) have considered further the division of interactions into those of 'direct' and 'indirect' character, emphasizing the importance of polaron formation and the fact that the relatively long wavelength of the subexcitation electron tends to average out the electron's response to its interaction with the numerous particles in its vicinity. Knipp (1988) has recently used the Shockley model of the 'deformation potential' in a condensed matter system to represent the coupling of low-energy electrons to phonons via the short-range, or direct, interaction.

A good general survey of the state of theory of the structure of water and the question of electron localization in this medium has been given by Kestner (1987). Voltz (1991) has discussed electron thermalization in dense molecular matter in a qualitative and illuminating manner.

LaVerne and Mozumder (1984) have recently made a general review of experimental and theoretical evidence about the processes of electron energy loss and thermalization in the low energy region and in gaseous and liquid media. Uncertainties in cross sections and in treatment of Coulombic effects make theoretical estimates of the thermalization process difficult. Early workers speculated that a subexcitation electron would reach a separation from its parent ion of ~ 15 nm in water before becoming attached to a molecule to form H_2O^-. Samuel and Magee (1953) used a semiclassical random walk method to calculate the probability of escape of an electron ejected from a positive ionic center; using cross sections from gas phase experiments and schematic assumptions about the energy loss process, they found the thermalization distance of 1.2–1.8 nm for a typical subexcitation electron. The theory of Frölich and Platzman (1953) seems not to have been used to estimate the thermalization distance of an electron liberated from a parent ion in water. Mozumder and Magee (1967) have developed a random walk model to calculate the electron thermalization distance, including the Coulombic field of the parent ion, but do not apply it to the case of water. LaVerne and Mozumder (1984) estimate the thermalization distance to be ~ 1.5 nm but do not describe how they arrive at this figure.

EXPERIMENT

The first experimental studies that were interpreted in terms of thermalization lengths for subexcitation electrons were carried out by measuring the yield of electron-ion pairs that escape recombination in the presence of an electric field. The dependence of this yield on the strength of the field was interpreted using the Onsager (1938) theory for the diffusion of electrons in the (weak) field of a positive ion and under the influence of an electric field. To find $\ell_{th}(E)$ is not at all straightforward by this approach; it is necessary to make several assumptions about the properties of the system.

More definitive electrochemical photoemission experiments have been used subsequently and extensively in this connection. Electrons are photoejected from a metal target into an aqueous medium; the current to an external circuit is collected by means of an applied field. After thermalization the injected electrons may return to the metal and decrease the collected current, or they may be captured by a scavenger, and, assuming the capture product undergoes a reduction process on the electrode, will not reduce the collected current. Thus the dependence of the measured current on the concentration of a scavenger may be interpreted to yield a thermalization length, $\ell_{th}(E)$, that depends on the initial energy of the photoelectron. Barker et al., (1966) seem to have been the first to analyze such experiments quantitatively. This technique has been used subsequently by many different experimental groups.

Experiments of this kind, interpreted in terms of subexcitation electron thermalization, have been carried out by many workers. For example, Rotenberg and Gurevich (1975), Neff, et al. (1980), Kreitus, et al. (1982), and Konovalov, et al. (1985), infer $\ell_{th}(E)$ from the dependence of the collected current on scavenger concentration. Konovalov, et al. (1986) have measured the variation of collector current with time using nanosecond lasers to the same end. It should be emphasized that direct inference of $\ell_{th}(E)$ from such measurements is fraught with uncertainties. The distribution in energy and angle of electrons photoemitted from the metal-aqueous-solution surface in poorly known, many assumptions about the diffusion of solvated electrons in the presence of scavenging agents must be made, as well as about the transport of subexcitation electrons in the neighborhood of the complex metal-

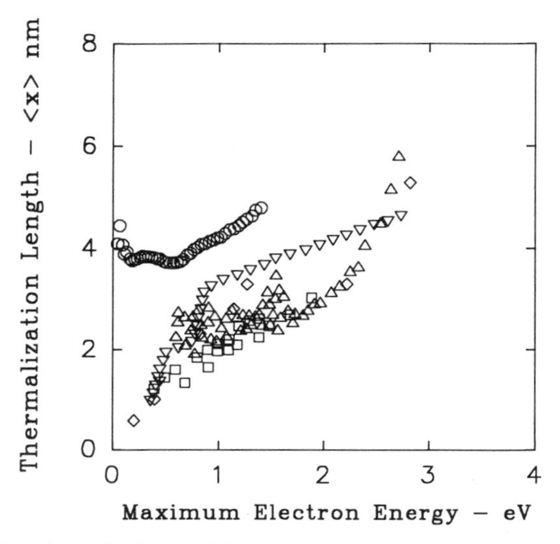

Figure 1. Experimental results on the first spatial moment of the thermalized electron distribution resulting from photoejection from metals into aqueous solutions. The data of Neff, et al. (1980) are shown as circles, the data of Konovalov, et al. (1985) as inverted triangles, that of Kreitus, et al. (1982) as triangles, that of Konovalov, et al. (1986) as diamonds, and that of Rotenberg, et al. (1975) as squares.

solution interface. Typically, the first moment, <x>, of the distribution of thermalized electrons in aqueous solutions, measured from the metal-solution interface, is obtained from the experiments referred to above. These data show substantial variations from one group to another when plotted as a function of the energy of the photons used minus the nominal work function of the metal employed. Figure 1 shows such data taken from the papers indicated in the caption. Considering the uncertainties attending this method, $\ell_{th}(E)$ cannot be regarded as well-known.

Here we consider in detail a model for losses by subexcitation electrons in water and estimate the corresponding thermalization length, accounting for Coulombic and transient energy loss effects when the ionization density along an ion or an electron track is small.

Comparison with experimental data is made. In addition, we study electron transport in the track of a high-LET ion where Coulomb forces are expected to be large.

INTERACTION CROSS SECTIONS

Existing ideas about elastic and inelastic interactions in liquid water are mainly qualitative in nature (Voltz, 1991, Magee, 1977, Fano and Du, 1991, Kestner, 1987). Electron transport at low energies in rare gas liquids is understood in terms of the Cohen-Lekner model (1967, Lekner 1967) in which condensed-phase, coherent scattering effects, arising from molecular order, are thought to be responsible for the observed high electron mobilities in these systems. Davis, et al. (1971) have shown that incoherent scattering effects may be important in liquid polyatomic systems and that mobilities may be correspondingly smaller than in the rare-gas liquids. As indicated above, inelastic interactions in the subexcitation range are understood to be due to the initiation of molecular vibrations and orientational motion, but accurate theoretical values for cross sections appropriate to these interactions are lacking. In view of these lacunae, and in view of the great difficulty of obtaining and interpreting relevant experimental data, we adopt a model for energy losses based on a phonon representation that has received much attention in the solid state literature. Negative ion resonance processes are ignored; note that Ibach and Lehwald (1980) saw no evidence for scattering due to such interactions for 5 eV electrons incident on layers of H_2O molecules adsorbed on Pt surfaces.

Experimental data obtained by the Sherbrooke group (Michaud and Sanche, 1987a, 1987b) on the transport of low-energy electrons in thin, solid films of water ice on metallic substrates have been used by them to infer elastic and inelastic cross sections for electrons in the subexcitation range. Although their cross sections are obtained from an approximate solution of the transport equation, Goulet and Jay-Gerin (1988) find them to be "self-consistent" when used in a Monte Carlo calculation of the transport of subexciation electrons in water ice. They also consider the possibility that the electrons may undergo dissociative attachment to water molecules in the thermalization process.

Elastic Scattering Cross Sections

Figure 2 shows the elastic transport Inverse Mean Free Paths (IMFP's), $\Lambda^{-1}{}_{el}(E)$, in H_2O vapor, at unit specific gravity, taken from Itikawa (1978) and Danjo and Nishimura (1985). In lieu of more definitive information, we follow Bolch, et al. (1988) in assuming that the energy-dependent cross section for electrons scattering elastically in liquid water is a constant, smaller than unity, times the corresponding quantity in the vapor at unit specific gravity. Also shown are data from the Sherbrooke group (Michaud and Sanche, 1987a, 1987b; Michaud, 1993) for amorphous water ice, indicated by the label "solid."

Inelastic Interactions

In order to represent inelastic processes in water we assume that the subexcitation electrons interact with phonon modes and with orientational motion of the water molecules in the liquid.

We first consider the possibility of interaction of an electron with longitudinal acoustical phonons. The treatment of Knipp (1988) yields an analytic expression for the stopping power of acoustical phonons in a medium for slow electrons, but is expressed in terms of the dynamical structure factor and the deformation potential E_1 of the system, where $E_1 \sim 1$–20 eV. E_1 for water does not seem to be known. In any case, the contribution

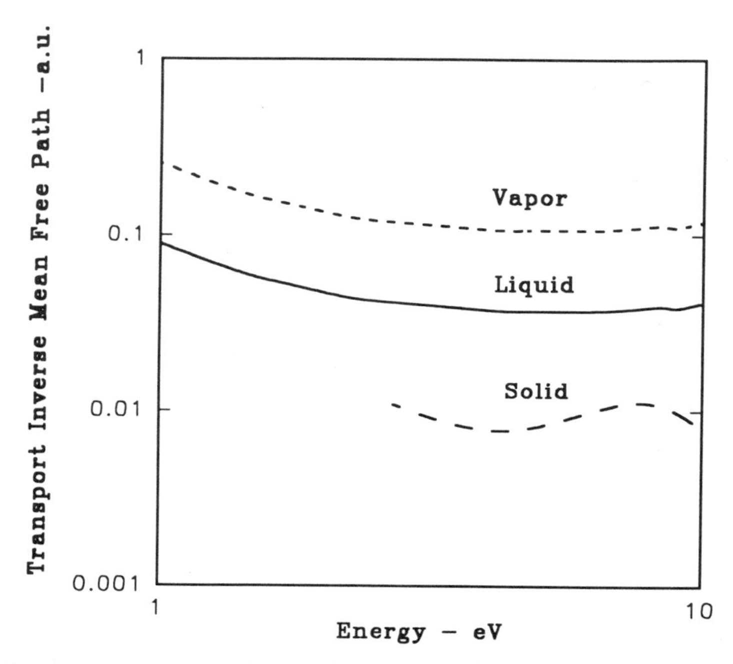

Figure 2. The transport inverse mean free path, in atomic units, for elastic scattering of electrons with various energies in water vapor, liquid water and solid H_2O. The sources of these data are described in the text.

to the stopping power of water on this model appears to be quite small compared with that due to other modes.

We employ a model for electron-acoustical phonon interaction based on the hydrodynamical analysis of Gras-Marti and Ritchie (1985). The constants entering this model come from measurement. We find that energy losses are negligible compared with losses to IR phonons in the energy region of interest; however, transport IMFP values for electron-acoustical phonon interaction may be an appreciable part of the total elastic scattering IMFP. The details of this treatment will be given elsewhere.

We now consider the interaction with longitudinal IR phonons. The Hamiltonian of the noninteracting phonon field may be written as

$$H_0 = \sum_i \sum_{\vec{q}} \hbar\omega_{i,\vec{q}} b^*_{i\vec{q}} b_{i\vec{q}} , \qquad (1)$$

where $\hbar\omega_{i,\vec{q}}$ is the energy of the ith mode in the water medium and $b_{i,\vec{q}}$ is the destruction operator for the ith kind of IR phonon with wavevector $\vec{q}$.

The interaction Hamiltonian between an added electron and the phonons may be written

$$H_1 = \sum_i \sum_{\vec{q}} \frac{D_{i,\vec{q}} e^2}{\Omega q^2} e^{i\vec{q}\cdot\vec{r}} \left(b_{i,\vec{q}} + b^*_{i,-\vec{q}} \right), \qquad (2)$$

where Ω is the normalization volume of the system and $D_{i,\vec{q}}$ is the coupling constant between an electron and a phonon. The coupling constants are evaluated from the dielectric response function of water.

Then the net stopping power of the medium for an electron with momentum $\hbar\vec{p}_0$ is given by

$$-\frac{dE}{dx} = 2\pi \sum_{i} \sum_{\vec{q}} \omega_{i,\vec{q}} |\langle H_1 \rangle|^2 \Big[\delta\big(E_{\vec{p}_0} - E_{\vec{p}_0 - \vec{q}} - \hbar\omega_{i,\vec{q}}\big) \cdot \big(1 - n_{i,\vec{q}}\big) - \delta\big(E_{\vec{p}_0} - E_{\vec{p}_0 - \vec{q}} + \hbar\omega_{i,\vec{q}}\big) n_{i,\vec{q}} \Big], \tag{3}$$

where energy loss and gain are accounted for in this equation, corresponding to the fact that when the medium is at temperature T, the probability of finding a given kind of phonon is

$$n_{i,\vec{q}} = \frac{1}{e^{\frac{\hbar\omega_{i,\vec{q}}}{kT}} - 1} \tag{4}$$

An approximation to the contribution to the effective mass due to phonon interactions can be found from the self-energy as,

$$\Delta E_{ph\,creation} = \sum_{i} \sum_{\vec{q}} \frac{|\langle H_1 \rangle|^2 \big(1 + n_{i,\vec{q}}\big)}{E_{\vec{p}_0} - E_{\vec{p}_0 - \vec{q}} - \hbar\omega_{i,\vec{q}}}, \tag{5}$$

with a corresponding expression for the contribution from phonon annihilation. Expressing the total energy in the form

$$E = \frac{\hbar^2 p^2}{2m} + \Delta E(p) \tag{6}$$

and expanding in a power series around the point $p = 0$ one finds a substantial correction in the mass of the subexcitation electron due to phonon interactions. The effect of this effective increase in the mass of the electron will be discussed elsewhere. For the present estimates, we take the electron to have its vacuum mass value.

Our estimate of the stopping power of water in the subexcitation region is shown in Fig. 3 as a solid line. Also shown are values of this quantity for water vapor at unit specific gravity (Ishii, et al., 1990), and for solid H_2O (Michaud, 1993). At room temperature the thermal effect on the stopping power is quite small; at 0.5 eV there is only a 10% increase over values found at T = 0 K, with much smaller increases at larger energies. At energies above 8.4 eV, stopping power values from our OREC code are used.

THE TRANSPORT OF SUBEXCITATION ELECTRONS IN WATER

Since the fractional energy losses experienced by subexcitation electrons in water are expected to be small, it should be possible to represent with good accuracy their spatial transport as a diffusion-type process. Numerical evaluation of $\ell_{th}(E)$ for the given interaction models could be carried out by means of a Monte Carlo approach (see Goulet and

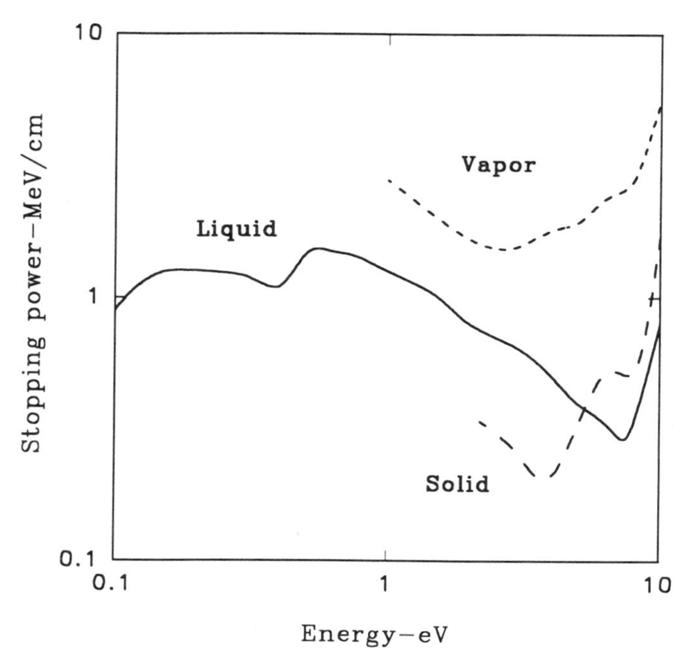

Figure 3. The stopping power of water for subexcitation electrons in water vapor, liquid water and solid H_2O. The sources of these data are described in the text.

Jay-Gerin, 1988). However, here we use an algorithm that is as accurate as the Monte Carlo procedure in the present application, as we have shown by numerical comparison.

Our interest here is to estimate the effect on the transport of subexcitation electrons of electromagnetic fields associated with the generation of those electrons in a water medium We approximate the electron spatio-temporal-energy distribution, resulting from transport in the Coulomb field of the track from which the electrons were ejected, by using an equation that represents their average behavior (Samuel and Magee, 1953). One may thereby include in a relatively straightforward way (a) the self-consistent motion of the electrons in the field of the positive ions, accounting for Coulomb interactions among these electrons, and (b) the time-dependent dielectric response of the medium.

Consider an electron with specified energy and position in the vicinity of a heavy ion track in water. It can (1) lose energy in inelastic collisions with the liquid, (2) experience elastic collisions that lead primarily to angular deflection, and (3) alter its energy in moving against the Coulomb attraction of core ions and the repulsion from other electrons in the system. Let a representative electron with energy E make n collisions and experience an average squared displacement of r^2 in moving from an (arbitrary) origin. If it experiences Δn collisions the change in the rms displacement $\Delta(r^2)$ will be given by $\Delta(r^2) = L^2\Delta n$, where L(E) is the transport mean free path for collisions in the medium. The kinetic energy of this electron will be diminished in inelastic collisions and in moving against the Coulomb force of the positive ion track; Coulomb interactions with other electrons tend to diminish this effect, as will dielectric polarization of the medium. If the electron's energy loss per unit

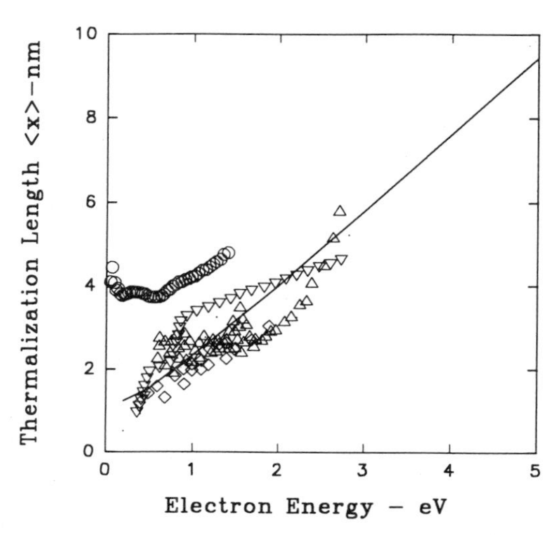

Figure 4. A comparison between the experimental data shown in Fig. 1 with theoretical results for the rms transport length, $\ell_{th}(E)/3^{1/2}$, found using the theoretical approach described in the text.

path length is S(E) then

$$\frac{dE}{dn} = -L(E)S(E) - F(\rho)\frac{dr}{dn} \tag{7}$$

where $F(\rho)$ is the Coulomb force on the electron. Passing to the limit where the increments become differential quantities and dividing both sides of the equation by $d(r^2)/dn$ one finds

$$\frac{dE}{dr} = -\left(2r\frac{S(E)}{L(E)} + F(\rho)\right) . \tag{8}$$

In keeping with the idea of diffusive transport, one replaces r^2 by $3\rho^2/2$, so that Eq. 9 becomes

$$\frac{dE}{d\rho} = -\left(3\frac{S(E)}{L(E)} + \left(\frac{3}{2}\right)^{1/2} F(\rho)\right) . \tag{9}$$

To account for the time-dependent polarization of the medium to the movement of charges it is important to keep track of the time. Thus the interval dt=ds/v and the pathlength $ds = Ldn = d(r^2)/L$, so one puts $dt = 3d(\rho^2)/2Lv$. Then the equations giving the time-dependence of the energy and radial position become, using atomic units in which $e = m = \hbar = 1$ and taking the electron to have mass m,

$$\frac{dE}{dt} = -\sqrt{2E}\left(S(E) + \left(\frac{2}{3}\right)^{1/2} \frac{\lambda L(E)}{\rho^2 \varepsilon(t)} \right) \tag{10}$$

and

$$\frac{d\rho}{dt} = \frac{L(E)\sqrt{2E}}{3\rho}, \tag{11}$$

where the Coulomb force is assumed to be only due to the positive ions of the track, assumed to be distributed in cylindrically symmetry about the track center. The time-dependent dielectric constant is taken to be

$$\frac{1}{\varepsilon(t)} = \frac{1}{\varepsilon_s} + \left(\frac{1}{\varepsilon_\infty} - \frac{1}{\varepsilon_s}\right) e^{-t/t_o}, \tag{12}$$

where $t_o = 1.033 \times 10^4$ a.u., $\varepsilon_s = 80$, $\varepsilon_{inf} = 2$ for water; the orientational response of the medium is assumed to determine the time dependence of dielectric screening here.

It is straightforward to generalize this scheme to describe multigroup diffusion in which each group of electrons propagates in energy, space and time and is coupled to the other groups by the self-consistent Coulomb force in the system.

To compare with experimental data pertinent to $\ell_{th}(E)$, we have computed the mean square distance to thermal energy of a single subexcitation electron that is ejected from a metal into an aqueous medium. We use forms of Eqs. 10 and 11 appropriate to propagation of such an electron under the influence of the transient, modified image potential that it experiences in this system, and losing energy to the medium at a rate that is both energy- and spatially dependent. We neglect loss of electrons in back-diffusion to the metal surface. Rotenberg and Gurevich (1975) have presented experimental evidence that this is a good approximation in such experiments. Figure 4 shows $\ell_{th}(E)/3^{1/2}$ from such a calculation compared with the several sets of experimental data on the projected diffusion length $<x>$ displayed in Fig. 1. Note that the factor of $1/3^{1/2}$ converts an rms diffusion length in three dimensions to its projected value along one dimension, appropriate for this comparison.

We have constructed a computer program that solves the multigroup set of equations describing transport of subexcitation electrons liberated in the vicinity of a dense track of ions generated by the passage of a high LET particle. The nonlinear motion of the electron cloud is quite complex.

With Coulomb interactions turned off we compute the rms distance traveled from the initial positions in the multigroup representation and find agreement with the results calculated for single electrons as shown in Fig. 4. With Coulomb interactions accounted for we find preliminary results for the fractional confinement $R(\lambda)/R(0)$. Here $R(\lambda)$ is the mean range of electrons when the linear charge density is λ. We have computed $R(\lambda)$ assuming that the electrons are released with an energy of 8 eV and that the initial spatial extension in space is 500 a.u. in an assumed Gaussian mode. One sees a substantial effect; for λ corresponding approximately to the ionization density created by a 5 MeV a-particle in water we find $R(\lambda)/R(0) \sim 0.75$. Further work along these lines is under way. We will employ a more realistic distribution of energies generated in the subexcitation region and will study the sensitivity of the results to the parameters entering the theory.

SUMMARY

We have described a scheme for estimating the transport parameters for subexcitation electrons of various energies in a water medium. Some experimental data relevant to these parameters are discussed. It is concluded that much more work is needed to elucidate these important quantities. We have constructed a program to solve a set of multigroup equations that describes approximately the transport and thermalization of subexcitation electrons in liquid water. Inelastic IMFP's for the transport process have been constructed using a phonon model. We have obtained preliminary results from this program that seem quite reasonable. We will extend these calculations to explore the sensitivity of the confinement function $R(\lambda)/R(0)$ to the assumptions of initial energy and spatial distribution as well as to the IMFP's that are used in the calculation. A still more accurate approach involves the use of particle simulation techniques accounting for the presence of Coulomb forces; we plan to adapt our computer code OREC to this end.

ACKNOWLEDGEMENTS

The authors thank Dr. M. Michaud for helpful communications about the his data relating to the several cross sections of subexcitation electrons in solid H_2O. Thanks are also due to Dr. J.A. Laverne for help in obtaining a copy of the paper of Kreitus, et al. (1982). This research was sponsored jointly by the Office of Health and Environmental Research, U.S. Department of Energy, under contract number DE-AC05-84OR21400 with Martin Marietta Energy Systems, Inc.

REFERENCES

Barker, G.C., Gardner, A.W., and Sammon, D.C., 1966, Photocurrents produced by ultraviolet irradiation of mercury electrodes, *J. Electrochem. Soc.* 113:1182

Bolch, W.E., Turner, J.E., Yoshida, H., Jacobson, K. B., Hamm, R.N., Wright, H.A., Ritchie, R.H., and Klots, C.E., 1988, Monte Carlo simulation of indirect damage to biomolecules irradiated in aqueous solution - the radiolysis of glycylglycine, *Oak Ridge National Laboratory Technical Report ORNL/TM-10851*

Cohen, M. and J. Lekner, J., 1967, Theory of hot electrons in gases, liquids and solids, *Phys. Rev.* 158:305

Danjo, A. and Nishimura, H., 1985, Elastic scattering of electrons from H_2O molecule, *J. Phys. Soc. Japan,* 54:1224

Davis, H.T., Schmidt, L.D., and Minday, R.M., 1971, Kinetic theory of excess electrons in polyatomic gases, liquids, and solids, *Phys. Rev.* A3: 1027

Fano, U. and Du, N.-Y., 1991, Dissipative polarization by slow electrons, *Appl. Rad. Isot.* 42: 975, and references given therein.

Frölich, H., and Platzman, R.L., 1953, Energy loss of moving electrons to dipolar relaxation, *Phys. Rev.* 92: 1152

Goulet, T. and Jay-Gerin, J.-P., 1988, Thermalization distances and times for subexcitation electrons in solid water, *J. Phys. Chem.* 92: 6871 (1988)

Gras-Marti, A, and Ritchie, R.H., 1985, Charged-particle excitation of ripplon fields, *Phys. Rev.* B31:2649

Hart, E.J., and Platzman, R.L., 1961, Radiation chemistry, *in*: "Mechanisms in Radiobiology," A. Forssberg and M. Errera, eds., Academic Press, New York, p. 63

Ibach, H. and Lehwald, S., 1980, The bonding of water molecules to platinum surfaces, *Surf. Sci.*, 91: 187

Ishii, M. A., Kimura, M. and Inokuti, M. 1990, Electron degradation and yields of initial products. VII. Subexcitation electrons in gaseous and solid H_2O, *Phys. Rev.* A42:6486

Itikawa, Y., 1978, Momentum-transfer cress sections for electron collisions with atoms and molecules-- revision and supplement, *Atomic Data and Nuclear Data Tables,* 21:69

Kaplan, I.G. and Miterev, A.M., Interaction of charged particles with molecular medium and track effects on radiation chemistry. *in:* "Advances in Chemical Physics," I. Prigogine and S.A. Rice, eds., Wiley, New York (1987)

Kestner, N.R., 1987, Theoretical studies of electrons in fluids, *in:*"The Liquid State and its Electrical Properties" E. E. Kunhardt, L. G. Christophorou, and L. H. Luessen, eds., Plenum, New York,
Knipp, P., 1988, Interaction of slow electrons with density fluctuations in condensed materials: calculation of stopping power, *Phys. Rev.* B37: 12
Konovalov, V.V., Raitsmiring, A.M., Tsvetkov, Y.D., and Benderskii, V. A., 1985, The thermalization length of low-energy electrons determined by nanosecond photoemission into aqueous electrolyte solutions, *Chem. Phys.* 93:163
Konovalov, V.V., Raitsmiring, A.M., and Tsvetkov, Y.D., 1986, Determination of the thermalization length of low-energy electrons in H_2O and D_2O solutions by photoelectric emission, *High Energy Chem.* 18:1
Kreitus, I.V., Benderskii, V.A., Beskrovnyi, V.M., and Tiliks, Yu.E., 1982, Thermalization length of low energy electrons in concentrated aqueous solutions of electrolytes, *Khimiya Vysosokikh Ènergii* 16: 112
LaVerne, J.A., and Mozumder, A., 1984, Energy loss and thermalization of low-energy electrons, *Radiat. Phys. Chem.* 23: 637
Lekner, J., 1967, Motion of electrons in liquid argon, *Phys. Rev.* 158: 130
Magee, J.L., 1977, Electron energy loss processes at subelectronic excitation energies in liquids, *Can. J. Chem.* 55: 1847
Michaud, M. and Sanche, L., 1987a, Total cross sections for slow-electron (1-20 eV) scattering in solid H_2O, *Phys. Rev.* A36:4672
Michaud, M. and Sanche, L., 1987b, Absolute vibrational excitation cross sections for slow-electron (1-18 eV) scattering in solid H_2O, *Phys. Rev.* A36:4684
Michaud, M, 1993, private communication.
Mozumder, A., and Magee, J.L., 1967, Theory of radiation chemistry VIII. Ionization of nonpolar liquids by radiation in the absence of external electric fields, *J. Chem. Phys.* 47: 939
Neff, H., Sass, J.K., Lewerenz, H.J., and Ibach. I., 1980, Photoemission studies of electron localization at very low excess energies, *J.Phys. Chem.* 84: 1135
Onsager, L., 1938, Initial recombination of ions, *Phys. Rev.* 54:554
Paretzke, H.G., Turner, J.E., Hamm, R.N., Ritchie, R.H., and Wright, H.A., 1991, Spatial distributions of inelastic events produced by electrons in gaseous and liquid water, *Radiation Research* 127: 121
Platzman R.L.1955, Subexcitation electrons, *Rad. Res*. 2:1
Ritchie, R.H., Hamm, R.N., Turner, J.E., Wright, H.A., and Bolch, W.E., 1990, Radiation interactions and energy transport in the condensed phase, *in*: "Physical and Chemical Mechanisms in Molecular Radiation Biology," W. A. Glass and M. N. Varma, eds., J. Wiley, New York.
Rotenberg, Z.A. and Gurevich, Yu.Ya., 1975, Photodiffusion phenomena stimulated by photoelectron emission in solutions, *J. Electroanal. Chem.* 66:165
Samuel, A.H., and Magee, J.L., 1953, Theory of track effects in radiolysis of water, *J. Chem. Phys.* 21: 1080
Voltz, R., 1991, Thermalization of subexcitation electrons in dense molecular matter, *in:* "Excess Electrons in Dielectric Media," CRC Press, Cleveland, OH, (1991), e.g., and references quoted.

DISCUSSION

Inokuti: What did you use for the phonon spectrum?

Ritchie: I took that from photon data.

Inokuti: I tried to do the same sort of thing but wasn't quite sure how well the phonon spectrum was known.

Ritchie: Well, it turns out that thermal effects are mostly important for the lower energy phonons.

Inokuti: I see.

Curtis: How much contribution do you think these effects would have on the magnitude of stopping power?

Ritchie: I think the main effect would be the quantal interference phenomenon. There is, of course, some uncertainty in the parameters that one uses in this calculation.

Curtis: What I mean is, on the order of five percent, ten percent, or what? For instance, if you see a number for dE/dx for water in the literature, how much would the effects that you're talking about modify that number if they had been taken into account?

Ritchie: The interference effect has never been considered before, I think it could amount to 20 percent effect in the stopping power for very low energy electrons.

Zaider: The Sanchez group has published ranges for thermalization for very low-energy subexcitation electrons. They are not direct measurements. They are obtained by making all kinds of assumptions, and that is where the differences are coming from.

Ritchie: Yes, and their data are for ice, not water.

Zaider: I wanted also to point out that one of the important domains where you need to know the range of these electrons is in radiation chemistry. We discovered that if you want to reproduce, say, the decay of the solvated electron as a function of time, then the actual position of this electron, at the time it becomes solvated, is crucial.

Paretzke: There are much larger contributions from hole migration in liquid water, which is not considered at all in this type of calculation.

Hamm: Aren't these distances considerably less than the thermalization distances?

Paretzke: No.

Ritchie: Yes, but we're talking here about 30 to 40 angstroms.

Varma: The question that we need to ask is how important are the thermal electrons. I think Mitio mentioned that 60 percent of all secondary electrons eventually end up as subexcitation electrons.

Inokuti: In liquid water, it is much more, even though it contributes to 30% of energy, but the total number is closer to 100%.

Varma: And then the question is, if you look at the bulk quantities such as stopping power, W- value, etc., which are average quantities, how much difference do thermal electrons make in these quantities? I guess not significant. On the other hand, the mechanism by which the electrons transport energy, even though it is 30 percent, is very crucial if you want to understand how the damage is produced at the molecular level.

Inokuti: That's certainly the whole idea behind scavenger kinetics.

Varma: Another question I had was in terms of the comparison of your results with those of Ritchie's. Is there a difference between the two approaches?

Inokuti: Well, I mean for one thing, maybe it's a good time to raise one question. And, Rufus, you compared the calculation with the Sanche experiment, but I think some qualification is necessary.

Ritchie: But the relation is not clear because you have negative ion formation in the gas. But Ibach finds that negative ion formation is not detectable in water deposited multilayers on metal substrates, so things are very different.

Varma: The information this program is generating is very important and basic. But once you put this information into a Monte Carlo code, you end up with average spatial and temporal distribution of ionizing events, etc. How can you test your calculations from a biological point of view? I mean, it is very nice to say that the chemistry changes, or that chemistry might change. But how can you relate it to a specific end point that is measured as a biological effect. Quite often we say that information generated is important for biology, but fail to show where and how!

Nikjoo: Some investigators are carrying out calculations of initial distributions of clustered damage and looking to identify those types that could be correlated with DNA strand breaks.

Varma: So you would be the user of this kind of information?

Nikjoo: Yes

Chatterjee: When I think of biology I think of an electron in a cell, there would be many trapping centers and how difficult it is to put into your calculations some effect of trappings centers.

Ritchie: Well, of course it could be done, but there is not much guidance as to what the trapping cross-sections are.

Miller: I don't know of any data that specifically addresses that question. The data that I was talking about concerns electron adducts on DNA products by interaction with solvate electrons.

Chatterjee: These are dry electrons.

Miller: Yes

Zaider: Mitio, how actually different is liquid from ice?

Inokuti: Of course the biggest difference between liquid water and ice is the rotation. In ice, there is no rotation and the very fact that hydrated electron happens is a result of the rotation. Molecular orientation has to get just right for an electron to get solvated. That is the biggest difference. And indeed, the calculation by Froehlich and Platzman just use only the rotator, and then later on it has been generalized by Magee.

Ritchie: It turns out that rotational excitation does not contribute very much to the stopping power of water for subexcitation electrons, at least until they approach thermal energies.

Miller: Is there any experimental evidence that suggests that liquid water is able to trap electrons before they are solvated? Is negative ion formation possible in liquid water?

Chatterjee: Yes, but it's not very stable.

Inokuti: What is observed recently, I think you have to know these things before a hydrated electron is formed. It's the work of a group in Paris, and there, you see, they looked at the absorption in the pico second region, and they observed a different spectrum that peaked in the infrared. And that is some kind of a trapped state that with time eventually goes to hydrated state. And also the H_2O^-, of course, doesn't exist in water, it cannot exist in water. Actually, it immediately dissociates, and that is a primary mechanism of decay of a hydrated electron. The hydrated electron stays, so long as an electron goes around from one molecule to another. And eventually, an electron decides to localize on one molecule, then the dissociation occurs. That's how hydrated electron decays in the pure water.

Ritchie: But probably this state that you talk about has a smaller binding energy than that for the hydrated electron.

Inokuti: Yes.

Ritchie: This happens also at the very low energy end of the spectrum. The electron has to lose most of its energy before it becomes localized.

Inokuti: Yes. But it's probably not quite thermalized.

Miller: The reason I raised this question is because in discussing these matters with J. Laverne and Mozumder at the Rad Lab, they feel that if they take Sanche's cross-section information, that they get spatial distributions of hydrated electrons that are too large to satisfy the requirements of radiation chemistry. So they want to narrow these distributions down by letting the electron be captured before it thermalizes. Is there really any physical reason to suggest that this mechanism would make the Sanche data compatible with the decay rate of hydrated electrons? I think what you're telling me is no, you can't get out of the problem by this mode.

Inokuti: Right.

Varma: From the point of view of biology, a cell is not just all water and moreover it is not pure water. In addition, how will one take into account structural considerations?

Ritchie: If one has an idea of what the phonon spectrum for the structures are, one could make an estimate of the effect of water on hydration around DNA on thermalization lengths in such systems.

Varma: That can tie in with the type of work Mike Dillon is doing.

Inokuti: Those are essentially information on the electronic structure, which is necessary to make a general theoretical framework to yield quantitative results.

Miller: But is it not possible to measure the phonon spectrum? Aren't you just talking about infrared absorption and things like that? This is basically how you measure the phonon spectrum.

Inokuti: Yes.

Varma: I think it is the logical step to go forward with knowledge from water molecules, to realistic biological structures and this I believe can be done with developed theoretical framework.

Miller: Yes.

Ritchie: This is another case where we need more experimental data.

ELECTRON EMISSION RESULTING FROM FAST ION IMPACT ON THIN METAL FOILS: IMPLICATIONS OF THESE DATA FOR DEVELOPMENT OF TRACK STRUCTURE MODELS

R.D. DuBois and C.G. Drexler

Pacific Northwest Laboratory
Richland, WA 99352 USA

ABSTRACT

Experimental information useful in improving and testing theoretical models of energy deposition and redistribution in particle-condensed phase media is discussed. An overview of information available from previous, and ongoing, studies of the electron emission from thin foils is presented and the need for doubly differential electron energy and angular distribution data stressed. Existing data are used to demonstrate that precisely known, and controllable, surface and bulk conditions of the condensed-phase media are required for experimental studies of this type. Work in progress and improvements planned for future studies at the Pacific Northwest Laboratory are outlined.

INTRODUCTION

Energetic particles traversing matter undergo coulomb interactions with bound target electrons. These interactions deposit energy in the target media at localized sites having nanometer dimensions. Energetic electrons produced in ionizing collisions at a particular site can redistribute a portion of the deposited energy over a larger area via subsequent interactions with other bound media electrons. In each interaction, media atoms or molecules are excited and will therefore react with their neighbors in a different manner than they do when in their relaxed states. This can produce "chemical" changes within the media.

Modeling the effects evolving from energy deposition is therefore rather complex. It involves knowledge about the products generated in individual projectile-media interactions, about emitted electron-media scattering and ionization processes, and about how excited/ionized target atoms and molecules relax or interact with their neighbors. As both the impinging projectile and the emitted electrons lose energy and slow due to interactions with the media, information about these processes is required for a wide range of energies. An added complication is that in finite sized media some of the ionized electrons may reach

Computational Approaches in Molecular Radiation Biology
Edited by M.N. Varma and A. Chatterjee, Plenum Press, New York, 1994

the surface and escape into the vacuum. This removes some of the originally deposited energy from the media.

Understanding the energy deposition and redistribution processes is vitally important in the fields of radiation induced damage to biological tissue and materials. Information about electron transport and emission is essential for designing systems where plasma-surface interactions occur, e.g., in thermonuclear reactors or in gas discharges, or where surface charging effects play a role. Electron emission from surfaces also plays a significant role in areas where accurate measurements of beam currents are required or for situations where particle detectors must be designed or used.

Modeling Needs

The philosophy behind our studies at the Pacific Northwest Laboratory is that models used to describe energy deposition processes require two types of experimental information. First, and foremost, they require as detailed information as possible about individual particle-media interactions. These data constitute the input parameters upon which the model is constructed. Secondly, and of equal importance, they require accurate experimental measurements for testing assumptions required when "other than ideal" systems are modeled, i.e., data are required to ascertain the reliability of the assumptions and methodology used.

The type of information required as input parameters is illustrated in Fig. 1 where an impinging particle, P, possessing charge, q, and velocity, v_p, interacts with a target, T. Generally a single electron is ejected; its velocity and direction with respect to the incoming projectile velocity being v_e and Θ respectively. A target ion, capable of "chemically" interacting with its neighbors or recombining with other electrons or ions, is also produced. After the collision the ion charge is q', which may be the same, larger, or smaller, than q. The final ion velocity, v_p', is always smaller than v_p. This constitutes the original energy deposition process.

A partial redistribution of the deposited energy occurs when the ejected electron undergoes subsequent elastic and inelastic interactions with other bound electrons in the media. Elastic collisions only alter the direction of travel whereas inelastic collisions produce additional ionization of the media. These inelastic processes are identical to those diagrammed in Fig. 1 with, of course, the modification that $q' = q = -1$. Such processes

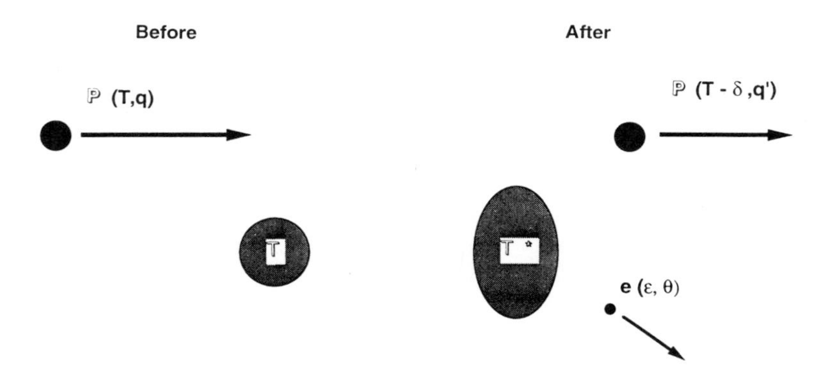

Figure 1. A schematic diagram or an individual ionizing interaction where initially the projectile, P, has energy, T, and charge, q. As a result of the interaction, the projectile energy decreases slightly is $T - \delta$, and its charge changes to q', where q' may be larger, smaller, or equal to q. An electron is ejected with energy E and direction T and an excited target, T^*, is produced.

continue until the original particle and all the ejected electrons come to rest or escape the media.

It should be mentioned that although energy deposition processes typically are expressed by a single electron being ejected to the continuum, i.e., single ionization, gas phase studies[1] have shown that the ejection of additional electrons is sometimes quite probable. Gas phase studies have also shown[2] that electron emission from interactions involving an increase or decrease of the projectile charge can also be quite important. For modeling purposes, multiple ionization or q changing interactions may be particularly relevant since they deposit different amounts of energy at localized sites and most likely produce different target ion species. This emphasizes that energy deposition models should treat the process in a stochastic rather than in an "average" manner.

Thus modeling the energy deposition may be cumbersome, but it is a very straightforward procedure once one has sufficient information about the individual particle-media interactions. Unfortunately, within a condensed-phase media these individual interactions simply cannot be experimentally investigated in such detail. Therefore a common method has been to use information obtained from gas-phase energy deposition measurements and to add appropriate modifications to account for bulk properties that are absent in the gas-phase data, e.g., to use condensed-phase differential oscillator strength information to modify the gas-phase cross sections in the low-energy regime. (see, for example, ref. 3). These modified data can then be used as input parameters and the modeling process proceeds as usual. But this requires various assumptions and the validity of those assumptions needs to be tested.

Testing the assumptions requires information about electron production and transport in condensed-phase media, which implies that measurements of the electron emission from the media can be used for this purpose. However the complication is that electrons are produced at varying depths within the solid and hence they suffer varying degrees of energy degradation as they travel toward the surface. Also they must be transmitted through the solid-vacuum interface in order to be observed. Hence the measured electron intensities are a convolution of the electron production, electron transport and the electron transmission probabilities associated with the media; all these pieces of information may not be available or easily obtainable for the particular media of interest.

A method of circumventing this problem is to test the model and its assumptions by studying "ideal" condensed-phase systems, meaning systems that can be modeled with sufficient accuracy and also be experimentally interrogated under well specified conditions. Sufficiently detailed measurements of the electron emission from such systems then provides a means of testing energy deposition models and may also provide insights into improving the models. To date we are aware of no detailed comparisons of this type, primarily because accurate experimental data are lacking.

Experimental work designed to supply these needed data is in progress at the Pacific Northwest Laboratory. However, prior to describing our studies, we will first provide a general overview of what is already known from previous electron emission studies. Due to limited space and available data, we will concentrate on electron emission resulting from fast ion (typically proton) impact on thin metal foils. We wish to emphasize that the examples cited are chosen primarily because we are more familiar with these studies; we do not intend to imply that they represent the only investigations in this field.

Previous Studies

Experimental studies of electron emission from condensed-phase targets are typically performed using thin, self-supporting metal foils in order to ensure that the impinging proton energy remains roughly constant during transit. As will be shown, the electron emission is extremely sensitive to the surface conditions, so only data involving reasonably clean or known surface conditions will be discussed here.

The most prevalent studies are of the total electron yield per incident particle, γ;[4–8] some studies of the forward versus backward emission, γ_f and γ_b have also been performed.[9–12] More detailed information is available from electron energy,[5,10,13–14] N(ε), and angular,[15–16] N(Θ), distribution measurements. However, for testing purposes we strongly believe that doubly differential energy and angular distributions, N(ε,Θ), are required since they will provide the most stringent tests. For example, total electron emission studies may provide certain insights into the overall process but using them for testing purposes is similar to evaluating various mathematical descriptions of a square and a triangle by only comparing their calculated surface areas with that of a test object. More highly differential tests are obviously required.

A few studies of N(ε,Θ) have been reported for "clean" surface conditions[8,16–23] but these rarely include full angular and energy information for the emitted electrons. From these and other studies, information about how the electron emission depends on the proton velocity, v_p, target z, target temperature and surface conditions is available.

a) Gaseous vs. Condensed-Phase Electron Emission. Since modified gas-phase data are sometimes used to provide the ionization probabilities in track structure codes, we begin by comparing the electron emission from gaseous and condensed phase targets. The only direct comparison of gas- and condensed-phase electron emission data of which we are aware is shown in Fig. 2. It is well known that bulk properties influence the production of low energy electrons and that impurities on the foil surfaces strongly influence the transmission of low energy electrons but these influences are believed to play minimal roles for higher energy electron emission. For that reason, Toburen[24] felt justified in comparing the electron emission from a thin carbon foil obtained using standard high vacuum conditions (approx. 10^{-6} Torr) with the emission obtained from a tenuous methane gas target.

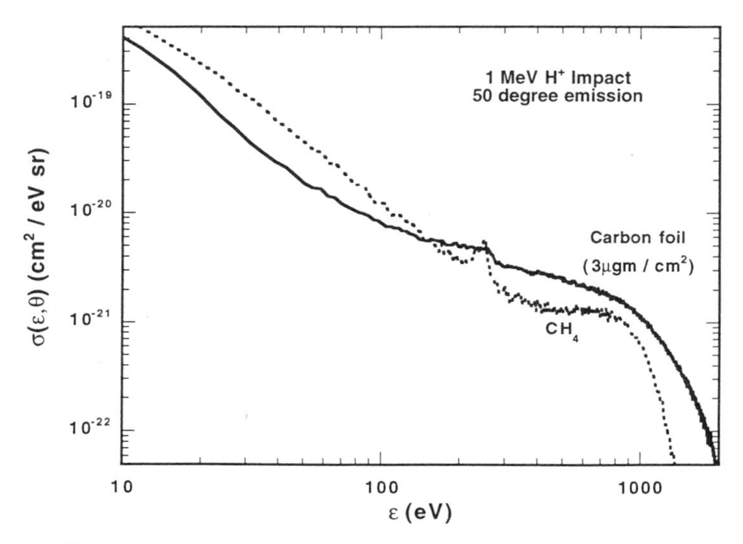

Figure 2. Doubly differential electron emission cross sections compared for 1 MeV protons interacting with a tenuous methane target and a thin carbon foil, ... and ———, respectively. The data are taken from ref. 24.

Because of uncertainties in the foil thickness, the carbon foil cross sections are uncertain within a factor of two. However the important differences between the gas- and condensed-phase data are a suppression of low energy electrons and a dramatic increase in the high energy electron production observed for the foil target. These are attributed to absorption of low energy electrons and scattering of forward directed high energy electrons to larger angles, respectively. Contamination of the foil surfaces and bulk effects undoubtedly also influence the electron emission.

b) Electron Emission from Thin Foils. We now discuss general features and what is already known about electron emission from thin foils. From an experimental viewpoint, the most important point in such studies is to provide a clean surface since the electron emission is extremely sensitive to surface contaminants residing on the surface. Surface contamination can be a combination of oxides chemically attached to the surface or adsorbed and occluded gases from the residual vacuum in the experimental apparatus. These contaminants tend to lower the work function of the surface;[26] thus as they are removed the emission of lower energy electrons decreases. (see Fig. 3) As can also be seen, removal of these impurities causes the maximum in the electron yield to shift to higher energies and the total yield of electrons to decrease dramatically. In addition, the width of the low energy maximum increases with the removal of impurities.[13] Therefore, unless the surface conditions are extremely well specified and controlled, comparisons between experimental data and theoretical predictions are meaningless.

With this in mind, the following generalities are based upon studies performed where the surfaces were quoted as being "clean." In many cases this was defined as "less than 0.3 monolayer."[8] Beginning with the total electron emission, for proton impact it has been shown,[4,7] as expected,[27] that the total yield of electrons is proportional to the stopping power of the primary ionizing particle. For modeling purposes, however, it is important to remember that γ is target and projectile dependent[9] and that γ increases in magnitude dramatically as the surface becomes contaminated.

Investigations of the forward and backward electron emission yields for fast protons impacting on thin carbon foils have shown[11] that approximately 15% more electrons are emitted in the forward (direction of exiting proton) than in the backward direction. Since

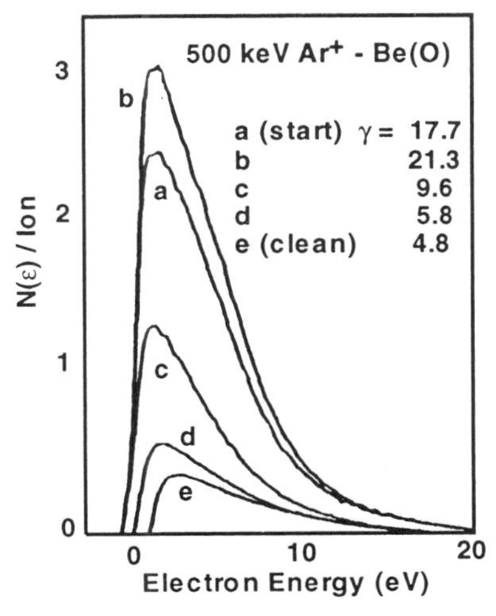

Figure 3. Electron yield, N(E)/incident ion, as a function of emitted electron energy and surface cleanliness for 500 keV Ar^+ - Be(O) interactions. Also listed are the total electron yields, γ. Data are taken from ref. 13.

low energy electron emission is the primary contributor to γ, these features may prove to be useful in testing assumptions about condensed-phase modifications to existing codes.

These forward/backward emission studies might also be used in a more unique fashion. For example, information about the range and ionization induced by electrons traversing condensed-phase media has been obtained from investigations of the forward and backward electron emission yields using fast ionic projectiles containing loosely bound electrons, e.g., H°.[11] Since the loosely bound electron is stripped from the projectile within a short distance after entering the foil, it can scatter back in the direction of the incoming beam or can travel in the forward direction causing additional ionization. If the foil thickness is less than the range of this electron, it, as well as any additional electrons produced along the way, will exit the foil. This increases γ_f as shown in Fig. 4. For thicker foils, these electrons cannot escape and the yields for H° and H^+ impact are identical. Since the stripped electron initially has the same velocity as the incoming beam, varying the beam energy offers a unique method of studying electron transport in foils.

Investigations of the low energy portion of the emitted electron energy spectra have shown that the shape is a function of the target material.[13] However for light ion impact, the position where the emitted electron yield maximizes and its width were found to be independent of the projectile impact velocity. For heavier projectile impact, the situation is different. The emitted electron energy distributions were wider and shifted to higher electron energies. Also the widths of the distributions increased with impact velocity for heavy ion impact. These changes as a function of projectile mass or velocity almost certainly are the result of differences in the initial energy deposition processes; hence they might be useful in testing this aspect of models.

The angular dependence of the emitted electrons might also be useful in testing the initial energy deposition processes since a cosΘ dependence (relative to the surface of the foil) is expected for <u>isotropic</u> source conditions. This dependence has been confirmed[17] for

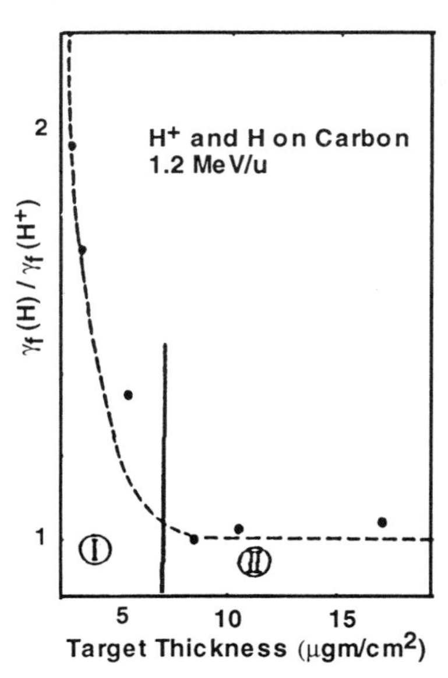

Figure 4. The forward electron emission, γ_f, for H atoms relative to that for protons passing through carbon foils. The curve serves only to guide the eye. Data are from ref. 11.

higher energy electron emission but, as will be shown later, it does not apply for low energy electron emission.

For testing theoretical models, we believe that doubly differential electron emission measurements should be used. However, from an experimental viewpoint, it is essential that these data be obtained under well specified conditions. The carbon foil data shown in Fig. 2 were obtained under standard high vacuum conditions where surface contamination was present but it was assumed that the higher electron energy data would be uneffected by any contamination. In Fig. 3, low energy electrons were indeed shown to be strongly influenced by surface contamination.

Figure 5 demonstrates this again for zero degree electron emission from sputter cleaned and untreated copper foils.[23] Figure 5 also demonstrates that removal of surface contaminants also leads to an increase in the yield for rather energetic electrons, electrons having velocities as large as the projectile velocity in the example used. This increase cannot be explained in a simple manner but may have ramifications concerning theoretical models.

The zero degree electron emission data in Fig. 5 also contain a strong convoy electron peak at approximately 650 eV. These convoy electrons might prove to be useful in providing information about electron transport and energy degradation in condensed-phase media since they are born within the media with rather well defined velocities and then suffer inelastic interactions prior to escaping the media.[19] Additional information about convoy electrons is contained in references 18–21.

Figure 5 also demonstrates that the differential electron emission depends on the target temperature. A close inspection of the data indicates the low energy electron emission decreases with decreasing target temperature, a feature that again emphasizes that experimental conditions must be extremely well defined if the data are to be useful for testing purposes.

Present Work

These studies have demonstrated that the electron emission is strongly influenced by surface and bulk properties and that low energy electron emission is particularly sensitive to bulk properties. Thus experimental conditions, particularly regarding low energy electron emission, must be extremely well known and controlled in order to provide data capable of testing and improving energy deposition models. The lack of doubly differential electron emission data of sufficient quality for testing track structure models is currently being

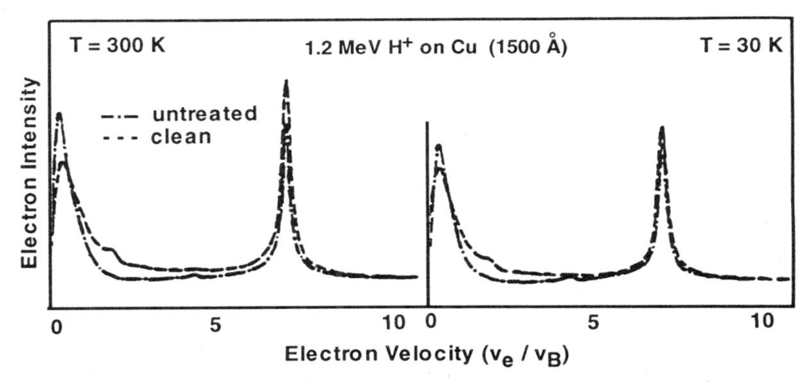

Figure 5. Zero degree electron emission spectra for protons impacting on sputter cleaned and untreated copper foils. The left and right sides of the figure are obtained for different target temperatures. v_B is the Bohr velocity. Data are from ref. 23.

addressed at PNL. We are studying the energy and angular distributions of electrons emitted from thin foils subjected to fast proton irradiation.

Initially emission from thin carbon foils is being studied since self-supporting carbon foils are readily available commercially and other experimental information for proton-carbon foil systems exists (see above). Electron time-of-flight (TOF) techniques are employed since they are particularly well adapted to studying low energy electron emission. Likewise extreme care is taken to avoid stray electric and magnetic fields within the target and spectrometer regions which may influence the low energy electron emission.

Unfortunately, our initial studies were hampered by surface cleanliness conditions. We had anticipated that the foils could be sputter cleaned in a similar manner to that used previously.[8] But we were unable to achieve reproducible emission spectra which means that either our technique was in error or the electron emission at a specific angle is more sensitive to surface contamination than is the total yield of electrons. Our initial studies also suffered from restricted base (and operating) pressures of $1\text{–}2 \times 10^{-9}$ Torr arising from a damaged sealing surface. These limitations are currently being addressed in the next phase of the work.

Data presented here were accumulated in the following manner. Energy and angular information were obtained by interacting a pulsed proton beam (intensity $< 10^{-11}$ A) with a thin (5 $\mu g/cm^2$) carbon foil and measuring the emitted electron flight times, i.e., energies, using a rotatable detector. Experimental parameters allowed us to measure electron energies up to 100 eV with reasonable accuracy and higher energies with ever decreasing accuracy. Emission angles relative to the proton beam that could be investigated were 15–75° and 105–155°, the foil surface being normal to the proton beam. Additional details are available in ref. 28.

Since our attempts to sputter clean the foils using the unpulsed proton beam were unsuccessful, we heated the foils to between 200 and 300 degrees C for approximately 24 hours in order to remove adsorbed gases. However, we had no direct method of confirming the surface cleanliness that resulted. Rutherford backscattered protons indicated a 2–3% oxygen contamination of the foil after baking; whether this represents surface or bulk contamination is uncertain.

After baking, the foil was removed from the oven and a TOF spectrum accumulation began within a few seconds. Sufficient statistics were achieved within approximately 100 seconds, the data were stored, and a new spectrum was measured. Using this procedure, we obtained "snapshots" of how the buildup of surface contamination influences the electron emission. Results are shown in Fig. 6.

Two major features are seen. One is that surface contamination causes the low energy electron emission to increase in magnitude which broadens and shifts the energy distribution to lower energies. The shift is in agreement with the electron emission data integrated over all angles as shown in Fig. 3 but the broadening is opposite to that shown. The second feature is the decrease in the yield for high energy electrons with increasing surface contamination. This is in agreement with data shown in Fig. 5.

Assuming unit sticking coefficients, the buildup of H_2 contamination on the surface was calculated and is shown in Fig. 6. The assumption used is that the surface was initially clean, although this could not be directly verified. Nevertheless, the data clearly indicate that doubly differential emission is sensitive to even a few percent of a monolayer of surface contamination. If subsequent experiments using foils where we are absolutely certain of the initial surface conditions confirm this, it casts doubt upon the surface conditions used in previous measurements. For example, Hasselkamp et al.[4] stated that they sputter cleaned their targets and interrogated them within 30 minutes. With their quoted base pressure of 5×10^{-9} mbar, this implies a surface contamination of roughly half a monolayer. Burkhard et al.[8] claim their surfaces to be clean to better than 0.3 monolayers of contamination.

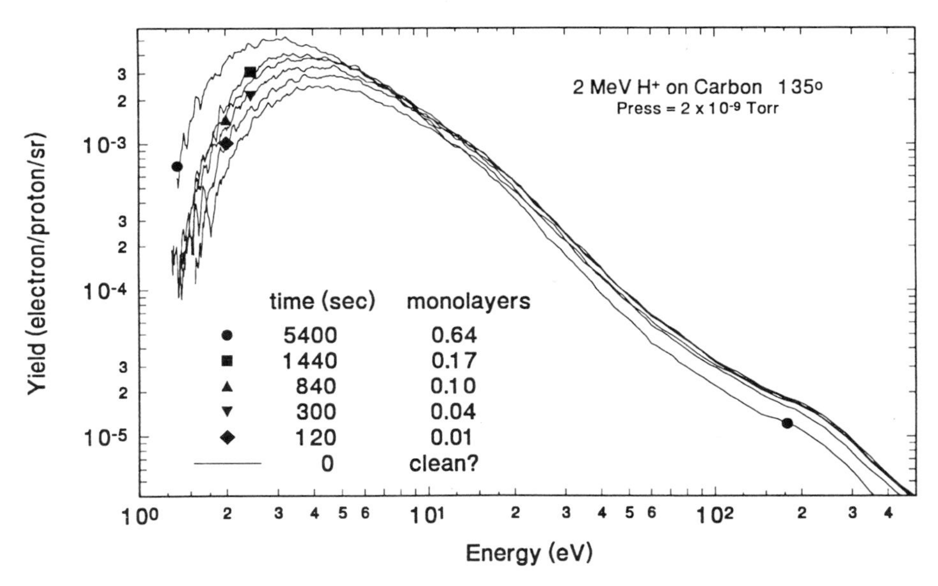

Figure 6. Electron yields measured at 135° with respect to the beam direction for 2 MeV protons interacting with a thin carbon foil. The spectra were taken as a function of time after the foil was removed from an oven used to remove surface impurities. Assuming initially clean surface conditions, the buildup of H_2 on the surface was estimated.

Figure 6 demonstrates that such levels of contamination seriously alter the low energy electron emission yields which should also effect γ. We stress, however, that our findings may be subject to our initial surface conditions which we are unable to specify. We also point out that we may be observing a temperature dependence similar to that shown in Fig. 5 since the foil was cooling from oven to ambiant temperatures during spectra accumulation. We doubt that temperature plays a role, however, since temperature dependent features extend[8] to higher electron energies than where we observe changes in the spectra.

In Fig. 7 we show the electron emission yields as a function of angle of emission. The foil was again baked as described above, but a comparison of the 135° data in Figs. 6 and 7 shows differences in the time zero spectra. If the initial surface conditions in Fig. 6 are indeed contamination free, the comparison implies that approximately 0.17 monolayer of (presumably) hydrogen contamination is initially present in Fig. 7. Assuming this to be the case, the buildup of contamination from background gases is again calculated and the total contamination is tabulated in Fig. 7.

Also shown are the relative intensities as expected from a $\cos\Theta$ angular dependence. For electron energies above 30 eV this $\cos\Theta$ dependence appears to be correct; however lower energy electron emission exhibits a much stronger dependence. This may arise in part from the suppression of low energy electrons emerging more parallel to the foil surface.

Thus, the present data provide two new pieces of information. They indicate that surface contamination levels above a few percent of a monolayer strongly influence the low energy electron emission yields and that the electron emission stabilizes when the surface contamination is between approximately 1/3 and 1 monolayer. Secondly, they show that a $\cos\Theta$ angular dependence does not apply for low energy electron emission.

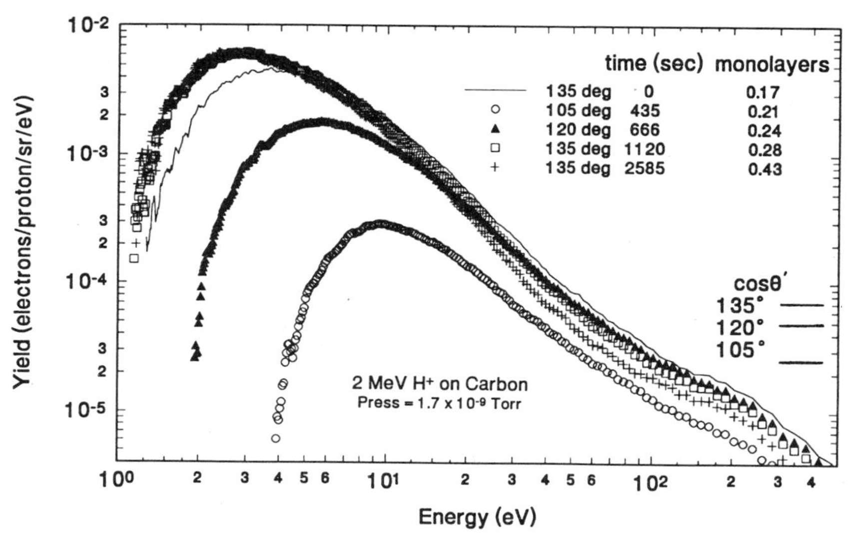

Figure 7. Electron yields as a function of angle of emission for 2 MeV proton-carbon foil interactions. The spectra were obtained at different times after removal from an oven used for cleaning the surfaces. Comparing the initial spectra with that in the previous figure indicates the initial level of surface contamination is approximately 0.17 monolayer. Additional H_2 buildup from the residual vacuum is calculated and shown. Also shown are the relative intensities expected from a cosΘ' dependence where Θ' is measured normal to the foil surface.

Summary and Suggestions for Future Studies

Experimental data required for modeling energy deposition and transport in condensed-phase media have been discussed. For testing purposes, measurements of the electrons emitted when thin metal foils are bombarded by energetic particles are desirable. Past studies were discussed in order to show what is already known from such studies and the need for detailed electron energy and angular distribution information is stressed. Experimental work in our laboratory designed to provide such information was discussed.

The data presented emphasize that electron emission measurements useful for testing energy deposition models must be obtained under extremely well controlled, and defined, surface and bulk conditions. Hence, additional studies are in progress in our laboratory where improved vacuum and surface cleaning procedures will provide known, and reproducible, surface conditions. A combination of electron time-of-flight and electrostatic energy analysis will be used in order to investigate the entire emitted electron spectrum. Simultaneous multiple-angle detection of the electrons is planned in order to investigate the angular and energy distributions under identical target conditions.

Future studies will employ condensation of gases on a cooled surface or in situ vaporization of metals on a clean surface to investigate electron production and transport in precisely know condensed-phase media. These improvements are intended to provide data having sufficient quality and details that should provide the first real test of track structure models.

ACKNOWLEDGEMENTS

This work was supported by the Office of Health and Environmental Research, U.S. Department of Energy under contract No. DE-AC06-76RLO 1830.

REFERENCES

1. RD DuBois and ST Manson, "Multiple-ionization channels in proton-atom collisions," Phys Rev A 35, 2007-25 (1987).
2. RD DuBois, "Multiple ionization of He^+ - rare-gas collisions," Phys Rev A 39, 4440-50 (1989).
3. JA LaVerne and A Mozumder, "Effect of phase on the stopping and range distribution of low-energy electrons in water," J Phys Chem 90, 3242–7 (1986).
4. D Hasselkamp, KG Lang, A Scharmann and N. Stiller, "Ion induced electron emission from metal surfaces", Nucl Inst and Meth 180, 349-356 (1981).
5. D Hasselkamp, A Scharmann and N Stiller, "Ion induced secondary electron emission as a probe for adsorbed oxygen on tungsten," Nucl Inst and Meth 168, 579-583 (1980).
6. A Clouvas, H Rothard, M Burkhard, K Kroneberger, C Biedermann, J Kemmler, K-O Groeneveld, R Kirsch, P Misaelides and A Katsanos, "Secondary electron emission from thin foils under fast-ion bombardment," Phys Rev B 39, 6316-20 (1989).
7. RA Baragiola, EV Alonso and A Oliva Florio, "Electron emission from clean metal surfaces induced by low-energy light ions," Phys Rev B 19, 121-9 (1979).
8. M Burkhard, H Rothard, J Kemmler, K Kroneberger and K-O Groeneveld, "Surface characterization of thin solid foil targets by ion impact," J Phys D 21, 472-7 (1988).
9. S Hippler, D Hasselkamp and A Scharmann, "The ion-induced electron yield as a function of the target material," Nucl Inst and Meth B34, 518-520 (1988).
10. W Meckbach, B Graunstein and N Arista, "Secondary-electron emission in the backward and forward directions from thin carbon foils traversed by 25-250 keV proton beams," J Phys B 8, L344-9 (1975).
11. K Kroneberger, A Clouvas, G Schlüssler, P Koschar, J Kemmler, H Rothard, C Biedermann, O Heil, M Burkhard and K-O Groeneveld, "Secondary electron yields from the entrance and exit surfaces of thin carbon foils induced by penetration of H^+, H° and H_2^+ projectiles (1.2MeV/u)," Nucl Inst and Meth B29, 621-6 (1988).
12. H Rothard, K Kroneberger, M Burkhard, J Kemmler, P Koschar, O Heil, C Biedermann, S Lencinas, N Keller, P Lorenzen, D Hofmann, A Clouvas, K-O Groeneveld and E Veje, "Secondary electron emission from entrance and exit surfaces of thin clean foils bombarded with H^+, C^+, and O^+ ", Rad Effects and Defects in Solids 109, 281-6 (1989).
13. D Hasselkamp, S Hippler and A Scharmann, "Ion-induced secondary electron spectra from clean metal surfaces," Nucl Inst and Meth B18, 561-565 (1987).
14. D Hasselkamp and A Scharmann, "The ion-induced low energy electron spectrum from aluminium," Surf Sci 119, L388-92 (1982).
15. M Burkhard, H Rothard, C Biedermann, J Kemmler, K Kroneberger, P Koschar, O Heil and K-O Groeneveld, "Heavy-ion-induced shock electrons from sputter-cleaned solid surfaces," Phys Rev Lett 58, 1773-5 (1987).
16. H Rothard, K Kroneberger, M Schosnig, P Lorenzen, E Vehe, N Keller, R Maier, J Kemmler, C Biedermann, A Albert, O Heil and K-O Groeneveld, "Secondary-electron velocity spectra and angular distributions from ions penetrating thin solids," Nucl Inst and Meth B48, 616-20 (1990).
17. J Mischler, N Benazeth, M Nègre and C Benazeth, "Angular distributions of secondary electrons emitted in Ar^+ -polycrystalline Al collisions," Surf Sci 136, 532-44 (1984).
18. R Rothard, M Burkhard, C Biedermann, J Kemmler, P Koschar, K Kroneberger, O Heil, D Hofmann and K-O Groeneveld, "The influence of solid surfaces on low-energy convoy electron emission," J Phys C 21, 5033-9 (1988).
19. P Koschar, A Clouvas, O Heil, M Burkhard, J Kemmler and K-O Groeneveld, "Convoy electron response to the charge pre-equilibrium of ions in solids," Nucl Inst and Meth B24/25, 153-8 (1987).
20. M Burkhard, H Rothard, C Biedermann, J Kemmler, P Koschar and K-O Groeneveld, "Strong convoy electron yield dependence on surface properties," Nucl Inst and Meth B24/25, 143-6 (1987).
21. RO Barrachina, AR Goñi, P Focke and W Meckbach, "Calculated convoy electron distributions due to electron loss collisions inside solid targets," Nucl Inst and Meth B33, 330-3 (1988).
22. S Suarez, GC Bernardi, P Focke and W Meckbach, "Double differential spectra of electrons emitted from the forward and backward surface of a carbon foil traversed by a beam of hydrogenic projectiles," Nucl Inst and Meth B33, 326-9 (1988).
23. H Rothard, "Ioneninduzierte elektronenemission von kontrollierten Oberflächen dünner Festkörper und von Hochtemperatur-Supraleitern," Ph.D. Thesis, Univ. of Frankfurt/M (1990).
24. LH Toburen, "Angular and energy distributions of electrons emitted from gases and thin foils during light ion bombardment," Scan Micro Suppl 4, 239-256 (1990).
26. D Hasselkamp and A Scharmann, ...Vak-Tech 31, 242 (1982).

27. J Schou, "Secondary electron emission from solids by electron and proton bombardment," Scan Micro 2, 607-632 (1988).
28. C Drexler, "Messung doppelt differentieller protoneninduzierter Sekendärelektronenausbeuten von Xenon-Gas und einer dünnen Kohlenstoffolie," Diplomarbeit, TU-München (1992).

DISCUSSION

Hamm: Are these absolute comparisons between gas phase and foil data?

Du Bois: The main point I want to stress is that the major difference for foil data is a suppression of low energy electron emission. This occurs partially because of electron transport through the medium. A second difference is the increased yield for higher energy electrons from the condensed phase. This is interpreted as scattering to a larger angle the high energy electrons emitted in the forward direction. That's all I really wanted to say about these data.

Ritchie: These are all electrons that come out?

Du Bois: Those are electrons emitted at zero degrees for 1.2 MeV protons impacting on a clean carbon foil 1,5000 angstroms thick.

Paretzke: You don't have the full surface covered. What are the surface effects?

Du Bois: Take a simple picture of a foil having a crystalline structure with 100 surface atoms. Now add four percent of a monolayer containment on the surface. Looking at the amount of surface covered you can probably convince yourself that this isn't going to influence the emitted electrons too much. But increase the coverage to 13 percent of a monolayer and it's not too hard for me to convince myself that the electron emission is going to be strongly influenced.

Ritchie: Have you seen shock electrons in any of your work?

Du Bois: I have not; the Frankfurt group has. We don't have the experimental capabilities in our lab. For those who do not know what shock electrons are, if you inject a fast heavy ion into the surface, you produce a shock wave, and an electron can be emitted perpendicular to this wave cone. Thus you have electrons created with well-defined energies and directions, and you can study those emitted from the surface. Note, they're born in the bulk. The unique thing is their energy, which is related to the velocity of incoming projectile in the few eV range, not hundreds of eV, so adjusting the velocity of the incoming beam may provide a method for studying electron transport of low energy electrons.

Ritchie: Are you familiar with the work of Pijper and Kreuit in Amsterdam?

Du Bois: No.

Ritchie: They have taken thin carbon foils and used kilovolt electrons in a coincidence experiment. They look at the secondary electrons in coincidence with the fast electron, peak in the electron spectrum at the surface plasmon energy, which is even stronger than the peak at the volume plasmon.

Paretzke: How did they clean the surface?

Ritchie: I think that they worked in the very high vacuum region, but I don't know the details.

Paretzke: Is a vacuum of 10^{-9} torr good enough?

Du Bois: It is not good enough.

Paretzke: Not for this study, but many people call this UHV, and that's right. What kind of problems can one run into with these experiments?

Du Bois: Now, I will state that there is an inherent problem with these type of studies. I'm a firm believer they would be done with thin, self-supporting structures, but the trouble is the variet of thin self-supporting structures is limited. Rupert in Stockholm, provided me with thin aligned DNA--I'll call them foils -- I attempted to take data with those but it was not very promising. It was very difficult to support the DNA as a foil. It appeared that the beam vaporized the DNA. I could obtain no reproducible results. I do not hold much hope for following that technique. Our best hope may be freezing, I'll say DNA, but let's just say freeze large molecules on a cold surface.

Zaider: Why do you have to freeze them?

Du Bois: To hold them in a condensed phase.

Nikjoo: What thickness were they?

Du Bois: If I remember, roughly, 1,000 to 2,000 angstroms. They were thin enough for our projectile beam to pass through easily. But the results were not promising. Because I showed data for metal foils, another major problem that we encounter for insulators is how does surface charging affect the emission? We are really interested in the low energy electron emission. That's where I say that what one really needs to do, particularly in a group like this, is select an "ideal" system to model and one that is experimentally feasible. Generate the best data one can and use that for testing the various codes.

Varma: But what I'm saying is that you will get a clean surface, if you continuously sputter the surface.

Du Bois: Surface charging may be a solved problem in the 10's of eV range but I don't agree in the few eV range. I don't know of any surface physics studies that investigate in the few eV range.

Varma: I don't know about few eV. I said only to the 10's of the eV range.

Du Bois: I will agree with that.

Varma: And the second point that you talked about was ideal system. What is an ideal system?

Du Bois: I will list one, it has good phonon structure.

Varma: How would that relate to the biology?

Du Bois: I admit that problem. However, for testing purposes it is ideal.

Varma: If you tell me that you can understand aluminum, then how do you translate that to the understanding of the biological system? That's a valid question to ask.

Du Bois: In my opinion, in testing the codes, there are two things you want to test (1) the assumptions and (2) the parameters used. Now, in an ideal system, if one has the input parameters, then what you're really doing is testing the assumptions. Those assumptions don't change when you transfer from aluminum to DNA. Then it's the input parameters that become an important factor.

Wilson: The primary attractiveness of aluminum oxide is that it's well-characterized from a physicist's standpoint; I would agree, it's not very biological.

Varma: It's like looking where the light is without looking at the problem. And I think we get ourselves into a problem by doing that.

Paretzke: That's why we plan to have the cold target test, but we can measure only reflections, but it's the only solution we have.

Ritchie: Getting back to physics again, are you familiar with a work of Yamazaki and Komaki on multiple emission of secondary electrons from thin carbon foils?

Du Bois: That's done with mostly heavy ions, if I recall.

Ritchie: With protons and also with He ions

Du Bois: I'm only familiar with studies using heavier ions.

Ritchie: Well, they've done that too, but mostly using protons. They collect all the electrons emitted from the foil. They collect one electron, two electrons at the same time, three electrons at the same time, etc. They find peaks in which many, many electrons are collected at the same time.

Du Bois: The work I'm familiar with along that line is done by Winter in Vienna.

Varma: In the past, in our program Powell did a lot of work with the foils of carbon and gold, and I think there was a tremendous amount of data generated. I don't know whether any of the data is useful or not. He has got a very big group on surface physics at the National Institute of Standard and Technology (NIST) where they are doing these measurements routinely.

Du Bois: They are typically looking at the Auger spectrum.

VarmaThey are looking at all the electrons that are going out, they're doing the cross-section measurement. You may want to use some of that information. Now, about this ideal system, you say that aluminum is the one. Can we have some kind of help from the chemists and biologists to identify an ideal system that should be looked at? Such as DNA?

Du Bois: I saw no hope for self-supported DNA.

Varma: But you could put it on a cold finger and do the experiment.

Du Bois: The intention is to put in a cold finger. The questions here would be, what one freezes on and is it better to use a cold finger and condensed vapor or vaporization of some metal?

Nikjoo: Optically flat and very high quality uniform polymer films can be produced by 'dip costing' method. The film is created by passing a vesicle substrate through a solution of the polymer. A thin film of the polymer solution coats the substrate and dries to form the film. Film thickness can be varied with the speed of withdrawing the slide from the solution and depends on viscosity and temperature. The uniformity of the films can be examined by "Copacitana method" VAMFO (Variable Angle Monochromatic Fringe Observation), Michelson Interferometer and electron microscopy.

Du Bois: Are they ultimately self-supporting?

NIKJOO: Yes

Chatterjee: In biology you're talking about thick target. You're not talking about thin target, and the targets are highly structural.

Paretzke: Conformational changes wouldn't influence secondary electron cross-sections to a large extent. Large means within five percent or so.

Chatterjee: Not conformational. I'm talking about the compactness of the chromosome. The DNA intertwining with each other and it's influence on electron spectrum. I wouldn't expect that this would change the ejection cross-section to a large extent.

Varma: I think this point is that after doing what you suggest, how does it help us in terms of biological understanding?

Paretzke: We can argue that even with a thick target or a thin target, it doesn't matter as much. Polyatomic molecules which have been studied to a large extent in the gas phase.

Chatterjee: An electron will go many places before they come out.

Zaider: Yes, but that's a transport problem, that's not a cross-sectional problem.

Hamm: You can get a self-consistent situation here where you measure the emission and then you do your own transport in that thin film. And if you get self consistency, then you are happy about the emission.

Paretzke: We can make errors which compensate each other, but it helps you at least to be consistent.

Chatterjee: So the conclusion is that if you have a thick target, you can reconstruct by bringing in the transport aspect of the electron.

Paretzke: Yes.

Du Bois: But from an experimental viewpoint you have to specify conditions extremely well.

Chatterjee: I think I know what you are trying to say about the aluminum. Aluminum is an ideal system where you can check your theoretical assumptions and things of that nature. However, if you can go into a DNA type of system, it would be much more direct and helpful in terms of the applications in biology.

Du Bois: It appears the closest we come to that is freezing.

Curtis: What is known experimentally about oscillator strengths in DNA?

Hamm: The one I was involved with was a reflection measurement on a dry film of DNA.

Paretzke: It was a few years ago.

Hamm: It was fairly thick.

Paretzke: It would have to be.

Hamm: It would have been optically thick.

Inokuti: Certainly, Linda Painter and others were very good at that time, but it has all been since '74. And after that, you see many things in techniques have developed, and I would think much more should be done now.

DIRECT IONIZATION OF DNA IN SOLUTION

J.H. Miller and W.E. Wilson

Pacific Northwest Laboratory
Richland, WA 99352, USA

R.H. Ritchie

Oak Ridge National Laboratory
Oak Ridge, TN 37831, USA

ABSTRACT

Most of the energy absorbed in the cell nucleus from a radiation field goes into the aqueous medium that surrounds macromolecules, like DNA, which are critical to the normal function of cells. This part of the energy deposition produces numerous reactive species that can diffuse to DNA sequences and induce chemical changes. The average diffusion distance of the free radicals that mediate this indirect mode of DNA damage is only a few nanometers because the cellular medium contains a high concentration of molecules that rapidly scavenge the radiation-induced species. Under these conditions, direct interaction of the radiation field with DNA can not be neglected as a potential mode of damage induction. Two aspects of the direct effect will be discussed in this paper: (1) screening of the interaction between DNA and charged particles by the dielectric response of the aqueous medium and (2) the impact-parameter dependence of these interactions.

INTRODUCTION

Technical advances in molecular biology have made it possible to construct chemically well-defined DNA targets (oligonucleotides, plasmids, mini-chromosomes, etc.) for radiation studies. Use of these targets in traditional radiolysis experiments, (dilute aqueous solutions exposed to low LET radiation) has the potential to reveal many interesting aspects of the interaction of macromolecules with homogeneously distributed free radicals that escape from the tracks of charged particles slowing down in the aqueous medium. As the radical scavenging capacity of the medium increases, the contribution to DNA damage from homogeneously distributed free radicals decreases and a component of damage emerges that is more strongly dependent on the structure of particle tracks. *In vitro* model DNA systems are believed to mimic the cellular medium when the scavenging capacity is of the order of 10^9 sec^{-1}. Under these conditions, DNA damage is generally considered to arise from three

Computational Approaches in Molecular Radiation Biology
Edited by M.N. Varma and A. Chatterjee, Plenum Press, New York, 1994

sources: (1) attack by homogeneously distributed free radicals, (2) spatially correlated free-radical attack, and (3) direct damage.

This conceptual basis for understanding the induction of DNA damage under conditions that may be similar to those that exist in the cellular medium has motivated some effort to resolve the three pathways experimentally. The contribution to DNA damage from attack by homogeneously distributed free radicals can be estimated by fitting a model of the competition between the DNA and radical scavengers for reaction with randomized free radicals to experimental yields of a radiolysis product (e.g., single strand breaks) as a function of the scavenging capacity of the medium. Application of this procedure by Schulte-Frohlinde and coworkers[1] suggests that about 60% of the single-strand breaks induced in naked plasmid DNA by exposure to low-LET radiation under radical-scavenging conditions comparable to the cellular medium are from randomized OH radicals. If this interpretation of the data is correct, then traditional radiolysis experiments coupled with advanced target preparation are highly relevant to radiation biology because they permit a detailed investigation of DNA damage by spatially uncorrelated free-radical attack under conditions where other modes of damage induction can be neglected. However, the only track information needed for interpretation of these experiments is the yield of water radicals that escape from track entities.

Schulte-Frohlinde and coworkers[1] used calculations by Schuler[2] of the probability to scavenge radicals from spurs to estimate the fraction of single-strand breaks induced by spatially correlated OH radicals under conditions that mimic exposure of DNA in cells to low LET radiation. This approach suggests that OH radicals in spurs and direct action are about equally effective in producing DNA damage with each responsible for about 20% of the strand breaks observed at high radical scavenging capacity. These semiempirical results need to be tested by comparison with calculations of the direct effect that are based on cross sections for impact ionization of DNA by charged particles in the radiation field.

Recent experiments by Wheeler and coworkers[3] on the yield of gamma-ray-induced base release as a function of DNA hydration suggest that ionization of water in the first solvation shell does not lead to hydroxyl radical formation with the same efficiency as is the case for ionization of bulk water. Some water molecules of the first hydration layer appear to be able to transfer charge to DNA when they are ionized. Thus, it appears that interpretation of experimental data from *in vitro* model systems designed to mimic DNA damage *in vivo* would benefit not only from cross sections for direct ionization of DNA but also from a semiclassical representation of these cross sections that gives the probability of energy transfer as a function of impact parameter. The latter will allow track simulation procedures to select which of the components of the heterogeneous system (bulk water, water of hydration, or DNA) is to be ionized in a given collision event. From the semiclassical theory developed by Bohr,[4] one expects that components that are close to the particle trajectory and that have low-energy excitation modes with high oscillator strength will have the greatest probability of receiving energy from a fast charged projectile.

In this paper we will describe calculations of cross sections differential in the energy of secondary electrons for direct ionization of DNA in solution and the semiclassical representation of these cross sections as an integral over impact parameters. Calculations of this type, tested whenever possible by experimental data, should provide the input data needed for track-structure simulations of DNA damage by direct interaction with charged particles in a radiation field.

METHODS

The first Born approximation for the differential cross section to eject electrons from absorption band i of a homogeneous condensed medium by energy transfer $\hbar\omega$ from a fast particle with nuclear charge Z and kinetic energy E is[5]

$$\pi a_o n_i E \frac{d\sigma_i}{d(\hbar\omega)} = \int_{k_{\min}}^{k_{\max}} \mathrm{Im}[\varepsilon_i(k,\omega)] \frac{Z^2}{|\varepsilon|^2} \frac{dk}{k} \tag{1}$$

where a_o is the Bohr radius (0.0529 nm) and n_i is the density of electrons involved in the transition that gives rise to the ith absorption band. The integral is over the range of momentum transfer allowed by the conservation of energy and momentum in the collision. $\mathrm{Im}[\varepsilon_i(k,\omega)]$ is the contribution from the ith band to the imaginary part of the complex dielectric response function ε and is analogous to the generalized oscillator strength of an electronic subshell of an atom or molecule in the gas phase.

The extension of this theory to a heterogenous medium is trivial when the macromolecular solute is present in such low concentration that its effect on the bulk dielectric response function is negligible. In this case, i denotes an impurity band in the medium, $\mathrm{Im}[\varepsilon_i]$ is proportional to the generalized oscillator strength of the band, and $|\varepsilon|^2$ is the square modulus of the complex dielectric response function of the medium in the absence of the impurity. At this level of approximation, the impurity is treated as a gas-phase molecule with a solvation structure.

The effects of collective modes of excitation in the medium, if they are present, are contained in $|\varepsilon|^2$. The primary role of the bulk medium in the collision process can be interpreted as an effect on the electromagnetic field of the projectile. Localized electronic polarization of the medium by the projectile may give rise to the condition $|\varepsilon|^2 > 1$; then screening of the impurity site from the projectile occurs in a manner that is somewhat analogous to the effect of bound electrons on a projectile in the gas phase. Collective modes of excitation of the bulk medium by the projectile may be characterizes roughly by the condition $|\varepsilon|^2 < 1$. This anti-screening effect can be interpreted as an electronic polarization wave excited in the bulk medium that may excite an electron of the impurity site.[6]

The general approach to obtaining an impact-parameter representation of collision cross sections involves the application of Fourier-transform analysis to replace the component of momentum transfer that is perpendicular to the projectile velocity by its canonically conjugate variable, interpreted as an impact-parameter.[7] This procedure is mathematically simple only in the limit of small momentum transfers where the recoil of the projectile can be neglected (i.e., the component of momentum transfer that is parallel to the projectile velocity is fixed by conservation of energy). This level of approximation is generally acceptable for the following reasons: (1) the effect of projectile recoil on the collision cross section can be included approximately by replacing $\hbar\omega$ in the energy-transfer dependence of ε and ε_i by $\hbar\omega - \hbar k^2/2m$, and (2) energy transfer in collisions with large momentum transfer will generally be very close to the projectile's trajectory. In the large momentum-transfer regime the effects of screening are negligible and the binary encounter approximation[8] may be used.

If the energy transfer transform[6] is used to obtain an impact-parameter representation of the differential ionization cross section, then

$$\pi a_o n_i E \frac{d^2\sigma_i}{d(\hbar\omega)d(b/b_c)} = \int_0^{\infty} f(\kappa)\mathrm{Im}[\varepsilon_i] \frac{Z^2}{|\varepsilon|^2} \frac{\kappa d\kappa}{k^2} \tag{2}$$

where $b_c = v/\omega$ is Bohr's maximum effective impact parameter and

$$f(\kappa) = \frac{b}{b_c} K_0\left(\frac{b}{b_c}\right) J_0(\kappa b) + \kappa b K_1\left(\frac{b}{b_c}\right) J_1(\kappa b) \tag{3}$$

where J_0, K_0, J_1, and K_1 are Bessel functions of integer order, $\hbar\kappa$ is the component of momentum transfer that is perpendicular to the projectile's velocity, and $k^2 = \kappa^2 + (\omega/v)^2$.

Ab initio quantum methods (see the chapter by Dr. Zaider) can be used to calculate the dielectric response functions needed to evaluate differential cross sections in the first Born approximation. Application of these methods has been greatly facilitated by rapid advancements in computer technology but still requires assumptions about the wavefunctions of the system (e.g., translational invariance) that we feel are not appropriate in the applications considered due to the high level of disorder in biological systems. Hence, we have adopted a semiempirical approach that makes use of photoabsorption data[9,10] to determine the parameters of the Orthogonalized Plane Wave Dielectric (OPWD) model that is based on a tight-binding approximation for occupied electronic states and orthogonalized plane-wave continuum states.[11] These approximations lead to a simple analytic form

$$\mathrm{Im}[\varepsilon(0,\omega)] = 1024\,\frac{\pi}{3}\,\eta\,\frac{\alpha^7}{\omega_B + \alpha^2/2}\,\frac{p^3}{(\alpha^2 + p^2)^6} \tag{4}$$

for the optical limit of the imaginary part of the dielectric function. Equqtion (4) with $p = [2(\omega - \omega_B)]^{1/2}$ contains all of the adjustable parameters of the OPWD model which are (1) the excitation threshold, $\hbar\omega_B$, (2) the orbital exponent of the occupied state, α, and (3) the electron density, η. If more than one band is present in the absorption spectrum, then a sum of terms like Eq. (4) is fitted to the experimental photoabsorption data. After the parameters of the OPWD model have been determined from the optical limit, the full dependence of $\mathrm{Im}[\varepsilon]$ on momentum and energy transfer can be calculated from a somewhat more complicated analytic function.[11] The real part of the dielectric response function is evaluated numerically using the Kramers-Kronig[12] relationship. It should be noted that the OPWD model is constrained to satisfy the Bethe sum rule.

RESULTS

Figure (1) shows the result of fitting the liquid-water photoabsorption data[9] by a sum of four terms like Eq. (4). Theoretical methods based on sum rules were applied by Painter and coworkers[13] to extend the experimental data from 25.5 eV to 100 eV. The optimum parameter values are given in Table I. The parameters of the model are strongly coupled and the optimum values given in Table I are not unique. Hence, little, if any, physical interpretation should be given to their value. A similar approach to obtaining the dielectric response function of liquid water has been carried out by others.[14] An alternative scheme for representing $\varepsilon(k,\omega)$ for water based on the same optical data[9] was incorporated early on into the Oak Ridge Monte Carlo code OREC.[15]

Photoabsorption data of DNA[10] were multiplied by the ionization efficiency[16] to obtain the experimental photoionization data for DNA that are shown by the open circles in Fig. (2). The solid curve in this figure is a fit to these data by Eq. (4) with an ionization threshold of 10.6 eV, $\alpha = 0.875$ au, and $\eta = 0.0289$ au. Using these parameters to evaluate $\mathrm{Im}[\varepsilon_i(k,\omega)]$ in Eq. (1) and the parameters in Table 1 to evaluate $|\varepsilon|^2$, we obtain the

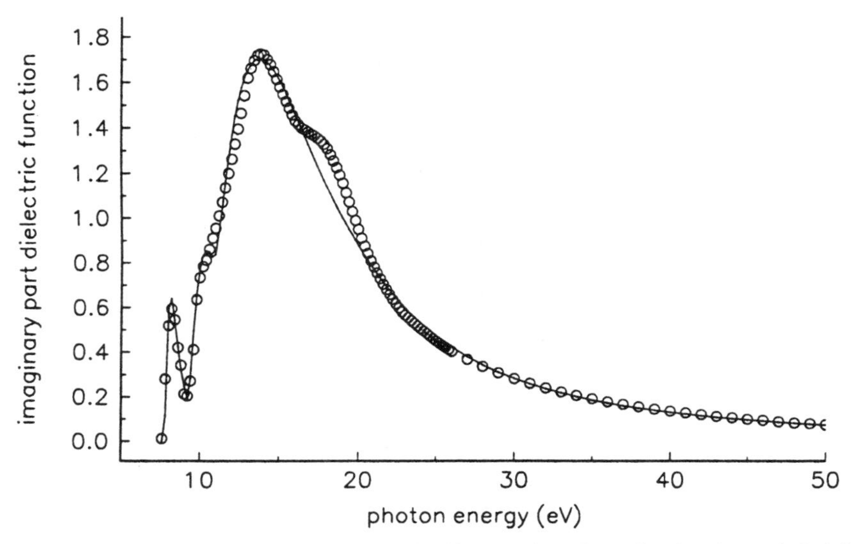

Figure 1. Fit of the OPWD model (solid curve) to liquid-water photoabsorption data (open circles) from Ref. 9.

differential ionization cross sections shown by the filled circles in Fig. (3) for a primary electron with a velocity of 3 au (122.5 eV). To illustrate the effects of the solvent, the open circles show the result obtained for the same primary- and secondary-electron energies but with $|\varepsilon|^2 = 1$. Clearly, the screening effect of the solvent dramatically reduces the cross section for ejection of low-energy secondary electrons and shifts the peak in the spectrum to a higher energy.[17] The small anti-screening effect for energy transfers between about 25 and 50 eV is consistent with the general finding that collective modes of excitation in liquid water have a broad energy spectrum and short lifetime.[18]

Impact-parameter probability densities calculated from Eq. (2) with the normalization $dP/db = (d^2\sigma_i/d\omega db)/(d\sigma_i/d\omega)$ are shown in Fig. (4) for ionization of DNA by 122.5 eV primary electrons. The following levels of energy transfer are illustrated: (1) 11 eV which is slightly above the threshold for ionization, (2) 22 eV which is near the maximum in the differential energy-transfer cross section (see Fig. (3)), and (3) 54 eV which is close to the energy transfer needed to produce the maximum secondary-electron energy in the usual convention of track simulation that the more energetic electron after an electron-impact

Table 1. Parameters for the OPWD model of liquid water.

Peak #	Threshold (eV)	Alpha (au)	Eta (au)
1	7.76	0.285	0.00047
2	9.25	0.531	0.0030
3	10.88	0.880	0.024
4	16.33	1.764	0.013

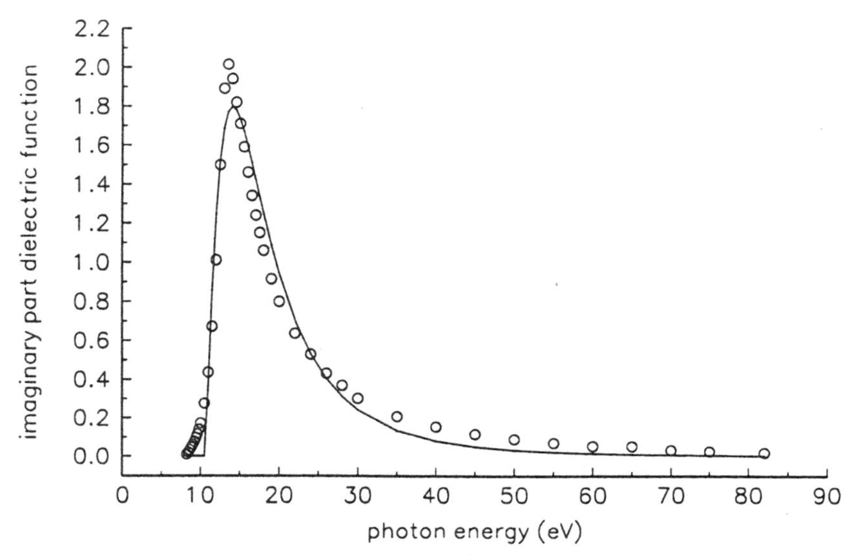

Figure 2. Fit of OPWD model (solid curve) to DNA photoionization data (open circles) from Refs.(10) and (16).

ionization is the primary. The probability distributions calculated in the second and third cases conform to the expectation that as Bohr's maximum effective impact parameter, v/ω, decreases, larger impact parameters become less efficient in collisional energy transfer. However, collisions with energy transfers near the ionization threshold appear to be an exception to this rule. Even though the maximum effective impact parameter is twice as

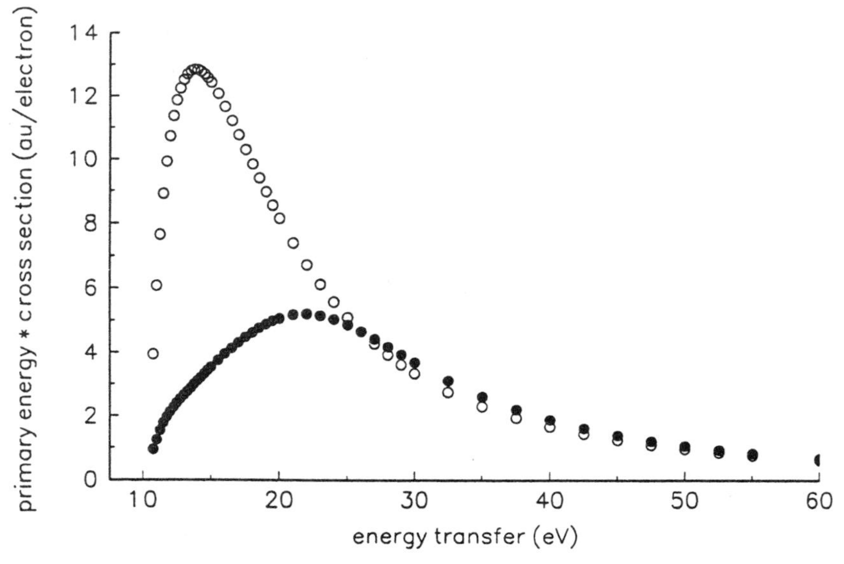

Figure 3. Differential cross sections for ionization of DNA with (solid) and without (open) screening by a water medium. The primary-electron energy is 122.5 eV.

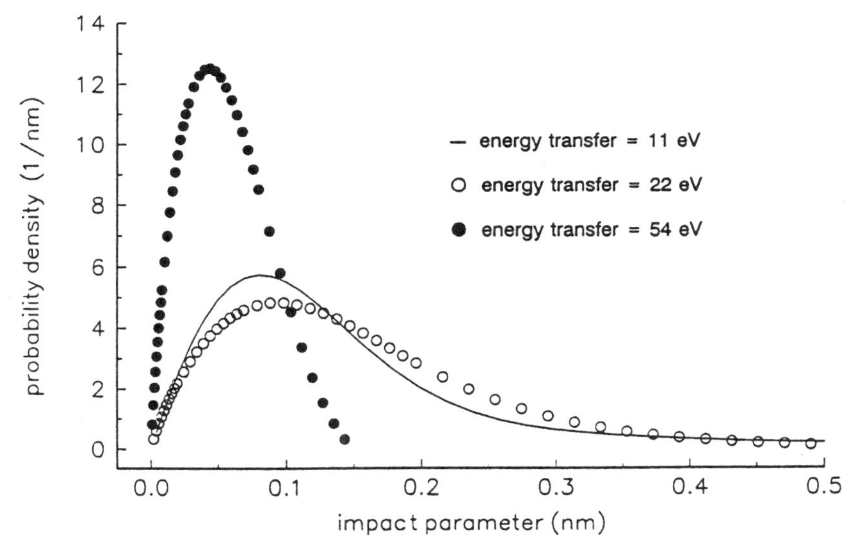

Figure 4. Impact-parameter distributions for three levels of energy transfer in ionizing collisions of DNA in water with 122.5 eV primary electrons.

large, the distribution of impact parameters in collisions where a 122.5 eV primary electron transfers 11 eV to the target is slightly more concentrated at small values than for collisions where 22 eV of energy is transferred. This effect is probably a consequence of the energy- and momentum-transfer dependence of the dielectric functions used in Eq. (2).

DISCUSSION

Our calculations show that Bohr's maximum effective impact parameter, $b_c = v/\omega$, determines the scale of the impact-parameter distribution in most cases. The probability to select an impact parameter much larger than b_c in a Monte Carlo simulation of inelastic collisions will be very small. Hence, it is useful to compare b_c with the characteristic size of the target of interest, the DNA double helix. This comparison is shown in Fig. (5) as a function of primary-electron energy for two levels of energy transfer. The filled circles show the ratio of the maximum effective impact parameter to the DNA radius (1.23 nm) for transfer of energy that is just sufficient to ionize the target (10.6 eV in our model of the DNA photoionization spectrum). The open circles show the ratio $b_c/1.23$ nm for the average energy transfer in an ionizing collision by a primary electron at the specified energy. Our results discussed above for 122.5 eV primary electrons suggest that the closed circles in Fig. (5) over estimate of the scale of the impact-parameter distribution in collisions that produce very low-energy secondary electrons. In general, Fig. (5) suggests that the scale of the impact parameter distribution is comparable to or smaller than the DNA radius for many of the electron energies that are encountered in radiation fields of radiological interest.

A cylinder with a 1.23 nm radius centered on the helical axis contains not only the atoms of DNA in the standard B conformation but also most of the water in the first two

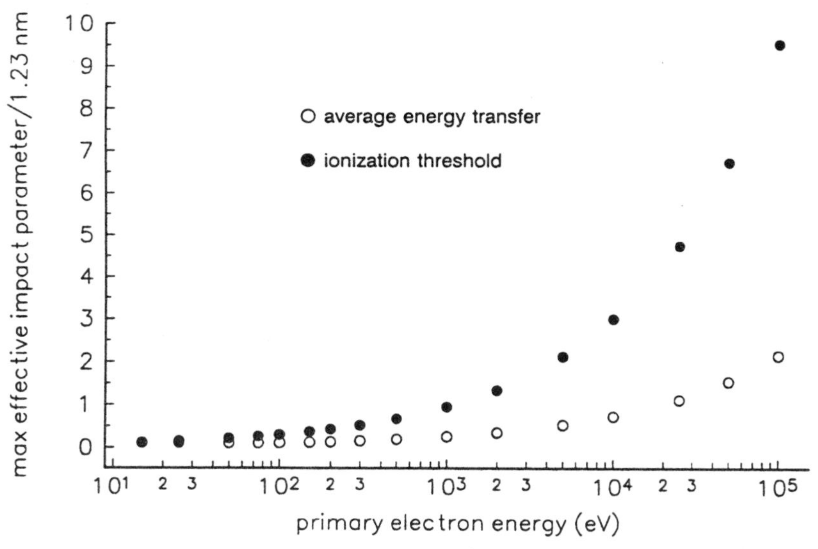

Figure 5. Ratio of the maximum effective impact parameter to the DNA radius for ionizing collisions with primary electrons of different energy.

hydration shells, about 80 water molecules per nucleotide pair.[19] The consequences of ionizing an atom in this volume are probably unaffected by the presence of radical scavengers in the medium; hence, the results presented in this paper that ignore the finite size and internal structure of the target may be useful in estimating the non-scavengeable component of DNA damage for comparison with experiments.[20] However, the atoms within this volume are not equivalent with respect to the probability that ionization leads to a specific product. For example, ionization of a sugar moiety is more likely to result in strand scission than ionization of a base. Furthermore, the recent data obtained by Wheeler and coworkers[3] with DNA at various levels of hydration suggest that ionization of water in the first hydration shell frequently leads to charge transfer to DNA bases while ionization of water in the second hydration shell appears to induce free radicals similar to those that result from ionization of bulk water.

Figure (6) illustrates the next level of theoretical development that we feel is needed to address questions concerning observable products that result from direct interaction between DNA and charged particles in a radiation field. This extension of the present model would include the internal structure of DNA by identifying the functional groups that make up the polymer (phosphate, sugar, and the DNA bases adenine, guanine, cytosine and thymine). We propose to develop an impact parameter distribution for each group by methods similar to those discussed above making use of optical data on the groups in isolation but with the constraint that the model remain consistent with DNA photoabsorption data.[10] This approach will allow us to include the internal structure of DNA in determining the impact parameter to a given functional group, which we argue on the basis of Fig. (5) will have a major effect on the probability to transfer energy to the group in many of the cases of current interest. Furthermore, this extension of the current theory will permit the different ionization thresholds of the components of DNA[16] to be included in the model.

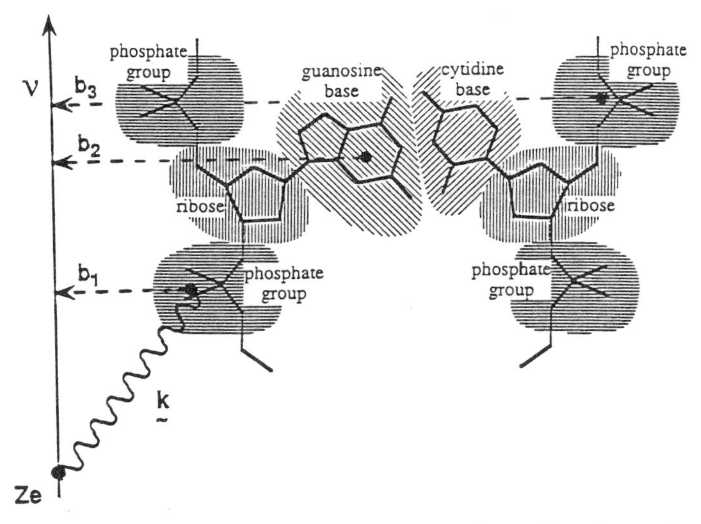

Figure 6. Spatial relationship between functional groups of DNA and a particle trajectory that passes close to the double helix.

SUMMARY AND CONCLUSIONS

A semiempirical model has been presented from which differential inverse mean free paths can be calculated for electron- and ion-impact ionization of DNA in solution. The model includes the screening effect that results from the dielectric response of the aqueous medium, which strongly influences the cross section for ejection of low-energy electrons. An impact-parameter representation of the differential inverse mean free path has been developed that will allow simulation of energy transfer to DNA when the trajectory of the charged particle slowing down does not penetrate the electron cloud of the target DNA molecule. Since the scale of the impact-parameter probability distribution is roughly equal to Bohr's maximum effective impact parameter, v/ω, this long-range mode of energy transfer is most important for ejection of low-energy secondary electrons by fast projectiles. In future applications of track-structure simulation to model direct induction of DNA lesions under cellular conditions, an extension of the current theory that allows for the internal structure of DNA will be required. An outline for future development of our model to meet this need has been presented.

ACKNOWLEDGEMENTS

This work was supported by the Office of Health and Environmental Research (OHER) of the United States Department of Energy.

REFERENCES

1. L. Udovicic, F. Mark, E. Bothe, and D. Shulte-Frolinde, Non-homogeneous kinetics in the competition of single-stranded calf-thymus DNA and low-molecular weight scavengers of OH radicals: a comparison of experimental data and theoretical models, *Int. J. Radiat. Biol.* 59:677 (1991).
2. R.H. Shuler, A.L. Hartzell, and B. Behar, Track effects in radiation chemistry. Concentration dependence of the scavenging of OH by ferrocynide in N_2O-saturated aqueous solutions, *J. Phys. Chem.* 85:192 (1981).
3. S.G. Swarts, M.D. Sevilla, D. Becker, C.J. Tokar, and K.T. Wheeler, Radiation-induced DNA damage as a function of hydration: I. Release of unaltered bases, *Radiat. Res.* 129:333 (1992).
4. N. Bohr, The decrease in velocity of alpha rays, *Phil. Mag.* 25:10 (1913).

5. R.H. Ritchie, The interaction of charged particles with a Fermi-Dirac electron gas, *Phys. Rev.* 114:644 (1959). R.H. Ritchie, R.N. Hamm, J.E. Turner, and H.A. Wright, The interaction of swift electrons with liquid water, *in:* "Sixth Symposium on Microdosimetry," J. Booz and H.G. Ebert, eds., CEC EUR 6064, (1978).
6. R.H. Ritchie, R.N. Hamm, J.E. Turner, H.A. Wright, J.C. Ashley, and G.J. Basbas, Physical aspects of charged particle track structure, *Nucl. Tracks Radiat. Meas.* 16:141 (1989). R.H. Ritchie, A. Howie, P.M. Echenique, G.J. Basbas, T.L. Ferrell, and J.C. Ashley, Plasmons in scanning transmission electron microscopy electron spectra, *Scanning Microscopy Supplement* 4:45 (1990).
7. U. Fano, The formulation of track structure theory, *in:* "Charged Particle Tracks in Solids and Liquids," G.E. Adams, D.K. Bewley and J.W. Boag, eds., The Institute of Physics, London (1970).
8. J.S. Samson, Formulation of the binary-encounter approximation in configuration space and its application to ionization by light ions, *Phys. Rev.*, 8:822 (1973).
9. J.M. Heller, Jr., R.N. Hamm, R.D. Birkoff, and L.R. Painter, Collective oscillation in liquid water, *J. Chem. Phys.* 60:3483 (1974).
10. T. Inagaki, R.N. Hamm, and E.T. Arakawa, Optical and dielectric properties of DNA in the extreme ultraviolet, *J. Chem. Phys.* 61:4246 (1974).
11. R.H. Ritchie, C.J. Tung, V.E. Anderson and J.C. Ashley, Electron slowing-down spectra in solids, *Radiat. Res.* 64:181 (1975). C.J. Tung, R.H. Ritchie, J.E. Ashley, and V.E. Anderson, Inelastic interactions of swift electrons in solids, Oak Ridge National Laboratory Report ORNL-TM-5188, (1976).
12. H.A. Kramers, *Collected Works*, North Holland Publishing Co. Amsterdam (1955), p. 333. R. Kronig, *J. Am. Opt. Soc.* 12, 547 (1926).
13. M.W. Williams, R.N. Hamm, E.T. Arakawa, L.R. Painter, and R.D. Birkoff, Collective electron effects in molecular liquids, *Int. J. Rad. Phys. Chem.* 7, 95 (1975).
14. J.C. Ashley, Stopping power of liquid water for low-energy electrons, *Radiat. Res.* 89, 25 (1982). D.J. Brenner, Stochastic calculations of the fast decay of the hydrated electron in presence of scavengers: Tests of model consistency, *Radiat. Phys. Chem.* 32:157 (1988).
15. See ref.(5) above and R.H. Ritchie, R.N. Hamm, J.E. Turner, H.A. Wright and W.E. Bolch, Radiation Interactions and Energy Transport in the Condensed Phase, *in*: "Physical and Chemical Mechanisms in Molecular Radiation Biology," W.A. Glass and M.N. Varma, eds., Plenum Press, New York (1991). The scheme described in ref.(5) and used in OREC for representing the response function of water also includes relativistic and exchange corrections. In the present application of the OPWD model, such corrections are not important.
16. W. Sontag and K.F. Weibezahm, Absorption of DNA in the region of vacuum-uv (3–25 eV), *Rad. and Environm. Biophys.* 12:169 (1975).
17. We have carried out calculations of the ejection of an electron from a charged center immersed in a water medium using the OREC scheme for representing the water dielectric function together with a somewhat different method for obtaining the impact parameter representation of the cross section. The results found are entirely consistent with those shown in Fig. (3).
18. W. Brandt and R.H. Ritchie, Primary processes in the physical stage, *in*: "Physical Mechanisms in Radiation Biology," R.D. Cooper and R.W. Wood, eds., National Technical Information Center, Springfield, VA (1974).
19. C.J. Alden and S.-H. Kim, Solvent-accessible surfaces of nucleic acids, *J. Mol. Biol.* 132, 411 (1979).
20. J.F. Ward, Mechanisms of radiation action on DNA in model systems - Their relevance to cellular DNA, *in*: "The Early Effects of Radiation on DNA," E.M. Fielden and P.O'Neill, eds., Springer-Verlag, Berlin (1991).

DISCUSSION

Paretzke: What's the ionization potential of DNA in cells?

Miller: When you multiply the photoabsorption data by the data obtained in Germany on the ionization efficiency of the DNA, you get something that starts about 8 eV.

Paretzke: DNA should be around eight eV.

Miller: Yes, but to reproduce the very rapid rise in the photoionization cross section between 8 and 10 eV, I have used an effective threshold that is a little higher than 8 eV

Varma: Can you say something about the fitting method that you used?

Miller: Mostly trial and error, but some non-linear least squares.

Varma: So you got a family of curves from the solutions to those equations and that gave you a set of parameters and you fit those parameters.

Miller: In the insulator model, you get an analytic expression for the imaginary part of the dielectric response function. You take that expression in the limit of zero momentum transfer and you get even a simpler analytic expression with three parameters. The ionization potential, the orbital exponent, and the electron density. For the water photoabsorption data, I've taken one of these functions for each band that is present. There is little shoulder in the photoabsorption spectrum near 18eV that I was never able to reproduce by this fitting procedure.

Zaider: Why not?

Miller: The deviation of our best fit from the data in this region is at most 25%. I did not feel that this difference justified a more exhaustive search of the parameter space or introduction of additional terms into the model. I should add that the 4 terms in the OPWD model for liquid water can be interpreted as being the result of 2S and 2P electrons in the water molecule. The ionization potentials come out to be roughly what one would expect for molecular orbitals made up from 2S and 2P atomic states. I think this is putting too much interpretation on this fitting process, which is really just procedure to reproduce the experimental data.

Zaider: How far was the other DNA from the target DNA for the screening purposes?

Miller: In my calculation of the screening of DNA by other DNA molecules, I took the density that Painter and co-workers measured for their DNA films, 1.35 g/cm^3

Zaider: So how far was it?

Miller: I'm sorry, I didn't calculate the average intermolecular separation from the measured density.

Inokuti: Could you tell a little bit about how you go to finite momentum transfer from zero momentum transfer?

Miller: That aspect of the calculation is a consequence of the model that Ritchie and co-workers constructed. Basically, this is a model that has the electron initially in a tight binding orbital. On ionization, it's raised to a conduction band, represented by plane waves which have been orthogonalized to the initial wave function of the electron. These model assumptions determine the momentum transfer dependence of the dielectric response function.

Inokuti: So essentially you used a lorentzian.

Miller: No. I don't think so. It's more complicated than that. I'm sorry I didn't bring the formulas to show you.

Ritchie: It also has the feature that the sum rules are built into it.

Miller: Yes, that's right. The sum rule is automatically satisfied.

Hamm: Wouldn't you say right offhand, though, that from what you were showing earlier, that if you just took water as a model, you would be pretty close?

Miller: No. If I want to answer questions like "what's the probability that we're ionizing the phosphate group relative to the probability that we're ionizing the base?", then the difference in the distance to these groups from the trajectory of the projectile will be important in most cases. With respect to the imaginary part of dielectric response function, if I want to calculate a cross section for ionizing a base, I've got to have the thing that is analogous to the generalized oscillator strength for that base, not for the whole molecule. It is certainly true that we should expect some interaction between these functional groups of DNA during an ionizing collision. I want to explore the approximation that we can decompose the photo absorption spectrum of the whole DNA molecule into absorption by its functional groups: base, sugar and phosphate.

Varma: John, these groups, they are really not static, you know, impact parameters are a function of time. They're all moving around and you end up with an average impact parameter.

Miller: The process of ionization is faster than the conformational fluctuations of the molecule so the projectile won't see the average conformation of the molecule in any one ionizing collision. It will respond to the instantaneous conformation that it sees.

Zaider: That will be different in each instance.

Varma: That's what I'm saying.

Miller: Right. Hence in principle, one should choose a different sample from conformation space for each collision, but the change in the relationship between the projectile velocity and the helical symmetry axis in each collision will be much more important than the fluctuation in the internal structure of the molecule.

Inokuti: I have only one question. I think putting the spectrum of DNA and then the spectrum of water is a very original idea, that never occurred to me.

Miller: Thank you, but is it correct?

Inokuti: Yes, I have been trying to think, maybe Rufus can explain.

Ritchie: Well, it's the same thing that we had built into our code many years ago.

Hamm: We have water, but we treat each type of excitation or ionization separately and put it over the same total water.

Inokuti: I see. So you use the same procedure in interpreting imaginary part of $-\varepsilon^{-1}$ in this way, is completely new to me.

Zaider: I didn't think you can separate this thing.

Inokuti:Well, one has to go back to the Maxwell equation and for two component systems.

Curtis: Most people talk about breaks on the sugars leading to strand break, and clearly, you're right. A breaks on the phosphate group will also lead to a strand break. What is the relative probability of such breaks?

Miller: I don't think we understand that very well at the present time. Studies of DNA fragmentation by monochromatic photons are beginning to provide information on DNA fragmentation pathways as a function of energy transfer. However, much more work will be needed before we can assign a branching ratio for the chemical alternatives that may result from direct ionization of DNA.

Chatterjee: Generally, I think it's becoming clear that base damage in double stranded DNA's (including ionized bases) do not cause strand breaks Now, in the backbone, it looks like either phosphate or sugar damage can lead to strand breaks.

CHARGED-PARTICLE TRANSPORT IN BIOMOLECULAR MEDIA: THE THIRD GENERATION

M. Zaider and A. Fung

Center for Radiological Research
Columbia University
New York, NY 10032

M. Bardash

Department of Radiation Oncology
Columbia University
New York, NY 10032

ABSTRACT

We describe Monte Carlo codes that simulate, event by event, the interaction of energetic electrons with a double-stranded DNA molecule and with the condensed water surrounding it. Both direct and indirect effects are treated explicitly. The cross-sectional input necessary in the transport codes was obtained via quantum-mechanical calculations of the dielectric response function, $\varepsilon(\mathbf{q},\omega)$, of polycytidine. For each inelastic event on DNA we score the energy deposited locally, the position of the event and the moiety that underwent that event. This information provides a detailed picture of the spatial disposition of molecular alterations for DNA exposed to ionizing radiation.

I. INTRODUCTION

Understanding the effects of ionizing radiation on cellular systems, with the potential for extrapolating this information to man, starts at the physical stage of energy deposition in biomolecules (e.g., DNA, proteins) and surrounding medium. A (necessarily incomplete) list of questions of concern must refer to: a) the kinds of damage inflicted by radiation on biomolecules (directly or indirectly), b) the radiation-type-dependent probability for inducing each kind of damage, c) the spatial configuration of damaged molecules, e.g., at a given dose the distribution of distances between, say, damaged bases, d) the degree to which such factors as the higher-order intracellular organization of DNA (e.g., chromatin), the base

Computational Approaches in Molecular Radiation Biology
Edited by M.N. Varma and A. Chatterjee, Plenum Press, New York, 1994

sequence in specific genomic targets, or the proximity of histones - to quote just a few - influence the outcome of radiation damage.

Questions a,b and d are, at least in principle, amenable to empirical determination and indeed much information, albeit often plagued by methodological uncertainties, has accumulated on the yield of different kinds of DNA damage. Point c however is *par excellence* a subject of theoretical inquiry as little or no empirical data exist on the nanodosimetry of radiation damage. The importance of examining this particular point comes from the fact that the spatial distribution of radiation products is expected to account for the relative biological effectiveness of different radiations directly or—as has been hypothesized recently[1]—through the agency of saturable repair processes. These have been, and remain, the main reasons for actively seeking information, with the aid of charged-particle transport simulation, on the spatial pattern of molecular damage.

The **first generation** of transport codes[2,3,4] dealt with interactions in water vapor, with density scaled to that of the liquid. Water has been often seen as the archetype of tissue material. The use of the vapor phase means that energy depositions (ionizations, excitations) are well localized; on the other hand, condensed-phase-specific phenomena such as plasmonic excitations, thought to be important in DNA surrounded by structural water, could not be included. A **second generation** of codes[5,6] has seen the change from gas phase to the liquid or solid phase of water; and the inclusion of a geometric scaffold representing an oligomeric piece of DNA.[7] Implicit in this approach is the assumption that the physics of DNA can be well approximated with that of water (liquid or vapor). Progress in solid state physics, as well as the increased availability of high-speed super-computers, makes the prospect of obtaining a quantum-mechanical description of DNA, and its inclusion in the transport code quite realistic. We term this stage "the **third generation** of transport codes."

This paper reports (to our knowledge, for the first time) results on the transport of electrons in condensed water containing a double stranded homopolynucleotide, polycytosine. This latter, taken in the same geometry as that of B-DNA, is a convenient model for the DNA duplex. The cross sections for the interaction of electrons with poly-C have been obtained from its frequency- and momentum dependent dielectric response function, $\varepsilon(\omega,q)$. To calculate the random-phase approximation (RPA) of $\varepsilon(\omega,q)$ we have determined the electronic structure of poly-(CSP) (wave functions and energy bands) with the aid of a semi-empirical Hamiltonian, the parameters of which were adjusted against optical data on various DNA-like biopolymers. This calculation allows us to obtain, on a per-electron basis, the probability of direct interaction with DNA, the interband or excitonic transitions involved, and—within quantum mechanical limitations—the moiety (base, sugar, phosphate) and the atom associated with specific transitions. We believe this to represent precisely the kind of information needed for the subsequent treatment of the chemical stage of radiation action that follows the energy deposition event. For completeness we also include results on the **indirect** action of electrons on this DNA model obtained with our stochastics chemistry code.[8]

The calculations reported here concern a DNA prototype made up of two polycytidine (cytosine-sugar-phosphate) strands. There are no serious restrictions to the extension of these calculations to a) paired A-T, G-C double helical strands, and b) to non-periodic DNA. Plasmids or even genes of known sequence may be used for this purpose, a good example being the 2.5 kb "house keeping" gene that codes for the APRT enzyme. However, the other important element, energy transport along the biopolymer following the energy deposition event, remains a challenge both conceptually and calculationally.

II. BASIC FORMALISM

Charged-particle transport in a medium requires information on the differential cross section for scattering off molecular targets. Within the first Born approximation (essentially for energetic particles) the double differential cross section, $d^2\sigma/d\omega d\Omega$, for scattering in the solid angle $d\Omega$ with energy loss in $[\omega,\omega+d\omega]$ is given by[9]:

$$\frac{d^2\sigma(\omega,\theta)}{d\omega d\Omega} = \frac{1}{(e\pi a_o)^2}\frac{1}{N}\frac{k_f}{k_i}\frac{1}{q^2} Im\left[-\frac{1}{\varepsilon(q,\omega)}\right] \tag{1}$$

In this equation $\varepsilon(\mathbf{q},\omega)$ is the wave-vector ($\mathbf{q}$) and frequency (ω) dependent dielectric response function of the medium, N is its electron density, a_o is the Bohr radius and k_i and k_f are, respectively, the initial and final momenta of the projectile.* The scattering angle, θ, may be calculated from:

$$\cos(\theta) = \frac{\vec{k}_i \cdot \vec{k}_f}{k_i k_f} \ . \tag{2}$$

Experimental data on $\varepsilon(\mathbf{q},\omega)$ is available from optical measurements[10] of the refractive index, n, and extinction coefficient, k, as a function of photon frequency; or from electron scattering. In both cases, because of the nature of the system (photon) or experimental limitations (electron scattering) one obtains data only at very low values of q (the so-called long wave length limit, $q \approx 0$). However, charged particle transport requires information at all kinematically-allowed values of q and one needs to resort then to theoretical evaluations of this quantity.

An exact expression for $\varepsilon(\mathbf{q},\omega)$ is[11]:

$$\frac{1}{\varepsilon(q,\omega)} = 1 - \frac{4\pi e^2}{q^2}\sum_n |<n|\rho_q|0>| \left[\frac{1}{\omega+\omega_n+i\delta} + \frac{1}{\omega_n-\omega-i\delta}\right] . \tag{3}$$

Here $|n>$ are **exact** many-body wavefunctions corresponding to eigenvalues E_n, ω_n are excitation energies ($E_n - E_0$) and ρ_q is the Fourier transform of the density operator. The limit $\delta \to 0$ is understood throughout. Because E_n and $|n>$ are not known (and not obtainable with currently available theoretical methods) the evaluation of $\varepsilon(\mathbf{q},\omega)$ is possible only if certain approximations are made. These are:

a) The single particle approximation: assume the wave function, $|n>$, describable in terms of Slater determinants made of single-particle wave functions.

b) The Hartree-Fock (HF) approximation: each electron is assumed to move in the field of the other ions and electrons.† The resulting HF wave functions depend—for a translationally periodic system—on the wave vector, $\mathbf{k}$, and band index, n: $\psi_n(\mathbf{k}) = |n\mathbf{k}>$; and the same for the single particle energies, $E_n(\mathbf{k})$. The main disadvantage of the HF approach (other than being impractical for large systems) is that the correlation energy is not taken into account and therefore the energies of the virtual (unoccupied) states—representing the excitations of the system—are too large by 50 to 100%. But these are precisely the quantities needed for calculating $\varepsilon(\mathbf{q},\omega)$.

c) The random phase approximation (RPA)[9]: by neglecting the phases of the virtual particle-hole pairs the following analytic expression for the dielectric response function

*We shall use whenever convenient atomic units.

†Another approximation (Born-Oppenheimer) assumes that the ions are fixed in space.

obtains:

$$\varepsilon(q,\omega)=1+\frac{4\pi e^2}{q^2\Omega}\sum_{n,l,k}\frac{|<l,k+q|e^{iqr}|n,k>|^2}{E_l(k+q)-E_n(k)-\omega+i\delta}\ . \tag{4}$$

Here Ω is the volume of the unit cell.

The excitation energies of the system need to be adjusted to include the correlation energy. The problem may be attacked with **ab initio** many-body theoretical techniques (the quasi-particle approach of Hedin[12]); or, if more approximate solutions could be tolerated, by applying the scissors operator: the idea here is to displace rigidly the HF energy bands until agreement with spectroscopic data is obtained. Clearly, such data need to exist in the first place for the application of this latter approach.

The calculation of $\varepsilon(\mathbf{q},\omega)$ lends itself well to corrections with the scissors operator. One could use for calibration available measurements of $\varepsilon(\mathbf{q}=0,\omega)$, and then extend the result to the $q\neq 0$ domain. We have adopted this particular technique for the polycytidine calculation. To further simplify the calculation, we have used a semi-empirical, tight-binding Hamiltonian, thus avoiding the need to iterate the solution of the Schrödinger equation to self consistency.

III. ADDITIONAL CONSIDERATIONS FOR POLYMERS WITH HELICAL SYMMETRY

Periodic polymers satisfy:

$$\rho(r+\nu a)=\rho(r)\ . \tag{5}$$

where $\rho(\mathbf{r})$ is the electron density and $\mathbf{a}$ is the translational vector of a unit cell. If furthermore the polymer has helical symmetry (i.e. a combined rotation by angle α and translation by $\mathbf{a}$ leaves the system invariant) one expects the Hamiltonian to commute with:

$$S^{\nu}(r)=D(\nu\alpha)\,(r)+\nu a, \tag{6}$$

where ν is the (integer) number of rotations and $D(\alpha)$ is the operator that rotates a vector around an axis through angle α. In analogy with the translationally periodic case, it becomes convenient to express the wave function as a linear combination of helical-symmetrized Bloch functions:

$$|nk>=\sum_j C_{nj}(k)\phi_j(k,r)\,,$$
$$\phi_j(k,r)=\frac{1}{\sqrt{2N+1}}\sum_{\nu}e^{i\nu k\cdot a}\chi_j\left(S^{-\nu}(r)-R_j\right)\ . \tag{7}$$

$\chi(\mathbf{r})$ are atomic or molecular orbitals and $\mathbf{R}_j$ is the position vector for unit j. The summation is over 2N + 1 cells (from –N to +N).

IV. THE SEMI-EMPIRICAL HAMILTONIAN

Under the tight-binding approximation one solves the Schrödinger equation by assuming a **local** crystal potential, V(**r**) (the potential satisfies, of course, all the symmetry operations of the system). Typically, one may use for V(**r**) a sum of **molecular** potentials calculated from self-consistent calculations. Implicit in this procedure is the assumption that the wave functions of the electrons are well localized near the nucleus. We take:

$$H = -\frac{1}{2}\nabla^2 + \sum_{\nu,\alpha} V_\alpha\left(S^{-\nu}(r) - R_\alpha\right) . \tag{8}$$

One needs to evaluate matrix elements of the form:

$$\left\langle \chi_i\left(S^{-\nu_1}(r) - R_1\right) |H| \, \chi_j(r - R_2)\right\rangle \tag{9}$$

and it can be shown[13] that by neglecting three-centered integrals one may express all the necessary matrix elements in terms of molecular products, $<\chi_i|\chi_j>$, $<\chi_i|T|\chi_j>$, and molecular eigenfunctions, E_i.

V. THE GEOMETRY OF THE POLYCYTIDINE CHAIN

A full Hartree-Fock calculation for a non-periodic segment of double-helical DNA chain may be quite involved. The selection of a **periodic** chain alleviates the computing effort substantially; this should be seen as a first step towards including in the calculation the long-ranged interaction of stacked nucleotides. Cytosine (see Table 1) is the simplest of the four DNA bases and was thus selected for this calculation.

Figure 1 shows schematically a poly-(CSP) chain. We have used the B-DNA geometry for this chain in order to reproduce a typical DNA configuration. A single polynucleotide chain can not have this configuration; however, a system such as poly(A-T) does have a helical structure. The assumption we are making is that, in base pairs, nucleotide-nucleotide interactions may be neglected. This is not unreasonable for transitions of, say, 10 eV or more (i.e., excluding excitonic transitions) given that hydrogen bonding is of the order of a fraction of eV. All calculations are performed for a single poly-(CSP) chain. In the Monte

Table 1. The atomic structure of DNA moieties.

Moiety	Chemical formula	No. of atoms	No. of electrons	No. of valence electrons	No. of orbitals
Sugar	C_5H_7O	13	45	33	37
Phosphate	PO_4	5	47	29	29
Cytosine	$C_4H_4N_3O$	12	57	41	44
Thymine	$C_5H_5N_2O_2$	14	65	47	50
Adenine	$C_5H_4N_5$	14	69	49	54
Guanine	$C_5H_4N_5O$	15	77	55	59

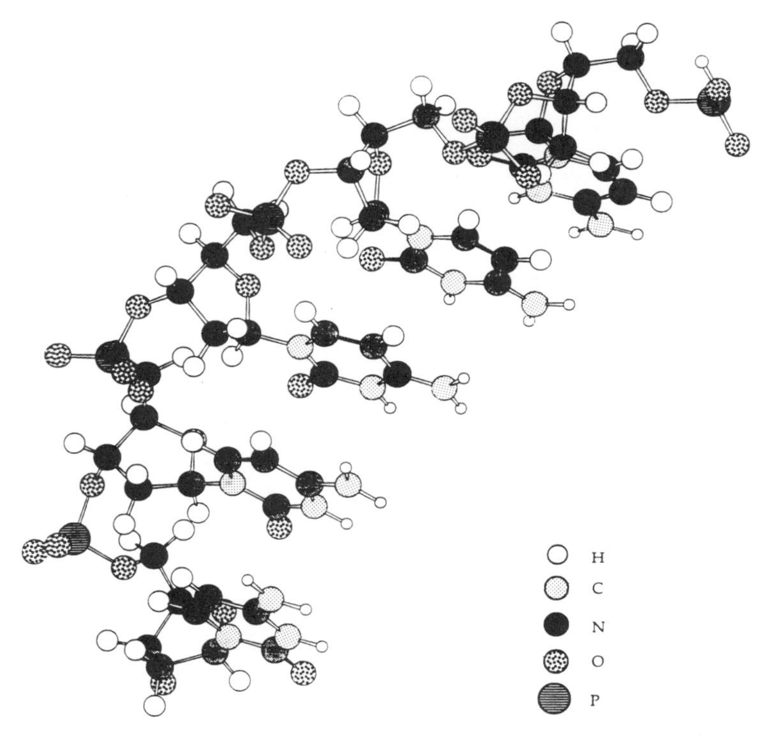

Figure 1. The structure of poly-(CSP).

Carlo code we use a double helical chain made of two poly-(CSP) polymers. As indicated in Fig.2, experimental data on the energy loss function above 10 eV for films of DNA or DNA bases indicate sufficiently small differences to justify—at least as a first approximation—our procedure. At a given momentum transfer, **q**, the area under the curve, Im[−1/ε], delimited by two values of ω is proportional to the probability of energy transfer in that energy interval; in this representation excitonic transitions amount to less than 3%.

The model depicted in Fig.1 has a pitch of 3.38 Å and a rotation angle of 36°. The cartesian coordinates of the individual atoms were taken from the Brookhaven Data Bank.[14]

VI. NUMERICAL RESULTS FOR ε(q,ω)

Calculations of the dielectric response function for poly-(CSP) have been reported by us in a recent publication. We summarize here the main results. Fig. 3 shows the band structure obtained: the gap has been set to 4 eV (i.e., conduction bands have been lowered by 7 eV) and one also notes the relative "flatness" of the bands, an indication of the strength of stacking interaction. The resulting energy loss function, Im[−1/ε] is shown in Fig. 4. Also shown here is the experimental determination of this quantity by Inagaki et al.[15] The jagged shape of the calculated result may be smoothed out by folding in a Gaussian-type resolution energy function (we have used a standard deviation of 3 eV). This is a standard procedure to account for the finite resolution of the detecting equipment, the finite life time

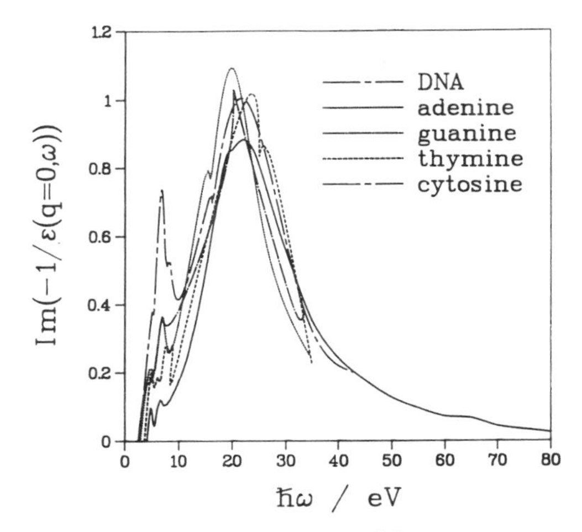

Figure 2. Experimental data for $\varepsilon(q=0,\omega)$ for DNA[10] and various nucleotide base stacks.

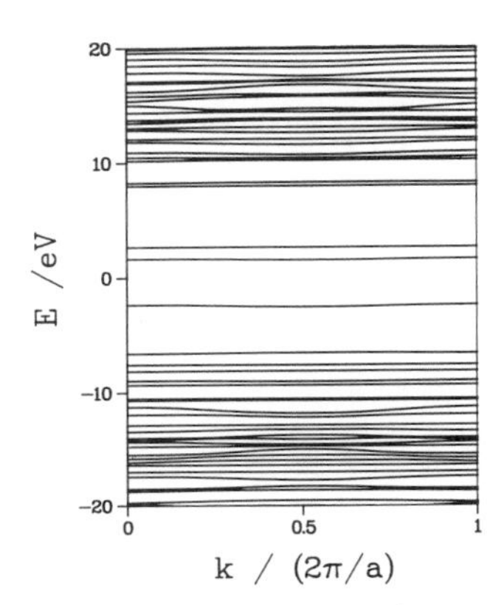

Figure 3. The calculated energy band structure of poly-(CSP).

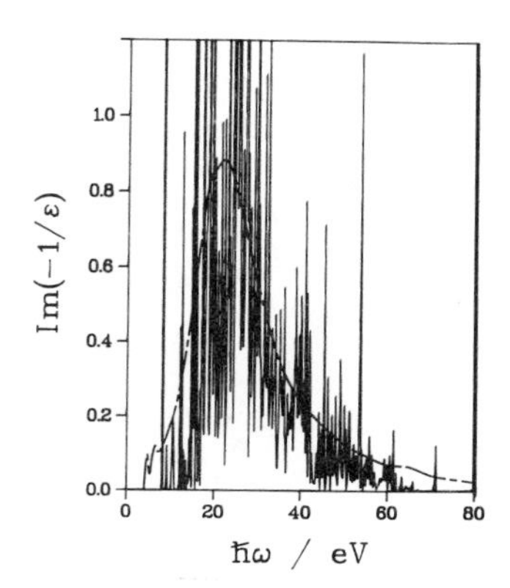

Figure 4. Calculated energy loss function, Im[$-1/\varepsilon(\mathbf{q} = \mathbf{0},\omega)$], for poly-(CSP).

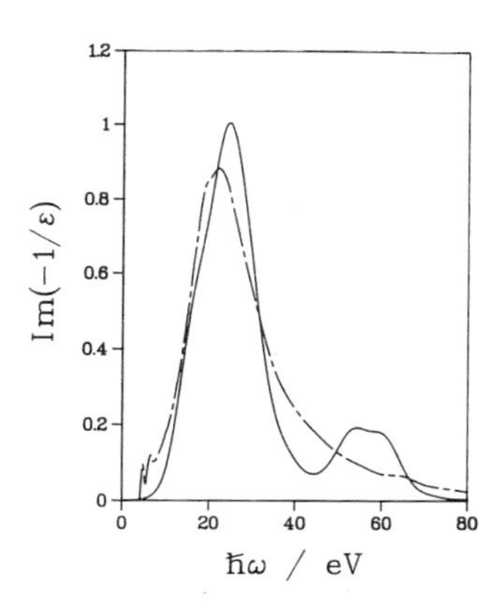

Figure 5. A "smoothed" version of the function represented in Fig. 4 (see text).

of the observed transitions, interchain and disorder effects in the film, and the fact that—because of the random orientation of DNA microcrystals—the experimental result is averaged over all directions of **q**. The smoothed result is shown in Fig. 5: reasonable agreement with the experimental determination may be noticed. With the energy bands thus set, we have extended the calculation at non-zero momentum transfer, a domain where no experimental data exist (see Fig. 6).

It is interesting to note that, in spite of the fact that a semi-empirical calculation is not so constrained, our results do satisfy 84% of the Thomas-Reiche-Kuhn sum rule.[11]

VII. THE MONTE CARLO CODE

The cross sections thus obtained have been used in the code MCDNA that transports charged particles in crystalline water containing a DNA double-helical chain. The mechanism of transport in water has been described recently[6] and is not repeated here. The mechanics of converting numerical results on Im[−1/ε] for DNA into analytical forms usable in Monte Carlo simulation follows a pattern similar to that described for water.[6] For instance, the single differential cross section, dσ/dω, may be obtained by integrating over q:

$$\frac{d\sigma}{d\omega} = \frac{2}{\pi N k_i^2} \int_{q_{\min}}^{q_{\max}} \frac{dq}{q} Im\left[\frac{1}{\varepsilon(q,\omega)}\right] \tag{10}$$

where (see Fig. 7):

$$\begin{aligned} q_{\max,\min} &= k_i \pm k_f \,, \\ k_i = \sqrt{2T}\,, \quad k_f &= \sqrt{2(T-\omega)} \,. \end{aligned} \tag{11}$$

T is the kinetic energy of the incident particle. For incident electrons the differential cross section needs to be corrected for exchange effects; we have followed the procedure suggested by Ritchie et al.[16] The differential inverse mean-free path (IMFP, $\Lambda^{-1} = \sigma N$) for 1 keV electrons, shown in Fig. 8, illustrates one of the results used in the transport code.

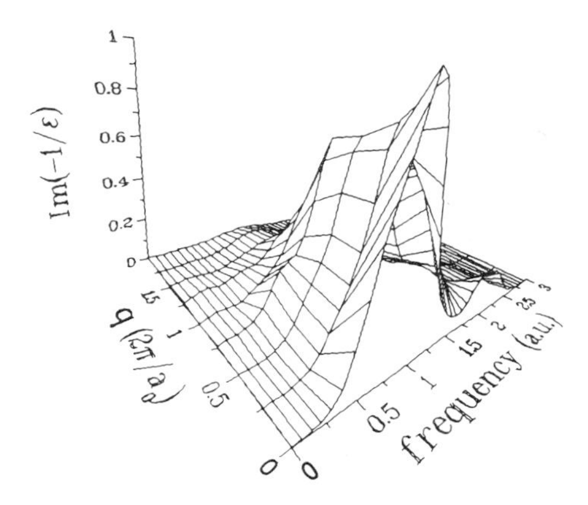

Figure 6. Three-dimensional representation of the calculated energy loss function for poly-(CSP).

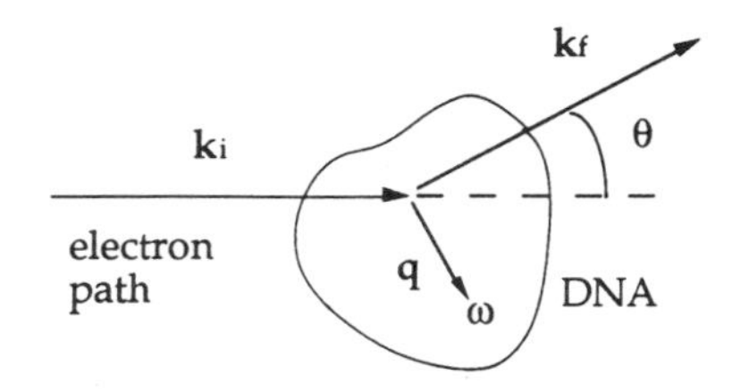

Figure 7. The kinematics of electron scattering.

A peculiar requirement of this calculation is the ability to identify (statistically) the moiety at which a certain electron-induced transition occured. This leads to a problem because from the energy band spectrum of Fig. 3 it is apparent that a given combination (q, ω) may be generated by more than one pair of occupied-unoccupied bands. It is necessary then to obtain, and include in the code, information on the energy loss function for individual pairs. As an example, we show in Fig. 9 the total IMFP for 1 keV electrons as a function of the valence band originating the transition. In the code, the moiety is selected at a rate proportional to the contribution to the wave function of the valence band of its orbitals.

The astute reader may also note that with the RPA expression, Eq. (4), for $\varepsilon(\mathbf{q},\omega)$ the energy loss function is **not** decomposable into contributions from separate pairs of valence-conduction bands (this is because of the minus sign). The **exact** expression [Eq. (3)], however, does satisfy this requirement.

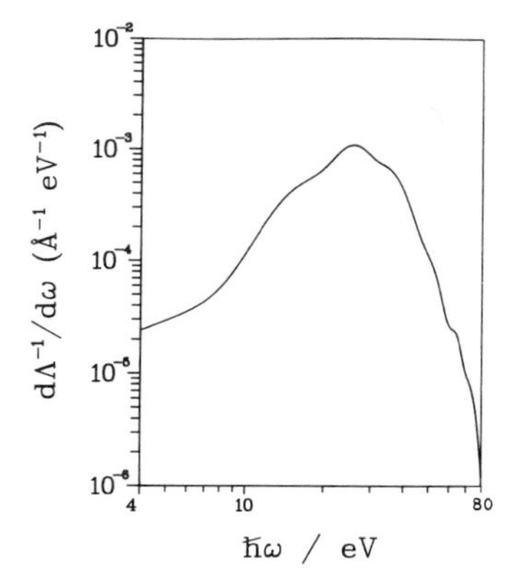

Figure 8. The differential inverse mean free path for 1-keV electrons incident on poly-(CSP).

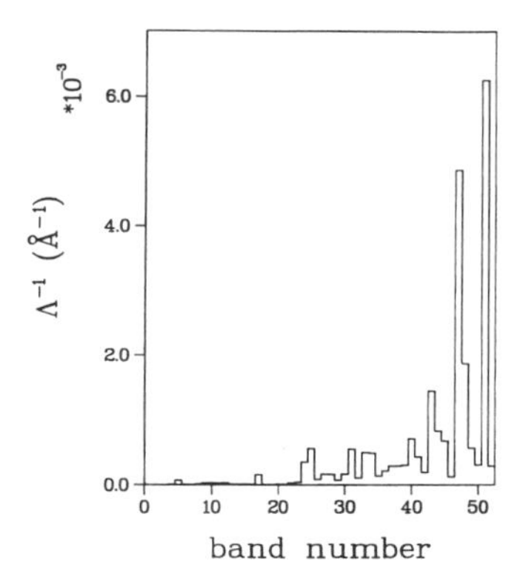

Figure 9. The total inverse mean free path for 1-keV electrons scattering off poly-(CSP) as a function of the valence band originating the transition.

To decide whether an electron in water will have its next interaction with DNA one has to establish: a) that the electron trajectory intersects the DNA "volume,"‡ and b) assess the probability of inelastic scattering off DNA. This latter is calculated from the following cumulative distribution (see Fig. 10a):

$$\begin{aligned} &\mathit{If}\ \xi \le F(a), \quad x\epsilon[0,a], \quad F(x) = 1 - e^{-\rho_1\sigma_1 x}; \\ &\mathit{If}\ F(a) < \xi \le F(a+b), \quad x\epsilon(a,a+b), \quad F(x) = 1 - e^{\rho_1\sigma_1 a - \rho_2\sigma_2(x-a)}; \\ &\mathit{If}\ \xi > F(a+b), \quad x\epsilon(a+b,\infty), \quad F(x) = 1 - e^{\rho_2\sigma_2 b - \rho_1\sigma_1(x-b)}\ ; \end{aligned} \tag{12}$$

and similarly for the other cases of Fig. 10. In the current version of MCDNA the (somewhat arbitrary) decision was taken to consider the range of excitations 4 to 10 eV as belonging to excitonic transitions. With probability of 0.03 (its cross sectional contribution) an exciton of energy 5 eV is produced. All other interaction are classified as ionizations. An ab initio study of the excitonic spectrum in DNA, and its corresponding energy loss function, is underway in our laboratory.[17]

VIII. INDIRECT DAMAGE ON THE DNA MOLECULE

Of all the radiation-induced free radical species present in aqueous solution containing DNA the only species that has significant reactions with DNA is the hydroxyl radical, OH.[18] For all other possible reactions the reaction rates are significantly smaller. Chemical

‡We have represented this volume as a cylinder of radius 11 Å This contains about 90% of the electron density of B-DNA.

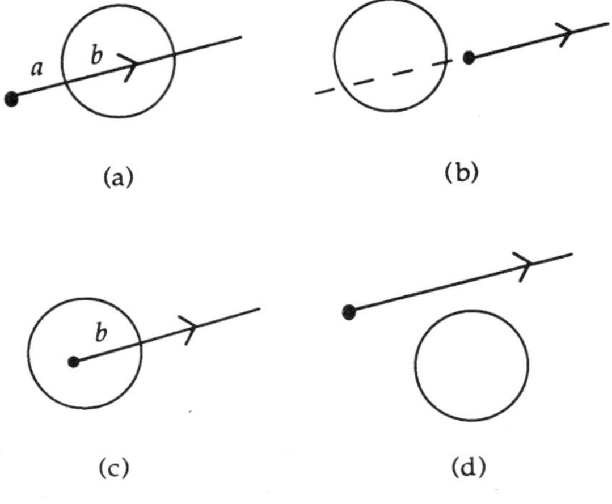

Figure 10. The geometrical representation used in Eq. (12).

mechanisms for specific kinds of damage have been studied quite extensively. For instance, DNA strand breakage may result from either the OH radical attacking a sugar moiety, or—as has been hypothesized[18]—damage to a base pair may migrate to the adjacent sugar. The precise rate at which base or sugar moieties are damaged by radiations remains largely unknown and experimental evidence—when available—is only indirect. To examine these questions we have developed a stochastic chemistry code[8] that models an oligomeric segment of DNA exposed to radiation. The DNA is in aqueous solution; the code also allows for different concentrations of radical scavengers. A full description of this code has been published.[8] Briefly, radiation-induced radical tracks (generated with the code DELTAI that simulates electron transport in crystalline water[6]) are superimposed on a three-dimensional representation of the DNA molecule consisting of nucleotides arranged in the B form. If needed, scavenger molecules are added at random spatial configurations but at a prescribed average concentration. The Monte Carlo technique is used to simulate the diffusion and pair-wise interaction of radicals among themselves or with target molecules (base or sugars). The code is always calibrated against experimental data on the time decay of e_{aq} and OH species.[19]

The results reported below concern energetic electrons; it is therefore necessary to consider only individual spurs, that is groups of radicals associated with single interactions of the primary particles.[§] Our previous studies[8] have indicated that it is necessary to follow the interaction of radicals with DNA only up to about 1 ms.

IX. RESULTS

To illustrate the capabilites of the codes described above we report here results on the **direct** and **indirect** interaction of 50 keV electrons with a double-stranded poly-(CSP) chain. In this simulation electrons are directed towards a 100-nucleotide-long helix set perpendicular to their initial direction (along the z axis) and at 50 nm distance. The oligomer is symmetric relative to the z axis. For each track we score the moiety damaged (if any), its

§The code can handle any length of track and not only spurs.

coordinates, the kind of damage (ionization, excitation, electron capture), and also—in the case of indirect effects—the distance from which the OH radical responsible for the damage arrives. It can be readily shown that, at the doses of concern in the radiobiology of DNA, individual tracks may be assumed to act independently. The results reported here correspond to the scattering geometry described above and should not be generalized to the situation where **random** electrons hit the DNA polymer. They are also not representative for exposure of DNA in cellular environment where scavenging material (e.g., proteins surrounding the DNA) will significantly diminish the indirect effect.[||]

Some 9% of the incident electrons result in direct effects on DNA. Of these, about 3% consists of electron capture by base or sugar, 0.2% are excitonic transitions, and 1.5 and 4% are, respectively, ionizations in sugar and bases. Within the statistics of this simulation the probability that one electron damages **directly** more than one moiety is less that 10^{-3}.

Indirectly, sugar molecules are damaged at the rate of 0.35 per incident track. Most of the damage (28%) consists of single nucleotides per track (see Fig. 11); in 5% of the tracks **two** nucleotides undergo sugar damage. One particular spur generated 7 damaged nucleotides. When damage occurs in the sugar moiety the possibility exists that the DNA strand becomes interrupted. When two strands are broken in close proximity, say within 10 nucleotides, a double strand break may occur. It becomes then important to know the distribution of intervals between pairs of damaged nucleotides (to be considered, a pair must be produced by the **same** electron track). The results shown in Fig. 12 indicate that 40% of the pairs are within 10 nucleotides with only a modest increase (an additional 10%) if one

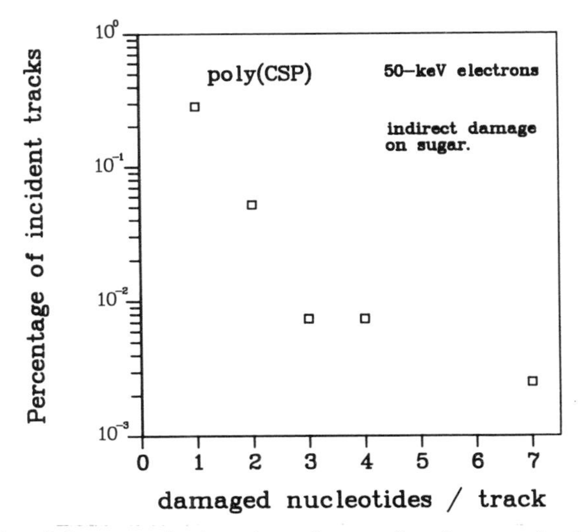

Figure 11. The fraction of electron tracks damaging a given number of sugar molecules by indirect action.

[||]In fact, experimental evidence has been presented to the effect that intra-cellularly radiation acts only via direct effect.

extends this range to 20 nucleotides. No significant difference was noticed between the distance distributions of pairs located on the **same** strand and pairs located on opposite strands.

An interesting question is whether one should consider the distribution of distances between **neighboring** damaged nucleotides or, alternatively, consider the distribution of distances between **any** random pair of nucleotides; this latter has been termed **proximity function of distances.**[20] The answer clearly depends on the ability of the cell to repair the damage. For instance, if repair of **single** strand breaks occurs fast relative to the repair of **double** strand breaks, proximity functions would be the relevant quantities.

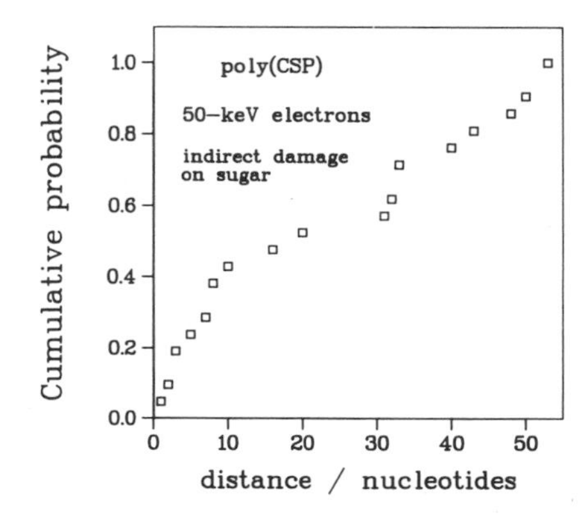

Figure 12. The cumulative probability distribution of distances between pairs of nucleotides damaged by indirect action.

Turning now to the other DNA moiety, the vast majority of hydroxyl radicals end up damaging a base. This result, which—again—may not be typical of the cellular environment, obtains simply because of the large number of available bases, at least compared with other competing species such as, say, OH radicals in the spur.

REFERENCES

1. J.F. Ward, Biochemistry of DNA lesions, *Radiat. Res.* 104:S103–S111 (1985).
2. H.G. Paretzke and M.J. Berger, Stopping power and energy degradation for electrons in water vapor, *in:* (ed): 6th symp. on microdosimetry, Brussels,London: Harwood Academic Publishers,CEC, pp. 749–58 (1978).
3. J.E. Turner, R.N. Hamm, H.A. Wright, J.T. Modolo, and G.M.A.A. Sardi, Monte Carlo calculations of initial energies of Compton electrons and photoelectrons in water irradiated by photons with energies up to 2 MeV, *Health Phys.* 39:49–55 (1980).
4. M. Zaider, D.J. Brenner and W.E. Wilson, The application of track calculations to radiobiology.-I. Monte Carlo simulation of proton tracks, *Radiat. Res.* 95:231–247 (1983).

5. J.E. Turner, H.G. Paretzke, R.N. Hamm, H.A. Wright, and R.H. Ritchie, Comparative study of electron energy deposition and yields in water in the liquid and vapor phases, *Radiat. Res.* 92:47–60 (1982).
6. M. Zaider, M.G. Vracko, A.Y.C. Fung, and J.L. Fry, Electron transport in condensed water, *Radiat. Prot. Dos.* 52:139-146 (1994).
7. D.E. Charlton, H. Nikjoo, and J.L. Humm, Calculation of initial yields of single- and double-strand breaks in cell nuclei from electrons, protons and alpha particles, *Int. J. Radiat. Biol.* 56:1–19 (1989).
8. M. Bardash and M. Zaider, A stochastic treatment of radiation damage to DNA from indirect effects, *Radiat. Prot. Dos.* 51:171-176 (1994).
9. P. Schattschneider, "Fundamentals of Inelastic Electron Scattering," Springer Verlag, New York (1986).
10. M.W. Williams and E.T. Arakawa, Optical and dielectric properties of materials relevant to biological research, *in:* "Handbook on Synchrotron Radiation," Elsevier, New York (1990).
11. D. Pines and P. Nozieres, "The Theory of Quantum Liquids," Benjamin, New York (1966).
12. L. Hedin and S. Lundqvist, Effects of electron-electron and electron-phonon interactions on the one-electron states of solids, *Solid State Phys*. 23:1-181 (1969).
13. F. Bassani, G. Pastori Parravicini, and R.A. Ballinger, Electronic states and optical transitions in solids, Pergamon Press (1975).
14. F.C. Bernstein, T.F. Koetzle, G.J.B. Williams, E.F. Meyer, M.D. Brice, J.R. Rodgers, O. Kennard, T. Shimanouchi, and M. Tasumi, The protein data bank: a computer based archival file for macromolecular structures, *J. Mol. Biol.* 112:535–542 (1977).
15. T. Inagaki, R.N. Hamm, E.T. Arakawa, and L.R. Painter, Optical and dielectric properties of DNA in the extreme ultrviolet, *J. Chem. Phys.* 61:4246–4250 (1974).
16. R.H. Ritchie and J.C. Ashley, The interaction of hot electrons with a free electron gas, *J. Phys. Chem. Solids* 26:1689–1694 (1965).
17. M.G. Vracko and M. Zaider, A study of the excited states in cytosine and guanine stacks in the Hartree-Fock and exciton approximations, *Radiat. Res.* 138:18–25 (1994).
18. C. Von Sonntag. "The Chemical Basis of Radiation Biology," Taylor & Francis, Philadelphia (1987).
19. C.D. Jonah, M.S. Matheson, J.R. Miller, and E.J. Hart, Yield and decay of the hydrated electron from 100 ps to 3 ns, *J. Phys. Chem.* 80:1267–70 (1976).
20. A.M. Kellerer and D. Chmelevsky, Concepts of microdosimetry. III. Mean values of the microdosimetric distributions, *Radiat. Environ. Biophys.* 12:321–335 (1975).

DISCUSSION

Chatterjee: You should compare how much direct ionization you're getting from your calculations as against damage by the water radical mechanism.

Zaider: Less than ten percent for radical mechanism.

Chatterjee: Another important checkpoint is to determine how many OH attack the sugar molecules and how many of them reach the bases? For a single stranded polymer 95 percent of the ·OH reaction is with the bases, and five percent with the sugar. Have you done these cross-checks?

Zaider: No.

Du Bois One thing you said you just did that for 50 keV electrons. If you would lower that electron energy, then I assume that your cumulative probabilities would all shrink in size; is that correct?

Zaider: I don't know.

Du Bois: Is that incorrect? I was just saying because of the shorter range.

Nikjoo: Low-energy electrons all get absorbed in the target, so you produce a lot of damage. But with the 50 keV, you've got a long-range electron, which misses the target.

Du Bois: Okay. But that's what he said. If you go to lower energy, then the range would be smaller. So on your cumulative damage where he's talking about 50 percent and ten nucleotides, would that be a hundred percent and five nucleotides or something of that order?

Varma: Marco, let me ask you a question. When I sit down with the biologist, they all tell me that the DNA repair is the most important aspect for them to understand the radiation effects. How does your work interest the biologists?

Zaider: We have no information on repair in our model. We have not addressed that yet.

Varma: I need to emphasize, this is only the chemical stage. That was not my question. My question is if you double up the effort in this area, would we be able to say something about the end result that biologists are measuring such as the DNA repair and misrepair. Do these studies tell you very much about the DNA repair or misrepair?

Chatterjee: I think we don't have enough experimental data yet to include repair in our models right now. However, some experimentalists are in the process of measuring repair of double strand breaks by faithful joining or unfaithful rejoining events. These measurements involve the use of restriction enzymes and specific probes for a pre-determined fragments. These data will certainly guide us to address the question of repair Varma has raised. However, one has to deal with solenoidal chromatin model for the DNA and determine whether the single strand breaks on the opposite strands are "staggered" or "blunt." It is very likely that the dsb of the former type will be rejoined faithfully. Hence, doubling the efforts in this area will certainly improve our rate of productivity. Eventually, we will be able to propose specific experiments where the models could be tested.

Nikjoo: The objective is to gain a better understanding of types of damage produced by high and low-LET radiation and to classify spectrum of damage in terms of their complexities, which may relate to their repairability. The information on strand breaks will be further enhanced by including the role of chemical environment to simulate the chemical processes involved following initial energy deposition. This type of information could help to understand the role of enzymatic repair.

Varma: This is very important because I think as I have mentioned earlier, that in the radiation protection field, track structure studies are used to relate very closely with biological data.

Zaider: What is it that makes certain damage repairable and others not? Is it the severity of the damage? I think that's a key question.

Ritchie: Marco, can I ask you a question? Did I understand that the charged particle has to go through this DNA molecule before it interacts with it in your model? So the kind of effect that John Miller just told us about would not be present, in your calculation? The next question comes, how different from your results would one be if one followed John's prescription and considered the effect of the medium; the screening effect of the medium to be that of water and then calculated the absorption in DNA using the experimental data.

Zaider: I don't know the answer to that. Of course, we can find it.

TRACK STRUCTURE, CHROMOSOME GEOMETRY AND CHROMOSOME ABERRATIONS

David J. Brenner

Center for Radiological Research
Columbia University
New York, NY 10032

John F. Ward

Department of Radiology
University of California at San Diego
La Jolla, CA 92093

Rainer K. Sachs

Department of Physics and Department of Mathematics
University of California
Berkeley, CA 94720

ABSTRACT

The joint role of radiation track structure and chromosome geometry in determining yields of chromosome aberrations is discussed. Ideally, the geometric models of chromosomes used for analyzing aberration yields should have the same degree of realism as track structure models. However, observed chromosome aberrations are produced by processes on comparatively large scales, e.g., misrepair involving two DSB located on different chromosomes or two DSB separated by millions of base pairs on one chromosome, and quantitative models for chromatin on such large scales have to date almost never been attempted. We survey some recent data on large-scale chromosome geometry, mainly results obtained with fluorescence in situ hybridization ("chromosome painting") techniques. Using two chromosome models suggested by the data, we interpret the relative yields, at low and high LET, of inter-chromosomal aberrations compared to intra-chromosomal, inter-arm aberrations. The models consider each chromosome confined within its own "chromosome localization sphere," either as a random cloud of points in one model or as a confined Gaussian polymer in the other. In agreement with other approaches, our results indicate that at any given time during the G_0/G_1 part of the cell cycle a chromosome is largely confined to a sub-volume comprising less than 10% of the volume of the cell nucleus. The possible significance of the ratio of inter-chromosomal aberrations to intra-chromosomal, inter-arm aberrations as an indicator of previous exposure to high LET radiation is outlined.

Computational Approaches in Molecular Radiation Biology
Edited by M.N. Varma and A. Chatterjee, Plenum Press, New York, 1994

INTRODUCTION

Radiation tracks have many aspects, and it has often been emphasized[1] that calculating and interpreting track structure should be done with an eye toward the ultimate biological effects of the radiation. Production of radiobiological damage may be thought of as occurring in three steps:

$$\text{energy deposition} \rightarrow \text{DSB} \rightarrow \text{chromosomal aberrations} \rightarrow \text{complex endpoints} \quad (1)$$

The yield of chromosomal aberrations is very closely correlated with complex biological endpoints, such as clonal inactivation[2] and oncogenic transformation[3]. Moreover, production of chromosomal aberrations is more sensitive to track structure than DSB formation, since low-dose RBE's for chromosomal-aberration yields can be higher than 20,[4] in contrast to RBE's ranging from ≈0.5 to 2 for DSB production.[5] In this article we shall focus attention on the second step in Eq. (1), chromosomal-aberration formation.

The dominant pathway for the production of chromosomal aberrations is often misrepair, specifically pairwise interactions between DSB to form exchange-type aberrations. Aberration yields depend not only on track structure, which determines the initial spatial pattern for ionization clusters of sufficient multiplicity and localization to produce DSB,[5] but also on chromosomal geometry and motion, which strongly influence the initial DSB patterns and subsequent DSB interactions. The probability that two DSB find each other and undergo an exchange depends not only on whether the DSB were made by the same radiation track or different tracks but also on whether the DSB are in the same chromosome or in different chromosomes. For estimating chromosomal aberration yields one needs spatial autocorrelation functions, both for ionization clusters induced by radiation of a given quality and also for chromatin density, as well as information about the spatial dependence of DSB interactions.[6,7]

From these considerations we suggest that, in terms of radiobiological mechanisms, models of chromosomal geometry are of comparable significance to models of radiation track structure. We here discuss recent experimental results, mainly obtained by fluorescence in-situ hybridization (FISH), which give more quantitative information than hitherto available about chromosomal spatial configurations on the very large scales of main interest in chromosomal aberration formation. The FISH techniques consist of "painting" specific chromosome homologues, or parts of chromosomes, with various fluorochromes.[8–13] Spatial configurations of chromosomes, or illegitimate recombinations between them, can then be recognized with comparative ease .[8–17] Using chromatin models based on these FISH experiments, we will here show how, in conjunction with track structure information, chromosomal geometry may be used to understand mechanisms of chromosomal aberration formation.

The standard chromosome aberration assays are not sensitive to small-scale rearrangements of the genome,[18] such as the deletion of a few base pairs in one chromosome. Compared to the situations analyzed in most other contributions to this volume, the processes and configurations considered in this article are slow and large. The time scale for DSB misrepair processes, which are believed to be enzymatic, is minutes or even hours, in contrast to the microseconds or less during which radiation- induced radicals remain spatially inhomogeneous, and during which, presumably, the initial spatial locations of DSB are determined. Once formed, the DSB retain their spatially inhomogeneous pattern for comparatively long periods of time, and thus provide the link between the early track structure and the later production of chromosomal aberrations. The length scales of interest in analyzing chromosome aberrations are often ≈1 μm rather than a few nm. We shall emphasize genomic separations ranging from 10^5 base pairs up to the length of a large human chromosome (≈3×10^8 base pairs), so that familiar configurations such as the basic

DNA duplex, the histone complexed "beads on a string" DNA configuration, and even the 30 nm chromatin fiber or the possible 50,000 base pair chromatin loops[19] are for our purposes only "fine-structure." An enormous amount of work has been done on quantitative models of this chromatin fine structure,[19,20] and models of its interrelation with radiation track structure during DSB induction have been analyzed,[21] but as yet there have, to our knowledge, been only very few attempts[17,22–25] to model quantitatively chromosome geometry on the scale of $10^5 - 10^8$ base pairs. We believe that such models are needed for radiobiology, and that the experimental and theoretical prerequisites for constructing them are now becoming available. This article describes some of the models and their radiobiological applications.

CHROMOSOME ABERRATIONS

We shall be concerned with chromosome aberrations produced in human cells, especially during the G_0/G_1 part of the cell cycle, prior to DNA replication, and typically scored at a subsequent metaphase. The scenario is that after DSB are produced by ionizing radiation the two ends of a DSB are held more or less together by proteins until *restitution* (i.e. correct repair, with at worst a few extra, missing or altered base pairs) or a pairwise DSB–DSB interaction (illegitimate recombination) occurs.

Aberration Types

It has long been known that many different kinds of chromosome aberrations can occur, and multi-color fluorescent in situ hybridization (FISH) techniques are now in fact revealing a rich spectrum of final products.[15,16,26] Each chromosome is divided into two arms by a centromere (Fig.1), which is required for the proper apportionment of chromosomes during cell division and is often located near the center of the chromosome. Aberrations resulting from DSB interactions with both DSB on the same chromosome arm (i.e., *intra-chromosomal, intra-arm* aberrations) occur but in this article we emphasize cases where the two DSB are on opposite arms (*intra-chromosomal, inter-arm* aberrations) or on different chromosomes (*inter-chromosomal* aberrations). Figure 1 shows some of the most frequently observed aberration types.

Suppose we assume complete randomness, i.e., that DSB induction, when averaged over stretches of millions of base pairs, occurs randomly in the genome, and that all DSB pairs produced are equally likely to interact. Then if we also take all 92 chromosome arms as equally long the measured ratio (F) of inter-chromosomal aberrations (Fig. 1C) to intra-chromosomal, inter-arm aberrations (Fig. 1B) should be $F = 90$. This can be seen in Fig. 1 by noting that, assuming complete randomness and equal arm-lengths, the end a' has equal likelihood of interaction with the opposite arm of the same chromosome or with any one of 90 other arms on other chromosomes. Correcting for the actual pattern of arm lengths in human chromosomes gives $F = 86$ as the prediction for complete randomness.[25]

Data on F Values

However it has long been known[27,28] that pairs of DSB on different chromosomes that are distant from each other within the cell nucleus are less likely to interact. If, as is now observed,[9,13,29–31] each chromosome is localized in a sub-volume of the cell nucleus even during interphase, this bias toward pairs which are nearby should result in a decrease of the F value.[22,27] In fact small F values, interpreted as evidence for proximity effects, chromosome localization and limited interaction range for DSB, have been observed for various kinds of cells.[22,32] Literature surveys show that for human lymphocytes subjected to

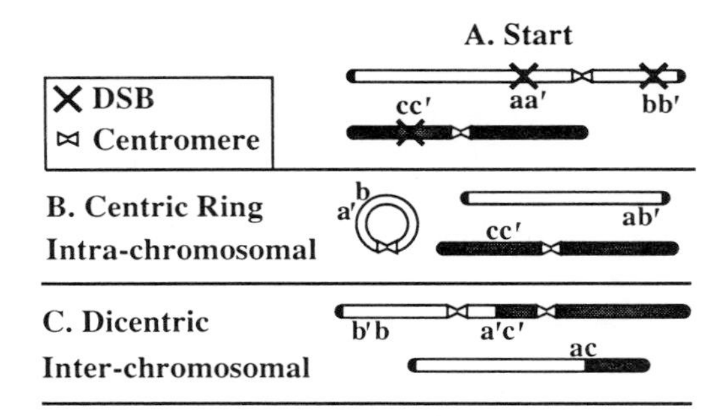

Figure 1. Suppose, as an example, the only damage to the genome after acute irradiation is the three DSB shown, i.e. the other 44 chromosomes are undamaged. The figure shows a few of the many aberrations that can result. In B the DSB *cc′* has restituted, the end *a′* has joined with the end *b* from the other arm of the same chromosome to produce a centric ring, and the end *a* has joined with the end *b′* to produce an "acentric fragment." In C the DSB *bb′* has restituted, while the joinings *a′c′* and *ac* have produced, respectively, a dicentric and an acentric fragment. The centric rings and dicentrics shown are typically lethal to a cell at mitosis. However, there are "stable" counterparts[32] (not shown), pericentric inversions and reciprocal translocations respectively, which can persist in cell clones for many cell divisions.

The probabilities that ends from two DSB find each other and rejoin depend on their initial separation and subsequent motion, which in turn depend on ionization patterns and on whether the DSB were formed on the same or different chromosomes. Consequently, comparing yields for intra-chromosomal interactions such as the one shown in B with yields for inter-chromosomal interactions such as the one shown in C gives information on track structure and on large scale chromosome geometry.

low LET the ratio is $F \approx 15$.[7,25] For human lymphocytes at high LET the ratio appears to be still smaller,[7,33-36] $F \approx 6$, presumably corresponding to the greater localization effects for ionization clusters made by densely ionizing radiations.

Measured F values are particularly sensitive to track structure and large scale chromosome geometry and we shall in this article, to be specific, focus attention on this assay. Most *in vitro* data on F values concern the ratio of dicentrics to centric rings (Fig. 1), but the ratio of translocations to pericentric inversions (Fig. 1 caption) is also of importance and is approximately equal to the ratio of dicentrics to centric rings.[7] Because the F value is a ratio it should be comparatively insensitive to various confounding factors. For example, interphase death complicates the interpretation of aberration experiments on human lymphocytes,[37] but there is no evidence, and no mechanistic reason to assume, that there is differential interphase death for cells with dicentrics (Fig. 1C) as compared to cells with centric rings (Fig. 1B), which indicates the F value should not be affected by interphase death.

Autocorrelation Functions

The considerations of the preceding section show that the DSB-DSB reactions which lead to chromosome aberrations are not well mixed spatially. In view of the pairwise nature of these interactions, the aspect of chromosome geometry most important for analyzing aberration formation is the chromatin spatial autocorrelation function.

Let $\sigma(\mathbf{R})$ be the density of chromatin within a representative cell nucleus centered at the origin, so that $\sigma(\mathbf{R}) = 0$ for $\mathbf{R}$ outside the nucleus. Then σ can be regarded as a random function, and for human cells during G_0/G_1,

$$\sigma(\mathbf{R}) = \sum_{j=1}^{46} \sigma_j(\mathbf{R}) \quad , \tag{2}$$

where σ_j is chromatin density for the jth chromosome. The autocorrelation function is

$$\langle \sigma(\mathbf{R})\sigma(\mathbf{R}') \rangle \tag{3}$$

where $\langle \ldots \rangle$ denotes an average (over cells experimentally and over an appropriate ensemble theoretically). Inserting Eq. (2) into Eq. (3) facilitates separating predicted aberration yields into the inter-chromosomal and intra-chromosomal types discussed in the previous section. Given that a particular location is occupied by chromatin it is of course true that there is an enhanced probability for other locations within a few nanometers to be occupied also, but this fine structure in the autocorrelation function is, we believe, only indirectly relevant to current assays of radiation-produced chromosome aberrations. What is highly relevant for F values is a larger-scale phenomenon: given that a particular location is occupied by chromatin from one specific chromosome, there is an enhanced probability that locations within, say, 1 μm are occupied by chromatin from that same chromosome.

One specific formalism for F values uses the *chromatin proximity function* $s(x)$, defined in terms of the autocorrelation function by[7]

$$s(x) = 4\pi x^2 \int d^3R \langle \sigma(\mathbf{R}+\mathbf{X})\sigma(\mathbf{R}) \rangle \quad , \quad \text{where} \quad d^3R = dR_1 dR_2 dR_3 \quad , \quad \text{and} \quad x = |\mathbf{X}| \quad . \tag{4}$$

Here, and throughout, integrals without specified limits go over all of 3-space (and have cutoffs automatically supplied by functions such as σ). In Eq. (4), it is assumed that, on average, chromatin distributions are spherically symmetric, so that s actually depends only on x, not the angles of $\mathbf{X}$. Inserting the prefactor $4\pi x^2$ in Eq. (4) gives a conventional normalization[6] of the functional form of $s(x)$.

In terms of the chromosome proximity function, the aberration yield Y can be approximated as[6,7]

$$Y \propto \int_0^{d_i} \frac{t(x)s(x)dx}{x^2} \tag{5}$$

Here the proportionality constant is irrelevant for an F value assay (which involves ratios, as discussed in the preceding section). In Eq. (5) d_i is the maximum initial distance for a DSB pair such that the pair can interact, and $t(x)$ is the familiar track structure proximity function. The function $t(x)$ can be interpreted by considering an ionization cluster sufficiently energetic and localized to produce a DSB in a chromosome.[1,5] Then $t(x)dx$ is proportional to the probability that there will be another such cluster at a separation between x and $x+dx$. Therefore if the radiation is on average spatially homogeneous the fact that different tracks are statistically independent implies

$$t(x) \propto 4\pi x^2 D \quad \text{(DSB induced by two different tracks)}, \tag{6a}$$

where the proportionality constant is again irrelevant for an F value assay. Moreover, in the "LET approximation" (chord length approximation),

$$t(x) \propto 2L \quad \text{(both DSB induced by same track)}, \tag{6b}$$

where L is the stopping power, and is independent of x. We have established by extensive Monte Carlo track simulations (to be discussed elsewhere) that, apart from spatial "fine structure" on scales less than 100 nm, this chord length approximation is applicable to ionization clusters of the kind suggested by Brenner and Ward[5] as being effective in producing DSB.

Using Eqs. (4)–(6) involves a number of idealizations and approximations.[7] One of these is neglecting cross correlations between chromatin density and radiation tracks, which will play a role to the extent that chromatin has significantly different properties as regards ionization cross sections and/or radical diffusion lengths from the surrounding milieu.[21] The formalism can be modified to include chromatin-track cross correlations, but for the F value assay we have in mind the resulting changes are expected to be minor, since the effects should not be biased for or against inter-chromosomal compared to intra-chromosomal, inter-arm aberrations. When considering intra-arm aberrations such cross-correlations could be more important. More generally, time-dependent, kinetic theoretical models more sophisticated than the approximation used here, given by Eqs. (4) and (5), will eventually be needed. However, it is clear, in view of the diffusion (or motion) limited nature of the DSB-DSB interactions, that the spatial chromatin autocorrelation function will continue to play a central role.

CHROMOSOME GEOMETRY

To compute $s(x)$ in Equation (4), and to distinguish between intra- or inter-chromosomal DSB-DSB reactions, we need a geometric model for chromatin. As before, we are concerned with interphase chromosomes considered on length scales of 100 nm or more, and with DNA stretches of $\approx 10^5$ base pairs or more.

Experimental Background

Some information on interphase chromosomes is more than a century old, but recently developed experimental techniques, especially FISH, have led to a considerable increase in our qualitative understanding of nuclear ultrastructure.[30] Importantly, it has been found that a single chromosome does not wind and twist its way throughout the entire interphase cell nucleus, as was once commonly assumed. Rather, at any instant, a chromosome in a mammalian cell seems to be confined to a sub-region comprising, very roughly, 10% or less of the nuclear volume[9] a finding which is in general agreement with the radiobiological evidence.[23,25] Enough data on nuclear ultrastructure is now accumulating to justify attempts to analyze, quantitatively rather than just qualitatively, overall chromosome structure in interphase mammalian cells.

A key question for a mathematical model is whether the higher order structure, beyond the level of the comparatively familiar 30 nm fiber,[19–21] is highly systematic and organized or has a considerable amount of randomness. In the latter case, a potentially useful approach is that of polymer physics, which models the behavior of a large, flexible linear molecule as a series of beads connected by links, with some of the bending angles at the beads partially or wholly random.[38] As long as random influences are dominant, many complications can be averaged out and surprisingly simple overall models can be obtained, even in situations where the molecule is subjected to a very complicated variety of forces.

Chromosomes do show systematic interactions, constraints, and structural motifs. It is known, for example, that certain specific portions of specific chromosomes are systematically associated in the nucleolus regions, where they cooperate in the biogenesis of ribosomes.[19] On the other hand, there is also evidence for considerable randomness in the ultrastructure of cell nuclei. For example, chromosome homologues are usually neither systematically associated nor systematically separated within the nucleus.[30] Moreover, such systematic geometric patterns as have been suggested on the basis of FISH observations often seem to be dependent on cell lines and/or location in specific sub-phases of interphase.[30,31] Radiobiologically, it is known that the number of inter-chromosomal aberrations involving one or the other chromosome of each chromosome homologous pair is approximately proportional to the length of that chromosome (see e.g., Ref. 39), as expected on a random model.

More quantitative information on large-scale interphase chromosome geometry comes from a recently developed technique for repeated measurements, in different cells, of the physical distance between two chosen points on one specified chromosome.[13,17] This technique directly gives data on the point-pair distance distribution function, closely related to the intra-chromosomal autocorrelation function. For genomic separations in the range $10^5 - 10^6$ base pairs, the distribution function was found to correspond to the distribution function predicted by a Gaussian polymer model.[17] In a Gaussian polymer, the angles at the beads are wholly random and the length of a link, rather than being deterministic, is given by a Gaussian-normal distribution. Any long, flexible molecule behaves like a Gaussian polymer on sufficiently large scales provided various complicating effects such as self avoidance can be neglected.[38]

In addition to giving a physical-length probability distribution for a given point pair, the Gaussian polymer model also relates average physical lengths to genomic distances. It predicts that the average physical distances between two points on a chromosome is proportional to the square root of the genomic separation. It was found, as indicated in Fig. 2, that this prediction works surprisingly well on scales of $\approx 10^5 - 10^6$ base pairs, at least for cells prepared by the methods of van den Engh et al.[17] But at still larger scales, instead of continuing to increase as the square root of the genomic separation, the measured

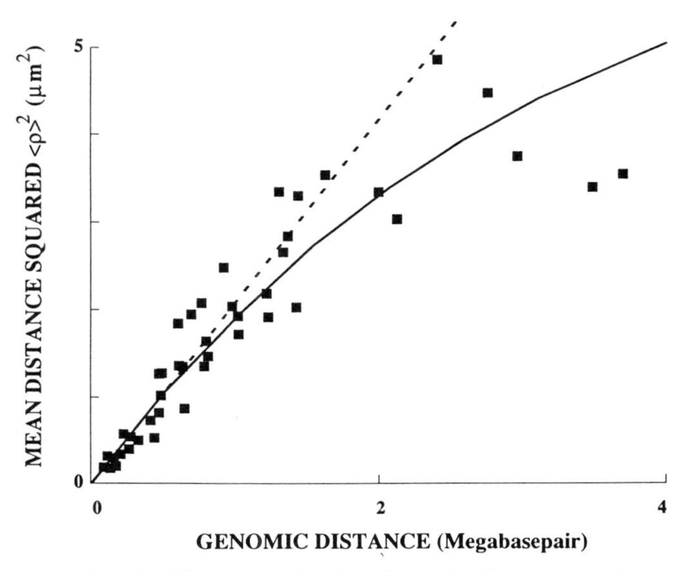

Figure 2. The square of physical distance as a function of genomic distance. Data, from van den Engh et al.,[17] for 45 point pairs in a six Megabasepair stretch near one end of human chromosome 4. The straight line shows the prediction of a Gaussian polymer model. The curved line is obtained from the confined Gaussian polymer model, model II of the text, by adjusting the parameters b and d_c.

geometric separations increase more slowly or become independent of genomic separations This leveling-off effect, which is visible for the larger separations in Fig. 2, may persist for separations up to the full length of a chromosome, from near one telomere to near the other.[13,40] The leveling-off effect indicates that at the largest scales a chromosome is subjected to some over-all localizing constraints not operative for a Gaussian polymer.

Mathematical Models

We now consider some formal models of chromosome geometry based on these experimental results. We shall assume, as is appropriate for human lymphocytes, that the cell nucleus is a sphere of diameter d, where $d \approx 6\,\mu\text{m}$. For convenience we normalize chromatin density using chromosome number, i.e.,

$$\int d^3R\, \sigma(\mathbf{R}) = 46 \quad . \tag{7}$$

As discussed above, the corrections needed to take into account the actual chromosome arm lengths in the human genome as compared to a model where all arm lengths are equal, are straightforward and lead to only minor numerical changes.[25] We thus take all chromosome centromeres to be in the center. We also take all the chromosomes in the genome to be independent, identically distributed, i.e., to have the same average geometry and no systematic interrelations. This assumption implies, in particular, that different chromosomes intertwine and overlap freely. Formally it also implies that for any $i,j = 1,\ldots,46$ with $i \neq j$:

$$\langle \sigma_i(\mathbf{R})\sigma_i(\mathbf{R}')\rangle = \langle \sigma_1(\mathbf{R})\sigma_1(\mathbf{R}')\rangle \quad , \quad \langle \sigma_i(\mathbf{R})\sigma_j(\mathbf{R}')\rangle = \langle \sigma_1(\mathbf{R})\sigma_2(\mathbf{R}')\rangle \quad . \tag{8}$$

We may define an inter-chromosomal chromatin proximity function $s_2(x)$, analogous to the function in Eq. (4) but contingent on the two DSB being on 2 specific, different chromosomes, as follows:

$$s_2(x) = 4\pi x^2 \int d^3R \langle \sigma_1(\mathbf{R}+\mathbf{X})\, \sigma_2(\mathbf{R})\rangle \quad , \quad \text{where} \quad x = |\mathbf{X}| \quad . \tag{9}$$

We shall take it that point pairs on different chromosomes are in effect random point pairs within the cell nucleus. Then for inter-chromosomal interactions we get

$$s_2(x) = (1/d)\; p(x/d) \quad , \tag{10}$$

where $p(y)dy$ is the probability that two points chosen at random within the unit sphere are within a small interval dy of being a distance y apart, i.e.,[41]

$$p(y) = 24y^2 - 36y^3 + 12y^5 \quad , \quad y \leq 1 \quad ; \quad p(y) = 0 \quad , \quad y \geq 1 \quad . \tag{11}$$

We now compare two different models. In model I, only over-all chromosome localization is taken into account. In model II, the linear connections implied by the actual DNA duplex as assembled into the 30 nm fiber, and the apparent polymer structure at length scales of $10^5 - 10^6$ base pairs, are also incorporated.

Model I.

We assume each chromosome is randomly dispersed within a chromosome localization sphere of diameter $d_c \leq d$, as indicated schematically in Fig. 3A. Possible physical

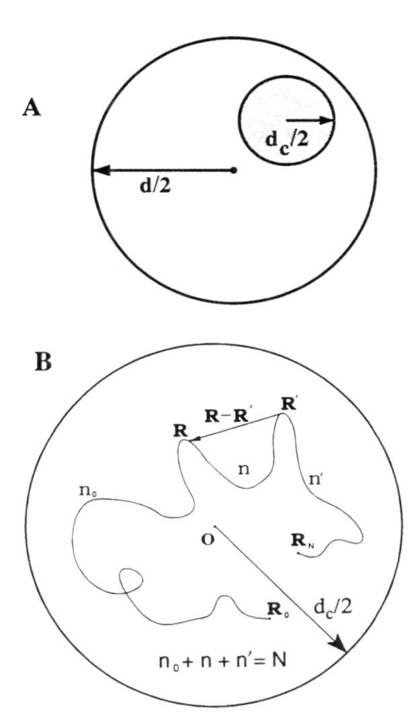

Figure 3. A. A chromosome localization sphere of diameter d_c inside a nucleus of diameter d. In model I, chromosome point pairs are considered to be random point pairs inside the localization sphere. B. Physical separations and base pair separations in model II. The cartoon shows a chromosome, a point **R**, which might be fluorescently labeled, separated by n polymer links from a point $\mathbf{R}'$ which might be labeled a different color. The two points are respectively n_0 and $n' = N - n - n_0$ links from the chromosome ends (i.e., the telomeres). As in model I the chromosome is confined in a spherical subregion of the cell nucleus.

mechanisms which might induce such localization include protein tethers to an organizing center for that particular chromosome or the confining influence of other macromolecules in the nucleus.

It now follows from a straightforward use of Eqs. (2), (4), and (7)–(10) that

$$s(x) = 46 \times 1/2 \times s_1(x) + 46 \times 1/2 \times s_1(x) + 46 \times 45 \times s_2(x) \quad , \quad \text{where} \tag{12}$$

$$s_1(x) = (1/d_c)p(x/d_c) \quad . \tag{13}$$

and p is defined in Eq. (11). In Eq. (12) the first term on the right refers to two DSB on the same arm of a chromosome, the second term refers to two DSB on opposite arms of the same chromosome, and the last term refers to two DSB on two different chromosomes.

Equations (4)–(6) and (9)–(13) are all that is needed to deduce F values from the model. Thus the model has two adjustable parameters, d_i and d_c. Even if we were to regard the nuclear diameter d as an unknown, instead of taking d=6 μm, the predictions for F values would depend only on the ratios d_i/d and d_c/d. For example we get from Eq. (4), (5), (6a) and (9)–(13) the low LET F value

$$F_{low} = 90(u^3 - 9u^4 + 2u^6)/g(w) \quad , \tag{14}$$

where $u = d_i/d$, $w = d_i/d_c$,

$$g(w) = (8w^3 - 9w^4 + 2w^6) \quad , \quad w \leq 1 \quad ; \quad \text{and } g(w) = 1 \quad , \quad w \geq 1 \quad . \tag{15}$$

Model II.

In model II, introduced by Hahnfeldt et al.,[24] the geometry of an interphase chromosome, analyzed on large scales, is taken as a Gaussian polymer constrained to lie within its own localization sphere, of diameter d_c, by forces localized at the surface of the sphere. Let N be the total number of links in the polymer, so there are $N + 1$ beads. In Fig. 3B, the chromosome is sketched on a scale so large the individual polymer links are not resolved. We could identify a bead with a matrix associated DNA region and assume a link is approximately the length of the putative chromatin loops between matrix associated regions,[19,20] i.e., is roughly 10^5 base pairs long, but these identifications are not essential features of the model because all the formal calculations are invariant under appropriate rescaling of link lengths.[38]

We consider configuration probabilities at thermal equilibrium, without explicitly analyzing polymer motions. To calculate averages such as $s_1(x)$ it is sufficient to obtain the polymer "Green function" $G(\mathbf{R_0},\mathbf{R_N};N)$, also sometimes referred to as the "statistical weight" or "distribution function" or "partition function."[38] In an appropriate continuum limit, which we henceforth adopt, N,n, and n_0 in Fig 3B are treated as continuous variables and the problem of finding the Green function $G(\mathbf{R},\mathbf{R}';n)$ reduces[38] to solving, for $n > 0$, a familiar initial-value, boundary-value problem for the diffusion equation, as follows:

$$\left[\frac{\partial}{\partial n} - \frac{b^2}{6}\frac{\partial^2}{\partial \mathbf{R}^2}\right] G(\mathbf{R},\mathbf{R}';n) = 0, \text{where } G(\mathbf{R},\mathbf{R}',0) = \delta(\mathbf{R}-\mathbf{R}') \text{and } G = 0 \text{ if } |\mathbf{R}| \geq (1/2)d_c \quad . \tag{16}$$

Here b is the thermal equilibrium rms length of one link in the absence of the spherical constraining surface, $\partial^2/\partial\mathbf{R}^2$ is the Laplacian, and δ is the three-dimensional Dirac delta function.

$G(\mathbf{R},\mathbf{R}';n)$ is interpreted both as the usual influence function for the diffusion equation and as the conditional probability that one end of an n-link polymer will be at $\mathbf{R}'$, given that the other end is at $\mathbf{R}$. The conditional probability interpretation of $G(\mathbf{R},\mathbf{R}';n)$ implies that the autocorrelation function $\langle\sigma_1(\mathbf{R})\sigma_1(\mathbf{R}')\rangle$ is, for the case that $\mathbf{R}$ and $\mathbf{R}'$ lie on opposite arms of the same chromosome,

$$\langle\sigma_1(\mathbf{R})\sigma_1(\mathbf{R}')\rangle =$$

$$\frac{1}{K}\int_0^{1/2N} dn_0 \int_{1/2N}^{N} dn' \int d^3R_0 d^3R_N G(\mathbf{R}_0,\mathbf{R};n_0) G(\mathbf{R},\mathbf{R}';N-n_0-n') G(\mathbf{R}',\mathbf{R_N};n') \quad , \tag{17}$$

where $K = N^2\pi d_c{}^3/24$ is a normalization constant. In Eq. (17) the integrals over n_0 and n' correspond to averaging over genomic locations on each arm and the integrals over $\mathbf{R_0}$, $\mathbf{R_N}$ correspond to the usual rule for expressing a probability as a superposition of conditional probabilities.

The solution of Eq. (16) is well known.[42]

$$G = \sum_{lmi} \frac{\psi_{lmi}(\mathbf{R})\psi_{lmi}(\mathbf{R}')}{A_{li}} \exp[-k_{li}n], \text{ where } \psi_{lmi}(\mathbf{R}) = Y_{lm}(\theta,\phi)\chi_{li}(R) \quad , \tag{18}$$

and $A_{li} = \int R^2 dR |\chi_{li}|^2$. Here Y_{lm}, ($l = 0,1...$, $|m| \leq 1$) denotes a spherical harmonic, the function $\chi_{li}(R)$ ($i = 1,2,...$) is a spherical Bessel function, and k_{li} is a corresponding eigenvalue. Specifically,[24,42] if $j_l(u)$ is the spherical Bessel function of order l and κ_{li} is the ith root of the equation $j_l(\kappa_{li}) = 0$, then $\chi_{li}(R) = j_l(2\kappa_i R/d_c)$ and $k_{li} = 2b^2 \kappa_{li}^2/3d_c^2$.

In the limit $nb^2/d_c^2 \to 0$, solving Eq. (17) or taking the appropriate limit in Eq. (18) gives for G the Green function of a Gaussian polymer, namely

$$G(\mathbf{R},\mathbf{R}';n) = \left[\frac{2\pi nb^2}{3}\right]^{-3/2} \exp\left[-\frac{3(\mathbf{R}-\mathbf{R}')^2}{2nb^2}\right] . \tag{19}$$

Thus for sufficiently small link number, or a sufficiently small ratio b/d_c, the spherical confining surface in Fig. 3B does not significantly influence the conditional probabilities.

In model II the intra-chromosomal, inter-arm chromatin proximity function $\hat{s}_1(x)$, is given by Eqs. (4), (17) and (18). For the experiment of Engh et al.,[17] and a nominal link length of 10^5 base pairs, $b = 0.05\ d_c$. Using $b = 0.05\ d_c$ gives, as in model I, equations for F values which depend on two parameters, d_i and d_c. We were not able to perform analytically all the integrals needed to compute $\hat{s}_1(x)$, so the final steps in its evaluation were done with Monte-Carlo methods.

NUMERICAL RESULTS

It was found that the two models give quite similar results for the F values. Thus when comparing pairs of DSB on different chromosomes with pairs on opposite arms of the same chromosome, the local chromatin connections, corresponding to the 30 nm fiber structure, which are incorporated into model II but omitted from model I, do not play a dominant role. We first discuss results from model I.

As an example, Fig. 4 shows the low and high LET F values as a function of chromosome localization sphere diameter d_c in a special case, namely the case where the

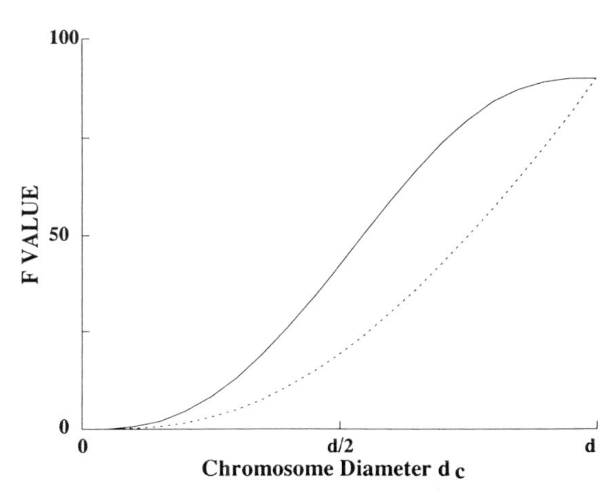

Figure 4. Predicted F values, according to model I, as a function of d_c in the special case $d_i = d_c$. The solid line is the low LET result, using Eq. (6A), the dotted line is the High LET result, using Eq. (6B).

maximum initial distance, d_i, for ultimate DSB-DSB interactions is taken equal to d_c. For lymphocyte data, the nuclear diameter d can be taken as $d \approx 6\ \mu m$. Since the model is scalable, in the sense that the F values are determined by d_i/d and d_c/d, the x axis in Fig. 4 is expressed in units of d. The graph shows that for $d_i = d_c \approx d/3$ the F values are approximately the values $F \approx 15$ and $F \approx 6$ observed at low and high LET respectively.

Similar graphs can also be obtained[7] for values of $d_i/d_c \neq 1$. In general the predicted F value is found to be a monotonic increasing function of both d_i and d_c. For d_c having its maximum value, equal to the diameter d of the cell nucleus, $F = 90$, the value for complete randomness, regardless of the value of d_i. This merely means that there would be no proximity effects deducible from F values if all chromosomes were equally, randomly and independently dispersed throughout the nucleus. Using the maximum value for interactions, i.e., taking $d_i = d$, also gives $F = 90$ at low LET irrespective of the value of d_c, but at high LET the F value for $d_i = d$, $d_c < d$ is less than 90. Thus, for localized chromosomes (i.e., $d_c < d$), the clustering of DSB induced by high LET tracks gives, on average, enhanced intra-chromosomal aberrations even if all DSB pairs within the nucleus are equally likely to interact (i.e., even if $d_i = d$). This result can be understood from the quadratic nature of the DSB-DSB interactions and considerations of the variance of the induced DSB number for one chromosome as follows.[7] High LET inflicts DSB in a localized way, so that, even if there are no spatial limitations on interactions, a localized chromosome will sometimes have many DSB, with a resulting large intra-chromosomal aberration yield quadratic in the DSB number, which more than compensates for the zero yield cases where the chromosome is missed entirely.

Figure 5 shows the predicted ratio of the low LET F value to the high LET F values. The values obey $F_{high} \leq F_{low}$, as would be expected from the greater clustering of high LET DSB. Notice in Fig. 5 that for $d_i \ll d_c$ the two predicted F values are nearly equal. The intuitive reason for this, at first sight surprising, result is the following. We may as well focus attention on a pair of DSB which are initially within the maximum distance d_i. When $d_i \ll d_c$ we may as well also assume that if one of the DSB is initially in the region occupied by a certain localized chromosome, the other DSB is also. Then the ratio of inter- to intra-

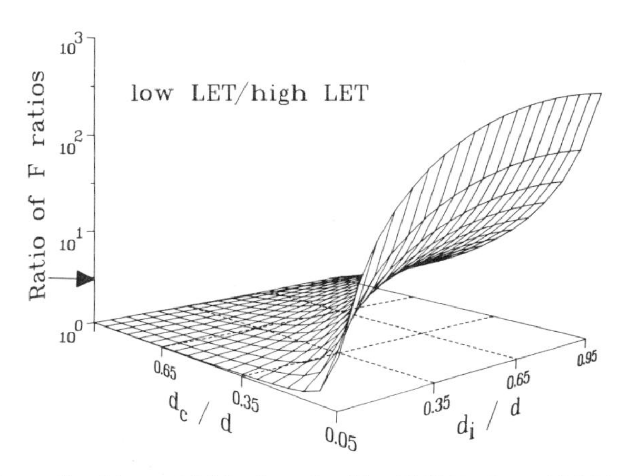

Figure 5. The results predicted by model I for the ratio of low LET F value to high LET F value as a function of chromosome localization sphere diameter d_c and of DSB-DSB maximum initial distance d_i in a spherical cell nucleus of diameter d. The arrow corresponds to the empirical value, ≈ 2.5.

chromosomal yields is simply given by a standard chemical concentration argument, involving the ratio of the volume of one chromosome to the volume of the cell nucleus, and this ratio is independent of the LET. It is true that in the case $d_i \ll d_c$ the high LET yields will tend to be much larger than the low LET yields, but the F values are the same.

If we assume, on the basis of the data quoted earlier, that $F_{low} \approx 15$ and $F_{high} \approx 6$, the equations of model I can be solved numerically for d_c and d_i. The result is

$$d_c \approx d_i \approx 0.32d \quad \text{(Model I)} \quad . \tag{20}$$

For the confined polymer model, model II, a corresponding analysis can be carried out, with qualitatively similar results. In particular, it is found that

$$d_c \approx 0.37d \quad . \quad d_i \approx 0.31d \quad \text{(Model II)} \quad . \tag{21}$$

However, perhaps the most useful way to compare the two models is to consider the rms chromatin radius for one chromosome, r, defined by

$$r^2 = \int d^3R \, \sigma_1(\mathbf{R})|\mathbf{R}|^2 \quad .$$

The rms radius is given by $r^2 = (1/8)d_c^2$ for model I and given by $r^2 \approx (1/12)d_c^2$ for model II. When expressed in terms of r instead of d_c the results given by Eqs. (20) and (21) are very similar, i.e., $r \approx 0.113d$ for model I and $r \approx 0.107d$ for model II.

CONCLUSIONS

Radiobiology needs information on large-scale chromosome geometry during interphase, and radiobiology can add significantly to this information.

Quantitative modeling of chromosome geometry on scales larger than 10^5 base pairs is in its infancy. The models presented here are conceptually simple and numerically tentative. But some progress has been made. The success of the polymer models in the range $\approx 10^5$ to $\approx 10^6$ base pairs[17] is evidence that over such scales there is indeed considerable randomness. The parameters inferred for the polymer model are consistent with the structure of DNA and chromatin on the level of the 30 nm fiber.[24] If polymer models are capable of predicting the main features of interphase chromosome geometry on still larger scales, the simplifications obtained by averaging out details will be very significant in a variety of applications, including radiobiology. Further experiments on point-pair distributions, of the kind developed by Trask and coworkers,[13,40] should help determine the appropriate mix of systematic and random structural motifs, leading to refinement of the models.

We have concluded from the radiobiological evidence, specifically from the ratio, F, of inter-chromosomal to intra-chromosomal, inter-arm aberrations, that chromosomes are localized, in regions with linear dimensions perhaps (1/3) of the nuclear diameter, i.e., $d_c \approx d/3$. The conclusion is not strongly model dependent, since models I and II gave comparable answers. The value of d_c is consistent with information obtained from FISH labeling of whole chromosomes during interphase.[30] Our additional result for the maximum initial distance that allows ultimate DSB-DSB interactions, namely $d_i \approx d/3$, is consistent with other types of radiobiological information.[23] Further geometric information on chromosomes should be available by considering ratios of restitutions to pairwise DSB interactions, ratios of radiation-produced intra-arm chromosomal aberrations, such as acentric rings, to intra-chromosomal, inter-arm aberrations, and ratios of 2-chromosome aberration types to the 3-chromosome aberration types which have recently been observed.[15,16]

Pairwise interaction of DSB is actually a dynamic process.[43–45] Models which have no explicit time dependence, such as the models presented in this article, are at best approximations, typically perturbation approximations,[45] to appropriate kinetic models. Kinetic models require some knowledge of chromatin motion, not just knowledge of chromatin geometry. An attractive feature of polymer models for radiobiological applications is that each model for average chromosome geometry at thermal equilibrium automatically implies a dynamic model with which to estimate chromatin motion.[38] Current Monte-Carlo computer codes for polymer motion[46] are in principle capable of handling 46 large polymers, moving with self and mutual avoidance, so a kinetic model of DSB-DSB interactions may be feasible.

Apart from their implications for the basic biology of cell nucleus ultrastructure, F values are also potentially significant in radiation dosimetry. As we have seen, F values for densely ionizing radiations appear to have a value $F \approx 6$, significantly lower than yet observed for any other clastogens including x-rays, γ-rays, or chemical carcinogens,[7] in agreement with mechanistic expectations. Therefore, F values have the potential to act as an indicator of prior high-LET radiation exposure. F values for stable chromosomal aberrations (compare Fig. 1, caption) in individuals who might have been exposed to a significant dose of high-LET radiation can be measured using Giemsa staining techniques and, importantly, much faster measurement techniques are now being developed based on FISH. The measured F values could provide stronger evidence than currently possible as to whether an individual was exposed to a significant dose of densely-ionizing radiation.

An example of the application of such a "chromosomal fingerprint" for high LET could be in regard to the question of the high- LET neutron exposure to A-bomb survivors at Hiroshima. For Hiroshima A-bomb survivors, an F value of 6.2 ± 0.7 has been measured.[39] This may be interpreted as direct evidence that a significant proportion of the equivalent dose to which the survivors were exposed was at high LET, in accordance with recent physical measurements,[47,48] but in contrast to dose reassessment calculations.[49]

A second application might be in the interpretation of "rogue" cells, i.e., aberrant cells in which an extremely large number of aberrations is observed, much higher than expected based on the mean yield of aberrations in other comparable cells in the same individual. These rogue cells have been observed in individuals exposed at Hiroshima or at Chernobyl.[50–52] Recent measurements of rogue cells in children exposed from the Chernobyl accident have shown a low F value of 4.6 ± 1.4[52]. One possible explanation of these rogue cells is that they are caused by "hot" particles, originating in the Chernobyl fire, which contain alpha emitters.[53] The anomalously low F value gives some credence to this explanation.

Finally, the most general application of this potential chromosomal fingerprint is in the field of radon. Currently, epidemiological studies to determine radon risks at low exposures are limited by the relatively large "background" lung cancer rates produced by carcinogens other than radon. It is possible that bronchial epithelial cells (basal cells and their progeny), taken from the normal-tissue margins around an excised lung tumor, could be assayed for F values in stable aberrations. These measurements might then provide a local history of high-LET exposure in the vicinity of the tumor.

Summary

Chromosomal aberrations are significant because they are an observable damage state intermediate between radiation induced DSB and the ultimate effects of medical interest. They are also significant for the light they can shed on nuclear ultrastructure and on enzymatic processing of DSB. Due to the advent of FISH, experimental information on chromosomal aberrations will no doubt continue to increase very rapidly. Although assaying chromosomal aberrations literally involves looking at individual DNA molecules, the main aspects of radiation track structure relevant to inter-chromosomal or intra-chromosomal, inter-arm aberrations, apart from those nm scale aspects relevant to DSB induction, are features on a comparatively large scale, which can be summarized in ionization cluster spatial autocorrelation functions for distances from 100 nm up to several μm. When intra-chromosomal, intra-arm aberrations are considered in detail, ionization cluster autocorrelation functions on scales from 10–100 nm also become relevant. Track features on these scales, from 10 nm to several μm, may be comparatively simple, but track structure modelers should not neglect them.

ACKNOWLEDGMENTS

Work supported by NIH grants CA-49062 and OH-02931 (DJB), DOE grant DE-FG02-88ER60631 (DJB), NIH grants CA-26279 and CA-46295 (JFW), DOE grant DE-FG03-88ER60-660 (JFW), and NSF grant DMS-9302704 (RKS). We are grateful to C. Geard, E. Hall and L. Hlatky for valuable discussions.

REFERENCES

1. D.T. Goodhead. Relationship of microdosimetric techniques to applications in biological systems. *Int. J. Radiat. Biol.* **56**:623–634 (1989).
2. R.B. Painter. The role of DNA damage and repair in cell killing induced by ionizing radiation. In *Radiation Biology in Cancer Research*, R.E. Meyn and H.R. Withers eds., pp. 59–68. Raven Press, New York (1979).
3. M. Sorsa, J. Wilbourn and H. Vainio. Human cytogenetic damage as a predictor of cancer risk. *Iarc Scientific Publications* **116**:543–54 (1992).
4. D.C. Lloyd and A.A. Edwards. Chromosome aberrations in human lymphocytes: effects of radiation quality, dose, and dose rate. In *Radiation-Induced Chromosome Damage in Man*, T. Ishihara and M.S. Sasaki, eds., pp. 23–29. Alan R Liss, New York (1983).
5. D.J. Brenner, and J.F. Ward. Constraints on energy deposition and target size of multiply-damaged sites associated with DNA double strand breaks. *Int. J. Radiat. Biol.* **61**:737–748 (1992).
6. A.M. Kellerer. Fundamentals of microdosimetry. In *The Dosimetry of Ionizing Radiation*, K. Kase, B. Bjarngard and F. Attix, eds., pp. 78–162. Academic Press, Orlando (1985).
7. D.J. Brenner and R.K. Sachs. Chromosomal "fingerprints" of prior exposure to densely-ionizing radiation. *Rad. Res.* **140**:134-142.
8. P. Lichter, T. Cremer, J. Borden, L. Manuelidis, and D.C. Ward. Delineation of individual human chromosome aberrations in metaphase and interphase tumor cells by in situ suppression hybridization using chromosome–specific library probes. *Human Genetics* **80**:R224–34 (1988).
9. H. Van Dekken, D. Pinkel, J. Mulliken, B. Trask, G. Van den Engh, and J. Gray. Three-dimensional analysis of the organization of human chromosome domains in human and hamster hybrid cells. *J. Cell Sci.* **94**:299–306 (1989).
10. J.N. Lucas, T. Tenjin, T. Straume, D. Pinkel, D. Moore, 2d., M. Litt, and J.W. Gray. Rapid human chromosome aberration analysis using fluorescence *in situ* hybridization. *Int. J. Radiat. Biol.* **56**:35–44, **56**:201 (1989).
11. T. Cremer, S. Popp, P. Emmerich, P. Lichter P and C. Cremer, Rapid metaphase and interphase detection of radiation-induced chromosome aberrations in human lymphocytes by chromosomal suppression in situ hybridization. *Cytometry* **11**:110–8 (1990).
12. J.W. Evans, J.A. Chang, A.J. Giaccia, D. Pinkel, and J.M. Brown. The use of fluorescence in situ hybridisation combined with premature chromosome condensation for the identification of chromosome damage. *British Journal of Cancer* **63**:517–21 (1991).
13. B.J. Trask, H. Massa, S. Kenwrick and J. Gitschier. Mapping of human chromosome Xq28 by two-color fluorescence in situ hybridization of DNA sequences to interphase cell nuclei. *American Journal of Human Genetics* **48**:1–15 (1991).
14. J.N. Lucas, A. Awa, T. Straume, M. Poggensee, Y. Kodama, M. Nakano, K. Ohtaki, H.-U. Weir, D. Pinkel, J.W. Gray and G. Littlefield, Rapid translocation frequency analysis in humans decades after exposure to ionizing radiation. *Int. J. Radiat. Biol.* **62**:53–63 (1992).

15. J.M. Brown and J.W. Evans. Fluorescence in situ hybridization: an improved method of quantitating chromosome damage and repair. *Brit. J. Rad. Supplement* **24**:61–4 (1992).
16. J.N. Lucas and R.K. Sachs. Using 3-colour chromosome painting to decide between chromosome aberration models. *Proc. Nat Acad. Sci. U.S.* **90**:1484–1487 (1993).
17. G. van den Engh, R. Sachs and B. Trask. Estimating genomic distance from DNA sequence location in cell nuclei using a random walk model, *Science* **257**:1410–1412 (1992).
18. J.F. Ward, G.D.D. Jones and J.R. Milligan. biological consequences of non-homogeneous energy deposition by ionizing radiation. *Radiation Protection Dosimetry* 52:271-276. (1994).
19. K.E. Van Holde. *Chromatin*. Springer Verlag, NY (1989).
20. A. Wolffe. *Chromatin: Structure and Function*. Academic Press, San Diego (1992).
21. A. Chatterjee and W. R. Holley. Early Chemical Events and Initial DNA Damage. In *Physical and Chemical Mechanisms in Molecular Radiation Biology*, W.A. Glass and M.N. Varma, eds., pp. 257–285. Plenum Press, NY (1992).
22. J.R.K. Savage, and D.G. Papworth. The relationship of radiation-induced yield to chromosome arm number. *Mutat. Res.* **19**:139–143 (1973).
23. D.J. Brenner. On the probability of interaction between elementary radiation-induced chromosomal injuries. *Radiat. Environ. Biophys.* **27**:189–199 (1988).
24. P. Hahnfeldt, J.E. Hearst, D.J. Brenner, R.K. Sachs and L.R. Hlatky. Polymer models for interphase chromosomes. *Proc. Nat. Acad. Sci. USA* **90**:7854-7858 (1993).
25. L. Hlatky L, R.K. Sachs, and P. Hahnfeldt. The ratio of dicentrics to centric rings produced in human lymphocytes by acute low-LET radiation. *Radiat. Res.* **129**:304–308 (1992).
26. L.G. Littlefield. Application of fluorescence in situ hybridization techniques in radiation cytogenetics. Radiation Research Meeting #41:126 (Dallas 1993).
27. K. Sax. An analysis of X-ray induced chromosomal aberrations in Tradescantia. *Genetics* **25**:41–68 (1940).
28. D.E. Lea. *Actions of radiations on living cells*. Cambridge University Press, Cambridge [Eng.] (1955).
29. L. Manuelidis. A view of interphase chromosomes. *Science* **250**:1533–4 (1990).
30. T. Haaf and M. Schmid, Chromosome topology in mammalian interphase nuclei. *Experimental Cell Research* **192**:325–332 (1991).
31. M. Fergusson and D.C. Ward. Cell cycle dependent chromosomal movement in pre-mitotic human T-lymphocyte nuclei. *Chromosoma* **101**:557-565 (1992).
32. J.R.K. Savage. Mechanisms of chromosome aberrations. In *Mutation and the Environment, Progress in Clinical and Biological Research* **340B**, M. Mendelsohn and R.J. Albertini, eds., pp. 385–396. Wiley-Liss, NY (1990).
33. M.A. Bender, and P.C. Gooch. Persistent chromosome aberrations in irradiated human subjects. II. Three and one-half year investigation. *Radiat. Res.* **18**:389–396 (1963).
34. S. Sasaki, T. Takatsuji, Y. Ejima, S. Kodama, and C. Kido. Chromosome aberration frequency and radiation dose to lymphocytes by alpha-particles from internal deposit of Thorotrast. *Radiat. Environ. Biophys.* **26**:227–238 (1987).
35. E.J. Tawn, J.W. Hall, and G.B. Schofield. Chromosome studies in plutonium workers. *Int. J. Radiat. Biol.* **47**:599–610. (1985).
36. J. Pohl-Ruling, P. Fisher, D.C. Lloyd, A.A. Edwards, A.T. Natarajan, G. Obe, K.E. Buckton, N.O. Bianchi, P.P.W. Buul, B.C. Das, F. Dashil, L. Fabry, M. Kucerova, A. Leonard, R.N. Mukherjee, U. Mukherjee, R. Nowotny, P. Palitti, Z. Polivkova, T. Sharma, and W. Schmidt. Chromosomal damage induced in human lymphocytes by low doses of D-T neutrons. *Mutat. Res.* **173**:267–272 (1986).
37. J.S. Prosser, A.A. Edwards and D.C. Lloyd. The relationship between colony forming ability and chromosomal aberrations induced in human T-lymphocytes after γ-irradiation. *Int. J. Radiat. Biol.* **58**:293–301 (1990).
38. M. Doi and S.F. Edwards. *The Theory of Polymer Dynamics*. Oxford Press, Oxford (1988).
39. R.K. Sachs, A. Awa, Y. Kodama, M. Nakano, K. Ohtaki, and J.N. Lucas. Ratios of radiation-produced chromosome aberrations as indicators of large-scale DNA geometry during interphase. *Radiation Research* **133**:345–350 (1993).
40. B. Trask, D. Pinkel, and G. van den Engh. The proximity of DNA sequences in interphase cell nuclei is correlated to genomic distance and permits ordering of cosmids spanning 250 kilobase pairs. *Genomics* **5**:710–17 (1991).
41. M.G. Kendall and P.A.P Moran. *Geometrical Probability*, pp. 53–54. Charles Griffin & Co., London (1963).
42. P.M. Morse and H. Feshbach. *Methods of Theoretical Physics*. McGraw-Hill, New York) (1953).
43. D.J. Brenner. Track structure, lesion development, and cell survival. *Rad. Res.* **124**: S29–S37 (1990).
44. S.B. Curtis. Mechanistic Models. In *Physical and Chemical Mechanisms in Molecular Radiation Biology*, W.A. Glass and M.N. Varma, eds., pp. 367–386. Plenum Press, NY (1992).

45. R. Sachs, P-L. Chen, P. Hahnfeldt, and L. Hlatky, DNA damage caused by ionizing radiation. *Mathematical Biosciences* **112**:271–303 (1993).
46. N. Madras and A. Sokal. The pivot algorithm: a highly efficient Monte Carlo method for self avoiding walks. *J. Stat. Phys.* **50**:107–186 (1988).
47. M. Hoshi, K. Yokoru, S. Sawada, K. Shizuma, K. Iwatani, H. Hasai, T. Oka, H. Morishima, and D.J. Brenner. Europium-152 activity induced by Hiroshima atomic-bomb neutrons. Comparison with the 32P, 60Co and 152Eu activities in Dosimetry System 1986 (DS86). *Hlth. Phys.* **57**:831–837 (1989).
48. T. Straume, S.D. Egbert, W.A. Woolson, R.C. Finkel, P.W. Kubik, H.E. Gove, P. Sharma, and M. Hoshi. Neutron discrepancies in the new (DS86) Hiroshima dosimetry. *Hlth. Phys.* **63**:421–426 (1992).
49. W.C. Roesch, (ed.), *US-Japan Joint Reassessment of Atomic Bomb Radiation Dosimetry in Hiroshima and Nagasaki*. Radiation Effects Research Foundation, Hiroshima (1987).
50. A.A. Awa, and J.V. Neel. Cytogenetic 'rogue' cells, what is their frequency, origin and evolutionary significance? *Proc Nat. Acad. Sci. USA,* **83**:1021–1025 (1986).
51. J.V. Neel, A.A. Awa, Y. Kodama, M. Nakono, and K. Mabuchi. 'Rogue' lymphocytes among Ukrainians not exposed to radioactive fallout from the Chernobyl accident, the possible role of this phenomenon in oncogenesis, teratogenesis, and mutagenesis. *Proc. Nat. Acad. Sci. USA* **89**:6973–6977 (1992).
52. A.V. Sevan'kaev, A.F., Tsyb, D.C. Lloyd, A.A. Zhloba, V.V. Moiseenko, A.M. Skrjabin, and V.M. Climov. 'Rogue' cells observed in children exposed to radiation from the Chernobyl accident. *Int. J. Radiat. Biol.* **63**:361–367 (1993).
53. A.E. Romanenko. *Medical consequences of the accident at the Chernobyl nuclear power station.* Medical Science Academy of the USSR, Kiev (1991).

DISCUSSION

Chatterjee: How does one include repair?

Sachs: The formalism doesn't include restitution explicitly, which dominates. But there is no preferential restitution for inter versus intra, so when we compute the ratio, the restitution cancels out.

Holley: How many base pairs are involved in the two points, the red and the blue?

Sachs: Somewhere between $\sim 10^5$ and $\sim 4 \cdot 10^6$, depending on the pair.

Holley: One other question. On these Gaussian chain models, what is the typical size of the link?

Sachs: In Gaussian models, everything is scalable so you can actually assign a nominal link size, which we chose to be a hundred thousand base pairs. It needs to be big enough.

Miller: Are you saying by that there's no natural scale for the link length?

Sachs: In a Gaussian polymer, there's no natural scale.

Miller: But in the genome, there's no natural scale?

Sachs: We tried for a long time to identify a link with a putative 50,000 or 100,000 basepair loop. That isn't working too well. At the moment, the right statement is: In the mathematics, there's no natural scale. Of course, the actual chromosome has various natural length scales.

Curtis: Did you look at the data from the survivors at Nagasaki?

Sachs: No, that should, of course, be on the agenda . When the data becomes available we will look at it.

Miller: Is the process by which double strand breaks evolve into these other entities a contact process?

Sachs: Sure. The idea is if pairs are made within a micron or two, then there's some chance that by the flexing, or reptation of the chromosome, eventually they will be brought within a nanometer or so and then they can with luck, or bad luck, make an aberration. I think this is possibly a glorious opportunity for Monte Carlo modeling because, of course, it's just nonsense to say that up to two microns, there's an equal probability as if they were initially half a micron apart, it must somehow fall off with the distance. That depends now, on 46 very large polymers, performing Brownian motions, presumably in a very viscous medium. I'm told polymer Monte Carlo methods can handle that problem. Part of what is hanging us up is we have no idea what the viscosity is. It must be enormous. The other part of what is hanging us up is we're worried about systematic restraints. A big Monte Carlo calculation of the actual motion should be made instead of the crude on/off assumption calculation.

Miller: The conceptual problem I'm having is that the process of making chromosome aberrations from DSBs intrinsically has to involve motion of the chromosome, but that motion must preserve the double strand break as an entity. Why are the fundamentals of the problem the double strand breaks and not the end groups of double strand breaks?

Sachs: Perhaps the ends are held more or less together by proteins, but perhaps end groups move independently. There's some recent data which, in fact, suggests that end independence is the case and our calculations could be repeated for this case without major changes. By the way, end independence is a very old model due to Lea and Sax around 1940, and others.

Nikjoo: You have based your model of tracks on ionizations only. Could you carry out your calculations with a more realistic model of track structure?

Sachs: You really have to ask Dave Brenner. I gather he looks for ionization clusters. A cluster, for example, is several ionizations within a few nanometers. Brenner and Ward got those numbers by a different RBE argument. Now, those clusters Dave takes as entities, then it is the autocorrelation function for clusters which is considered. I don't know any direct evidence that excitations are relevant to aberrations.

Terrissol: Do you think it is sufficient to use approximate proximity function? In the calculation, you used a "T" function.

Sachs: Let me say it this way — we're doing a naive time independent approximation to what is actually a time dependent process. I think one knows from the well-mixed case how the approximation is related to the whole process. It's a first order singular perturbation approximation. Now, for the time dependent process, it's of course completely wrong to look merely at these autocorrelation functions. You have to watch the time development, both of the chromosomes and of the track.

I believe that on a scale of a hundred nanometers and for the particular assay that we're looking at our model is a reasonable first approximation and, hopefully, will guide the big Monte Carlo calculations that, in fact, need to be done.

Paretzke: Do you have plans to take the nucleo skeleton and the normal Brownian diffusion of the molecules into consideration?

Sachs: Yes. The constraints I mentioned may be due to the nuclear matrix. Also working with John Hearst, Phil Hahnfeldt, Lynn Hlatky and others, we have some simple minded models where you put down a deterministic protein backbone, which is the same length in interphase and metaphase, and now you let the chromatin come out as loops from that, trying to combine the best features of the Laemmli model with the polymer models. Of course, sub nuclear structures are present and they play a crucial role.

Glaeser: You showed how fluorescent labeling demonstrates that the chromosomes are localized within a certain spatial domain. Is there any knowledge about whether two different chromosome domains interpenetrate, or are they completely separated? My impression is probably the latter, so the possibility of getting ligation between double strand breaks on two different chromosomes would be restricted just to those parts of the chromosome on the surfaces of their respective domains. That may really give rise to big changes in the theory.

Sachs: Absolutely. In fact, in a much older theory due to Savage and Papworth they get quite reasonable results on this kind of surface stuff. Dave Brenner subscribes to this surface picture. However, I believe in reality chromosomes interpenetrate freely. It's like a bowl of spaghetti, in my opinion.

Glaeser: What does Fluorescence in situ hybridization (FISH) say about whether the domains are separate or whether they interpenetrate?

Sachs: FISH, at the moment, says nothing except that they're so large that it's hard to see how they could fail to interpenetrate when you put 46 of them into the nucleus. I suppose one should color one chromosome red and one green, then look for the red/green overlap. That has not been done.

Glaeser: That would have to be done as a three dimensional reconstruction as well, and not just as a projection.

Sachs: If you look at the projections, you could maybe from the statistics, disentangle that much. If green was in front of red, would you see something different than if red was in front of green?

Glaeser: No, but you could use optical sectioning or tomography to look at special domains in three dimensions, and thereby separate the color, if they did not inter-penetrate.

Sachs: Yes. That's a much harder experiment. My opinion is they interpenetrate; yours is that they don't. I think it is a crucial question. We shall soon see experimentally.

Chatterjee: There are techniques which show that the centromere is attached to a nuclear membrane, and telomere is in the nucleus.

Sachs: That's both cell cycle location dependent and cell type dependent.

Glaeser: And maybe chromosome dependent.

Chatterjee: Could it be that the chromosomes are distributed randomly in the nucleus and also they are undergoing Brownian motion.

Sachs: Our hope is that you have some mixture of systematic and random motifs, each simple enough to sort out.

Zaider: I was quite interested to see that your calculations show support for an idea that Kellerer had some ten years ago, namely that cellular nuclear sensitive material was being made of these compact domains. That explains certain experimental data. Your calculations, in fact, bring further support to that idea.

Sachs: I agree that the "F" ratio data supports the other data for localization.

Zaider: In what way is your calculation different from that of Kellerer?

Sachs: First, we're trying to put in the polymer, though in the end it didn't make much difference. Second, we're focusing specifically on the "F" ratio, which was not done in those earlier calculations. In addition, when you introduce these autocorrelation functions, then in principle you have the possibility of introducing things like correlations between the chromatin and the track structure.

Zaider: There are two things that appear controversial in what you're saying. One is the ability to treat the track and the biological material as independent entities. This is probably the first assumption that will have to go. The other thing that appears to be implicit in your calculation is the assumption that double strand breaks are produced with an RBE of 1, independent of radiation quality.

Sachs: No, RBE is irrelevant, because using a ratio is a very powerful thing. It means we don't really need to think about DSB RBE's. Regarding your first point, the formalism can accommodate correlations between ionization and chromosome structure

Zaider: When you write a proximity function of ionizations, you basically assume that one ionization or a couple of ionizations will result in a double strand break, and that will immediately imply that the RBE for double strand breaks must be independent of LET.

Sachs: No, not if you look at the "F" value. Even if you're looking at yields, then this argument fails when looking at clusters. But we are looking at a ratio of yields so the RBE for double strand breaks doesn't matter.

Zaider: Wouldn't the efficiency with which the double strand breaks are produced come into the calculation?

Sachs: No.

Zaider: Now, as far as indirect effects are concerned, would those be unimportant for your argument?

Sachs: Again, I presume they're important, but I presume they ratio out because they should not bias for inter versus intra.

Inokuti: Just for clarification, I think you emphasized in your discussion with Zaider that the ratio cancels out, but isn't there implied an assumption that chemical nature of inter and intra is the same?

Sachs: Yes, but the chemical species is the double strand break locally. Now, it has to choose another strand break a hundred nanometers away. That other one doesn't know whether it's on the same chromosome by the local chemistry unless there's some signal. The local chemistry here and the local chemistry there do not know whether these are on the same chromosome; therefore, they should not bias for inter versus intra.

Inokuti: I still think it's a chemically significant assumption because the precise molecular environment at any time is not necessarily similar.

Sachs: What I'm saying is inter versus intra is a two double strand break consideration where the two double strand breaks are typically made a hundred nanometers apart and millions or tens of millions of base pairs apart if they're on the same chromosome. Chemistry is a much more local question.

Inokuti: That's an assumption. There may be a different amount.

Sachs: Well, are you saying the chemistry might know about things a hundred nanometers away and ten million base pairs away?

Inokuti: Well, I would say that micro dimension is not the precise enough description, but the real chemical arrangement, for example, how much water molecule is around, how much molecules are there?

Sachs: For a hundred nanometer scales or more?

Inokuti: Yes.

Holley: They may not be the same, but they're not going to be influenced by a break far away.

Sachs: Yes, the question is whether there's bias from one DSB due to a DSB that's a hundred nanometers away. If you say that's possible, then that's an interesting point.

Inokuti: There's an electric field changing which hasn't even been mentioned at all.

Sachs: Then that's certainly an interesting point.

MONTE CARLO AND ANALYTIC METHODS IN THE TRANSPORT OF ELECTRONS, NEUTRONS, AND ALPHA PARTICLES

Randall S. Caswell and Stephen M. Seltzer

U.S. Department of Commerce
Technology Administration
National Institute of Standards and Technology
Gaithersburg, MD 20899 USA

ABSTRACT

In this paper we discuss Monte Carlo calculational methods and analytic methods, and methods combining features of both, developed over the past decades at the National Institute of Standards and Technology. These include the Monte Carlo program ETRAN for electrons and photons developed by Berger and Seltzer; the neutron analytic method developed by Caswell and Coyne, and the incorporation into that program of the synthesis of Monte Carlo results for proton tracks of Wilson and Paretzke; and the modification of the analytic method for neutrons to the case of radon progeny alpha particles by Caswell and Coyne. Some comparisons with experimental results and with other calculations are given. Some applications of the calculational results to the prediction of biological effects using biophysical models are given.

INTRODUCTION

Monte Carlo methods have the advantage of being usable in principle for most any geometry or structural shape, and permit use of arbitrary cross sections. In practice Monte Carlo programs frequently contain analytic elements or methods of biasing to speed up the calculation and devote the calculational time to the important results sought. Analytic methods usually have the advantages of being faster calculationally and allowing better understanding of the elements affecting the outcome of the calculation.

In this paper we shall summarize the Monte Carlo program ETRAN for electrons (and photons) developed over the past two decades by Berger and Seltzer; the analytic method for neutrons and for radon alpha particles developed by Caswell and Coyne; and the incorporation into the neutron analytic program of the synthesis of Monte Carlo results for proton tracks of Wilson and Paretzke. In each case typical results are shown. Some

Computational Approaches in Molecular Radiation Biology
Edited by M.N. Varma and A. Chatterjee, Plenum Press, New York, 1994

comparisons with experimental results and other calculations will be given. Some applications of the calculations to biophysical models for the prediction of biological effects are given.

ELECTRONS

Over the years, work in our program at the National Institute of Standards and Technology has been done in numerous areas concerning electron transport, including both analytical and Monte Carlo methods. It is clear that the focus of this workshop is on electron Monte Carlo track structure calculations, by which is meant the direct analog simulation of each interaction in the histories of electrons and all progeny so as to allow one to extract complete information on all spatial and energy variables. Such information is used for further modelling of chemical and biological action, or processed to obtain physical distributions such as the spatial correlations of ionization events (cluster-size frequencies), the spectra of energy deposited in small sites (micro- and nano-dosimetry), and spatial distributions of energy deposition (radial distributions, point kernels). Our experience in such single-interaction Monte Carlo calculations is somewhat limited,[1–5] but our experience is much more established in electron Monte Carlo calculations based on multiple-scattering models. Therefore, we mainly discuss here the pertinence of our multiple-scattering electron Monte Carlo codes to problems associated with radiobiological modelling.

The ETRAN Monte Carlo model is based on a condensed-history random walk in which the effects of numerous Coulomb collisions along a path segment are sampled from relevant multiple-scattering distributions.[6] Collision energy-loss straggling is treated through the use of the Landau distribution,[7] with Blunck-Leisegang binding corrections,[8] truncated so as to have the correct mean collision loss (e.g., as given in Ref. 9). Net deflection angles due to elastic (with small corrections for inelastic) scattering are sampled from a Goudsmit-Saunderson distribution,[10] based on the screened Mott single-scattering cross section[11] at high energies and results from exact phase-shift calculations for elastic scattering[12–15] at low energies. The production of bremsstrahlung photons, and the corresponding electron radiative energy loss, is sampled from distributions derived from the synthesis of the best available theoretical information for the cross sections in the screened fields of the atomic nucleus and the atomic electrons.[16,17] Secondary knock-on electrons are sampled from the Møller cross section; and information on the electron-impact ionization cross section that governs emission of characteristic x rays or Auger electrons is obtained from various calculations.[18–23] Photon transport is treated according to a single-scattering Monte Carlo model that takes into account coherent and incoherent scattering with binding corrections, photoelectric absorption, and pair production. More detailed discussions of the methods used in the calculation can be found in references 24–28. The ability to handle a large variety of geometries, as well as a more complete treatment of atomic relaxation following ionization events, was incorporated in the development of the Integrated TIGER Series (ITS) codes.[29–32]

Such codes can handle all generations of the electron-photon cascade with energies from the GeV region down to about 1 keV. They have been used extensively during the last 30 years in a wide variety of applications. The demonstrated level of agreement found in numerous comparisons with measured data has established these codes as a reliable method for the calculation of electron transport quantities and has largely validated the cross-section information used over this energy range. The obvious advantage of this class of simulations is in the reduction of computer time required to trace the electron histories. For example, an electron slowing down from 100 keV to about 1 keV in a low-Z medium undergoes perhaps about 5,000 individual Coulomb collisions, and from 1 MeV perhaps about 50,000. The corresponding ETRAN calculations involve only about 100-200 steps in the random walk.

However, condensed-history models become unreliable at low energies. Contributions of this type of calculation to the needs of radiobiological modelling are then somewhat limited, but perhaps still valuable.

One area of application is in the calculation of the transport of relatively high-energy electron or photon beams to sites at some depth in a target or phantom. The calculation would provide fluence spectra to serve as source distributions for further calculations to low energy by other methods. Such an application is illustrated in Berger,[33] and in Berger and Seltzer,[34] for a beam of 20-MeV electrons incident on a water phantom. In that case, the ETRAN Monte Carlo calculations were carried out only down to 0.625 MeV, below which the spatial variable could be neglected, and the results used as a source term for a further energy-degradation calculation according to the method of Spencer and Fano.[35] In a similar vein, sample calculations for a ^{60}Co beam incident on a carbon phantom are given in Seltzer *et al*.[36] for the fluence of both the photons and the secondary electrons.

Cell nuclei of 1–5 μm diameter correspond to ranges in water of electrons with energies from about 6 to 15 keV. As shown in Figure 1, ETRAN calculations of absorbed-dose distributions in low-Z media for electrons in this energy range have been rather successful.[37] Thus it should be possible to use a condensed-history Monte Carlo model in microdosimetric calculations or in calculations of the radial distributions of absorbed dose from delta rays around proton tracks as used in Katz's radiation-effects models. Berger[38] did use ETRAN results for electron transport in his calculations of event size distributions from proton irradiation. Figure 2 shows event-size distributions for a 1-μm diameter water sphere irradiated by a broad beam of 20-MeV protons and associated secondary electrons. Cases include protons in the continuous-slowing-down approximation (no electron transport), protons including energy-loss straggling (no electron transport), and the full calculation including energy-loss straggling for the protons and energy transport by secondary electrons.

To some extent, the condensed-history and the single-scattering Monte Carlo models need equivalent cross-section information. Comparisons at the higher energies with the well-validated condensed-history models can be a valuable check on the newer Monte Carlo models which extend calculations down to final interactions in the few eV region. This is illustrated in Figure 3, which compares depth-dose distributions in water from 800-keV electron beams normally-incident on a semi-infinite water slab target. The points are from Turner *et al*.,[39] based on a sample of 5000 electron histories generated by an older version of the OREC single-scattering Monte Carlo code. The histogram is from ETRAN condensed-history Monte Carlo calculations,[40] based on the histories of 100,000 incident electrons. The poor agreement is ascribed to the use in OREC of defective elastic-scattering cross sections, which have since been improved.[41]

NEUTRONS

The objective of the overall scientific program for neutrons is to understand the "physical stage" of the interactions of neutrons with tissue and biological systems. A quantitative description of the processes and spectra involved in the transfer of energy from neutron radiation to the biological material should help in understanding the biological effects of neutron radiation. Physical information on the secondary particles produced by the neutron interactions, and on their interactions with (or energy transfer to) biological tissue can be used as input to biophysical models of neutron radiobiological effects.

Our original work with neutrons was based on an analytic model,[42] rather than Monte Carlo. When neutrons interact with the nuclei of tissue or a tissue-like medium (assumed infinite, homogeneous), a spectrum of secondary particles is generated which we call the **initial spectrum**, n(E),[43–44] where E is the energy of the secondary particle. n(E) has dimensions $MeV^{-1}g^{-1}$ for unit incident neutron fluence (cm^{-2}), and represents the particle spectrum generated per Mev per gram of tissue exposed to the unit neutron fluence. As the secondary particles with initial energies n(E) interact and slow down in the medium, on the average their spectrum becomes a lower-energy spectrum called the "particle fluence differential in energy," or the **fluence spectrum**, or the **slowing-down spectrum**. It is the same as the slowing-down spectrum used in electron physics, except for a question of normalization. The fluence or slowing-down spectrum for neutron-induced secondary charged particles has the dimensions of cm^{-2} MeV^{-1}, and is given by the relation

$$\varphi(E) = \frac{1}{S(E)} \int_{E}^{E_{max}} n(E')dE' \quad , \tag{1}$$

where S(E) is the stopping power in $MeV/(g/cm^2)$. The dimension of the integral in equation (1) is g^{-1}, and corresponds to the number of secondary particles per gram produced in the irradiated material by the incident neutron fluence. In electron physics the dimension of the integral in the equation corresponding to equation (1) is unity, and correspondingly the dimensions of the slowing-down spectrum Y(T) is g cm^{-2} MeV^{-1} which may be taken as a differential track length per unit energy. It is thus a matter of convenience whether one uses $\varphi(E)$ or Y(T). T is the electron kinetic energy.

Initial spectra have been calculated for secondary particles ejected by 9-MeV neutrons incident on tissue.[44] The predominant particles are protons produced by neutron elastic scattering. Other features are alpha particles produced by (n,α) nuclear reactions and heavy particle recoils such as C and O produced primarily by elastic scattering. At lower neutron energies the spectra are simple. At higher neutron energies such as 14 MeV, they become quite complex with many reactions energetically possible, and thus participating.

Slowing-down spectra for 14.5-MeV neutrons incident on tissue are shown in Figure 4. Since the slowing-down spectrum for a given secondary particle is inversely proportional to stopping power or LET (see Equation 1), the proton spectrum is much larger than the spectra for the other particles. This is somewhat compensated by the fact that the many protons deposit less energy per cellular interaction than the heavier charged particles. A similar slowing-down spectrum has been calculated for the charged particles produced in the interaction of ^{252}Cf fission neutrons with tissue. This broad spectrum of neutrons with average energy around 2 MeV leads to simpler charged-particle spectra than the charged-particle spectra shown in Figure 4 due to the lower neutron energy.

Early uses of this spectral information as input to a biophysical model were made by Katz.[45,46] Figure 5 shows survival curves of T-1 human kidney cells irradiated with 14-MeV neutrons under oxygenated and anoxic conditions as measured by Barendsen and Broerse,[47,48] compared to Katz's biophysical model.[45] This model which is a rather simple phenomenological model nevertheless yields excellent agreement with the experimental data. The Katz model has also been applied to irradiations of cells with mixtures of ^{60}Co gamma rays and 14-MeV neutrons under aerobic conditions. The model uses parameters for human kidney cells. It was compared to experimental measurements by Railton[49] for aerobic CHO (Chinese Hamster Ovary) cells. The agreement was excellent. However, a phenomenological model, however successful, gives us only a limited mechanistic understanding.

The next step in the analytic calculation is to calculate microdosimetric spectra. Secondary particles ejected by neutron interactions are incident on a spherical sensitive volume in a medium. In the simplest version[42] we assume random secondary particles travelling in straight lines and losing energy continuously (continuous-slowing-down approximation, or CSDA) as given by their stopping power. Secondary particles initiating within the sensitive volume, or cavity, start with the initial spectrum, n(E). Those starting outside the volume, "in the wall," start with the initial spectrum, but slow down and by the time they reach the cavity, on the average they have the slowing-down spectrum, $\varphi(E)$. Knowing the chord-length distribution for the cavity, it is simple to write down analytic equations for the energy deposition distributions for the four types of event which intersect the cavity.[42,50]

An example of the result of such a calculation is given in Figure 6 as a conventional plot in lineal energy, y, for a cyclotron neutron spectrum from 14-MeV deuterons on beryllium.[51,52] The curve marked "δ" is an energy deposition calculation corrected for the variation in energy per ion pair, W, and therefore corresponds to an ionization yield. This presentation is appropriate when comparing to a tissue-equivalent proportional counter measurement such as that of Ito. In this particular case we note generally excellent agreement between calculation and experiment.

In general the continuous-slowing-down-approximation works reasonably well for site sizes of about 1 μm and larger. In Figure 7 experimental and theoretical y spectra are compared for 14.5-MeV neutron energy. The major discrepancy is in the shape of the large "proton" peak on the left where the calculation is narrower and higher than the experiment. This is due to neglect of straggling (by definition) in the CSDA calculation. The differences in the higher peaks, corresponding to alpha particles, below about 400 keV/μm, and heavy recoils probably are due to the nuclear data used in the calculation. A comparison between a Monte Carlo calculation and an analytic calculation is shown in Figure 8 for 0.1-MeV neutron energy. Note that even such a detail as the oxygen cutoff (maximum energy which oxygen recoils can lose in the sensitive volume) around 36 keV/μm is reproduced both in the analytic and the Monte Carlo calculations.

To handle problems such as straggling or to go to sites with diameters on the scale of nanometers, one must do better than CSDA. The synthesis of Monte Carlo results for proton tracks of Wilson and Paretzke[54–56] allows the incorporation of Monte Carlo data into analytic calculations in a systematic, convenient way. For example, neutron energy deposition calculations on the micrometer and nanometer scales have been carried out for tissue including proton straggling and "passer" or "toucher" events (events where the main particle track is outside the sensitive volume, but which irradiate delta rays into the site). The heavier neutron secondaries are treated in the continuous-slowing-down-approximation (CSDA). In this way it is easy to see the effect of straggling on the microdosimetric spectra, as well as the contribution of passers. Results may be calculated for either energy deposition or ionization yield, the latter being important for comparison to tissue-equivalent proportional counter (TEPC) data.

Some calculations have been made for tissue[57] using the analytic code with the inclusion of the Wilson-Paretzke synthesis of Monte Carlo results. The effect of inclusion of straggling is to improve the agreement with the experimental data shown in Figure 7. If we move to a smaller site size, say 100 nm, the y distribution including straggling is still broader.

The next step is to include the "touchers" or "passers." In deciding how to include the "touchers" into the calculation, we chose to use "energy balance" to estimate the contribution of the touchers. The idea of "energy balance" is that the energy deposited in a given sensitive volume by the sum of the energy deposition events should equal the average energy deposition given, for example, by the Fano or Bragg-Gray theorems. In a CSDA calculation, since one is using the average energy deposition based on stopping power or unrestricted LET, "energy balance" should automatically be true, and we use this as a check on inherent calculational accuracy. In this work we use "energy balance" to determine the magnitude of the contribution of the passers. That is, we run the program in CSDA, and determine the total energy deposition. Then we run the program with straggling and delta-ray energy losses from the volume, which yields a smaller energy deposition. We use the difference between the CSDA result and the "straggling" result to normalize the passer contribution. Figure 9 shows a calculation including straggling and passers compared to the Menzel data of Figure 7. Note the improved agreement.

Figure 10 shows a comparison with a Monte Carlo calculation for 6 MeV neutrons, 50 nm site diameter by Morstin and Olko.[58] We note that the present calculation shows somewhat higher energy losses, and that the heavy particles appear to be missing from the straight Monte-Carlo calculation. Further study is needed to understand these differences.

Calculations have been carried down to 2 nm site size, relevant to DNA. At very small site sizes we have recently found a discrepancy between calculations and experiment[59] which needs further study.

Our calculations should serve as a suitable basis for studies of the hit-size effectiveness function (HSEF),[60,61] but this has not yet been done.

ALPHA PARTICLES

Epidemiological studies have demonstrated that sufficiently high levels of radon and short-lived progeny in uranium mines can cause lung cancer. The problem of a suspected carcinogenic risk at low radon levels in homes, however, is still unsettled and requires additional research. To investigate this problem, the neutron analytic code discussed above has been modified to determine alpha particle slowing-down spectra and lineal energy y spectra incident on the cells at risk (basal cells and secretory cells) in the bronchial epithelium.[62-64] This physical information is then used as input to a biophysical model due to Hofmann and collaborators.[65–67]

The geometry of the calculation is shown in Figure 11. We consider the alpha radioactivity of the radon daughters, ^{218}Po (6.0 MeV) and ^{214}Po (7.7 MeV), distributed uniformly through a 15-μm mucous-serous layer. We determine the fluence rate (slowing-down) spectrum at the depth of the cells. The nominal source strength is taken as 1 Bq cm^{-2}. To calculate the spectrum one integrates over the entire source up to the maximum ranges of the alpha particles involved (72 μm in tissue). Calculations have been performed for airways of diameters 1.130, 0.651, 0.435, and 0.198 cm (generations 2, 4, 6, 10 of the Yeh and Schum[68] morphometry).

Fluence-rate spectra for ^{214}Po are shown in Figure 12 for cell depths ranging from 10 to 60 μm for a generation 2 human airway. For the 10 μm depth, note the peak around 6 MeV resulting from alphas from the near side of the airway, and a lower broad peak around 4 MeV from alphas crossing from the other side of the airway. The corresponding microdosimetric spectra to Figure 12 are shown in Figure 13. Similar calculations have been made for ^{218}Po and for mixtures of the two daughters.

To apply the effect-specific track-length model,[65] one starts with an assumed ^{222}Rn concentration of 1 pCi l^{-1} (37 Bq m^{-3}), and a decay product ratio of ^{222}Rn:^{218}Po:^{214}Pb:^{214}Po = 1.0:0.9:0.6:0.4. Appropriate assumptions are made about attached and unattached fraction and normal breathing conditions to obtain deposition of inhaled radon progeny aerosols for each bronchial airway generation. The initial deposition pattern is subsequently modified by mucociliary clearance, diffusion into the bronchial epithelium and radioactive transmutations. This leads to resulting steady-state alpha particle emission rates per unit surface area and time for ^{218}Po and ^{214}Po. This is the input to the calculations described above of slowing-down spectra and microdosimetric (y) spectra at various cell depths in the bronchial epithelium.

Secretory and basal cells in the bronchial and bronchiolar epithelium have been identified as the primary progenitor cells of bronchial carcinoma.[69] Bronchial airway generations 2, 4, 6, 10 are considered since they are representative of the anatomic region of preferential tumor appearance. The track-length model presumes a constant probability per unit track length for damaging cellular targets by radiation of fixed LET. The probabilities per unit track length have been derived from the analysis of *in-vitro* survival, mutation, and transformation of V79 and C3H10T1/2 cells. The probabilities per unit track length for cell killing, $P_d(L)$, mutation, $P_m(L)$, and transformation, $P_t(L)$, have similar dependence on LET. Mutation probabilities, transformation probabilities, and inactivation probabilities are roughly in the ratio 1:20:1000. We use measured depth density distributions of the basal and secretory cells, and assume diameters of 8.8 μm for the cell nuclei. Calculations have been made of the transformation probabilities for the basal and secretory cells induced by ^{218}Po and ^{214}Po alpha particles as functions of depth in the bronchial epithelium of airway generation 2 for 1 Bq cm^{-2} alpha particle source strength of each radionuclide.

Next the distribution of biological effects among the bronchial airway generations is studied.[66] The result is shown in Figure 14. Basal and secretory cell doses are plotted as a function of the generation number. While secretory cell doses are generally higher than basal cell doses, transformation risk is highest in the basal cells. Although dose decreases strongly with depth, it appears that cancer probability is rather uniformly distributed throughout epithelial tissue.

ACKNOWLEDGEMENT

Work supported in part by the Office of Health and Environmental Research, U.S. Department of Energy. Not subject to copyright.

SUMMARY

We have discusssed Monte-Carlo and analytic radiation transport programs, and some programs having features of both. They are useful for radiation transport calculations for electrons, photons, neutrons, and alpha particles. The programs have been widely used, and verified with comparisons to experiment and to other programs. In some cases we are now exploring interactions at the few nanometer site size corresponding to interactions with DNA. Use of these calculations with biophysical models should be increasingly valuable in predicting the biological effects of ionizing radiation and in understanding mechanisms of biological effect through testing biophysical models.

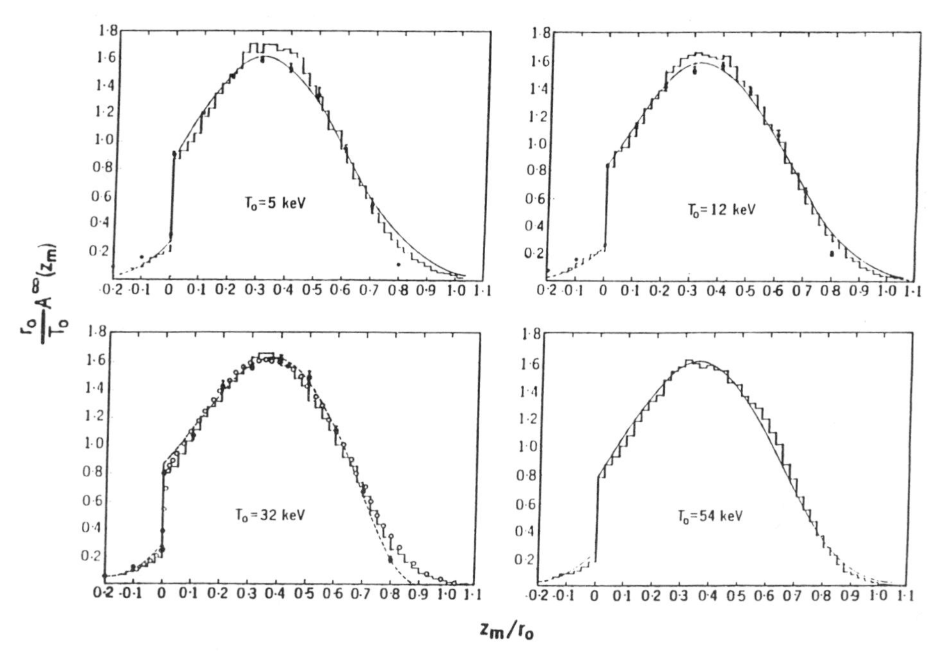

Figure 1. Distribution of absorbed dose as a function of depth in an infinite air medium for a plane-perpendicular electron source. Experimental results are represented by the solid curves for source energies T_0 of 5, 12 and 54 keV, and by open circles for 32 keV. The dashed curve for 32 keV is from the CSDA moments-method calculations of Spencer.[11] The solid points are from csda single-elastic-scattering Monte Carlo calculations. The histograms are from condensed-history ETRAN Monte Carlo calculations. Depth is expressed in units of the electron CSDA range r_0.

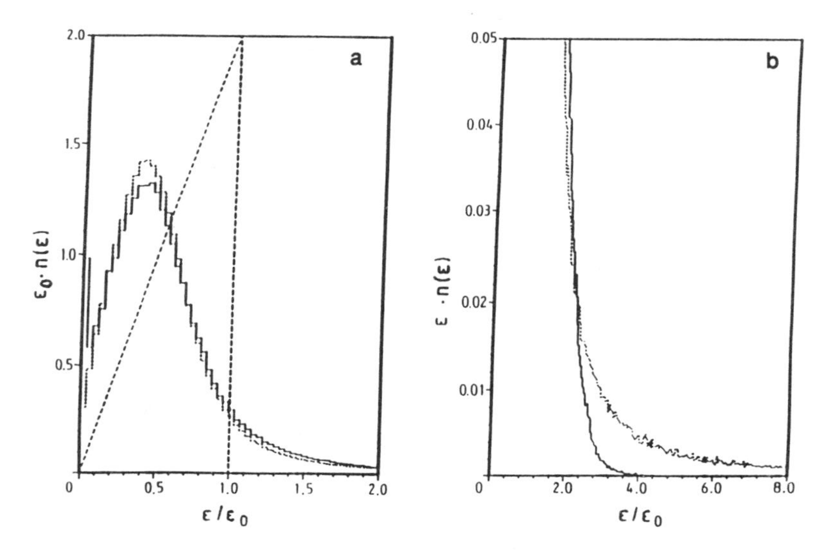

Figure 2. Event-size distributions $n(\varepsilon)$ in a 1-μm diameter sphere of water irradiated by a broad beam of 20 MeV protons and associated secondary electrons. The energy-loss scale factor ε_0 is 2.63 keV. Dashed triangle from a CSDA calculation ignoring secondary electrons. Dashed histogram from a Monte Carlo calculation taking into account proton energy-loss straggling but neglecting energy transport by secondary electrons. Solid histogram from a Monte Carlo calculation taking into account proton energy-loss straggling and including energy transport by secondary electrons through the use of ETRAN results. The expanded plot on the right shows the behavior in the large-event tail of the distribution.

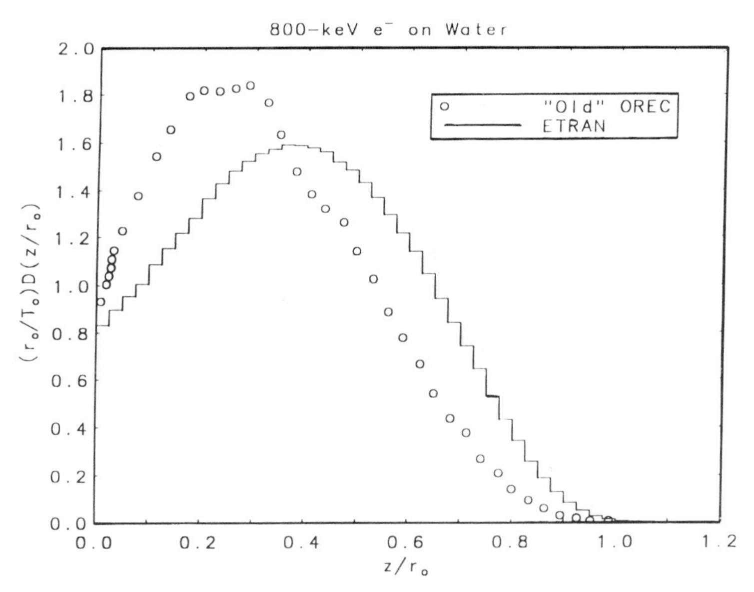

Figure 3. Distribution of absorbed dose as a function of depth in a semi-infinite water medium for electrons incident perpendicularly with an energy T_0 of 800 keV. Depth is expressed in units of the electron CSDA range r_0. The points are from a calculation with the single-scattering Monte Carlo code OREC.[39] The histogram is from calculations with the condensed-history Monte Carlo code ETRAN.[40]

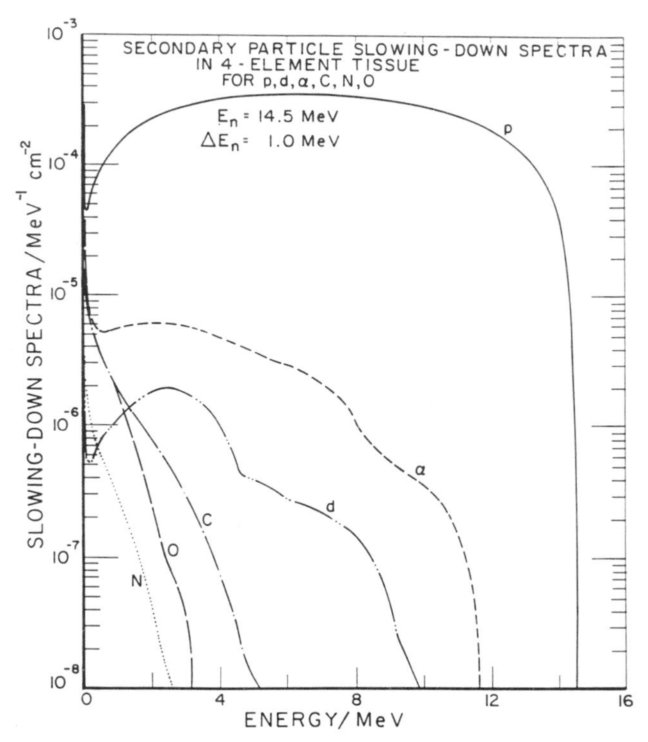

Figure 4. Secondary particle slowing-down spectra for 14.5 MeV neutrons incident on 4-element tissue. Note the strong predominance by protons. In terms of biological effect the large number of protons is somewhat compensated by the fact that each proton deposits less energy per cellular interaction than do heavy charged particles such as C, N, and O.

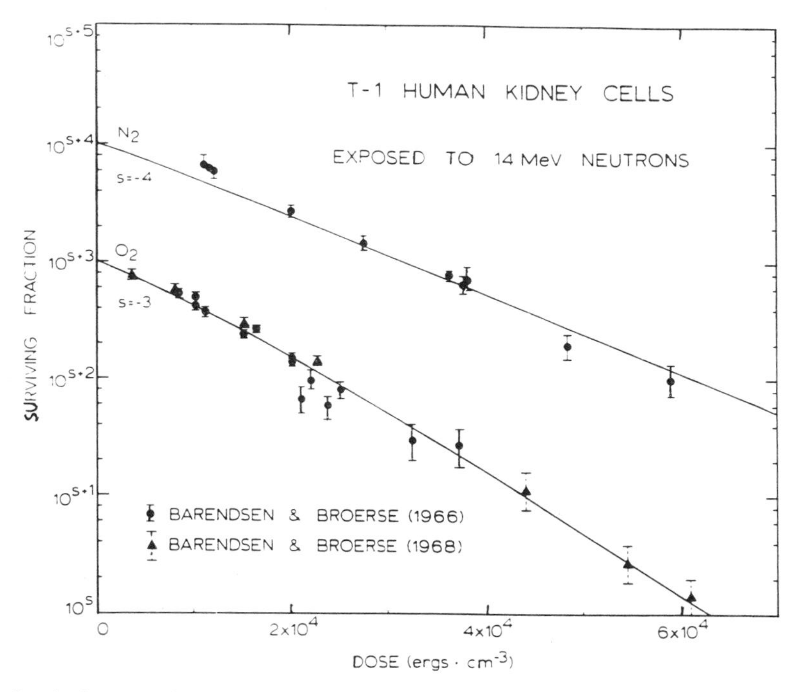

Figure 5. Survival curves of T-1 human kidney cells irradiated with 14-MeV neutrons under oxygenated and anoxic conditions as measured by Barendsen and Broerse[47,48] compared to the calculations of Katz and Sharma.[45] The calculations used the initial spectra of Caswell and Coyne.[43]

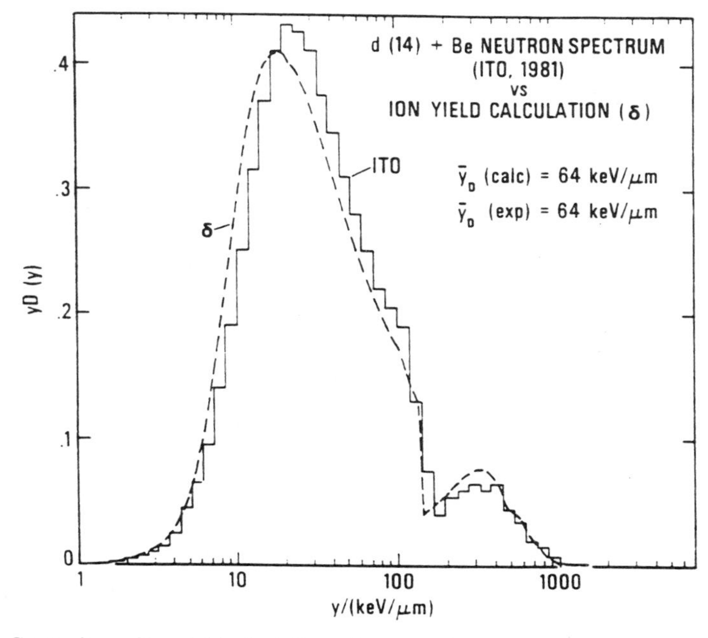

Figure 6. Comparison of ion yield calculation with an experimental y-spectrum measurement for a d(14) + Be cyclotron neutron source due to Ito quoted in Reference 51.

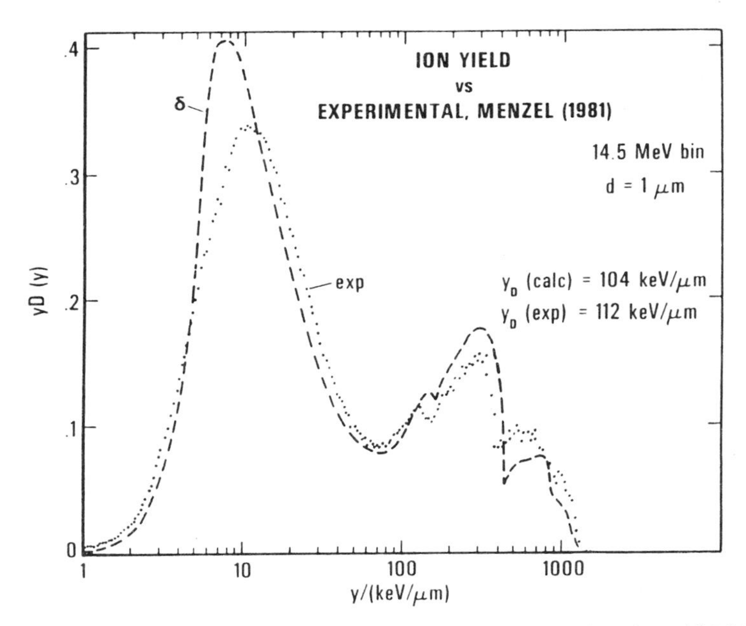

Figure 7. Comparison of 14.5-MeV ion yield calculation with the experimental results at 15.1-MeV neutron energy from Reference 53. The ion yield calculation is appropriate for comparison with experiment since the tissue-equivalent proportional counter response is proportional to ionization.

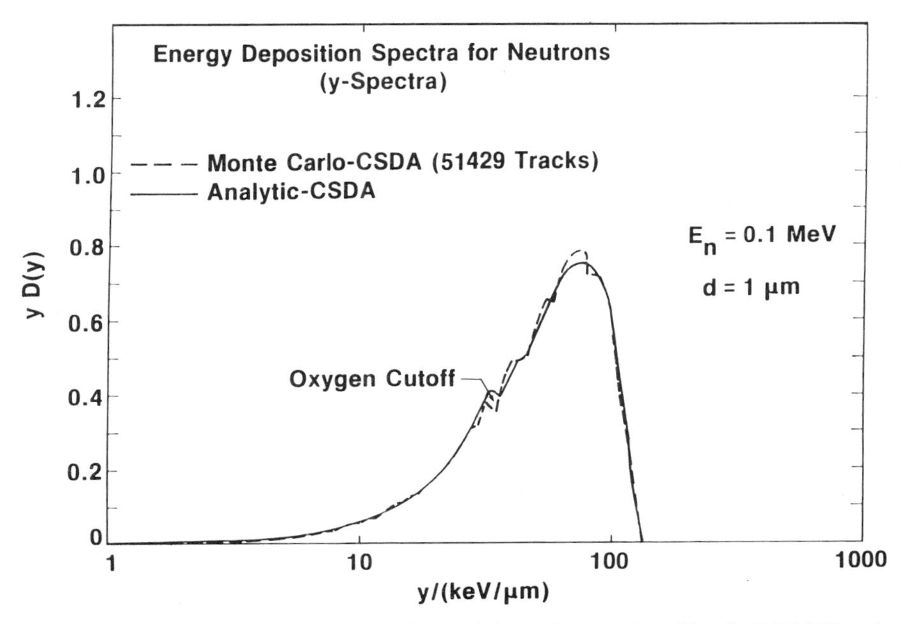

Figure 8 Comparison of Monte Carlo and analytic calculations of energy deposition for 0.1-MeV neutrons for a site diameter of 1 μm in the continuous-slowing-down approximation. Note the generally close agreement.

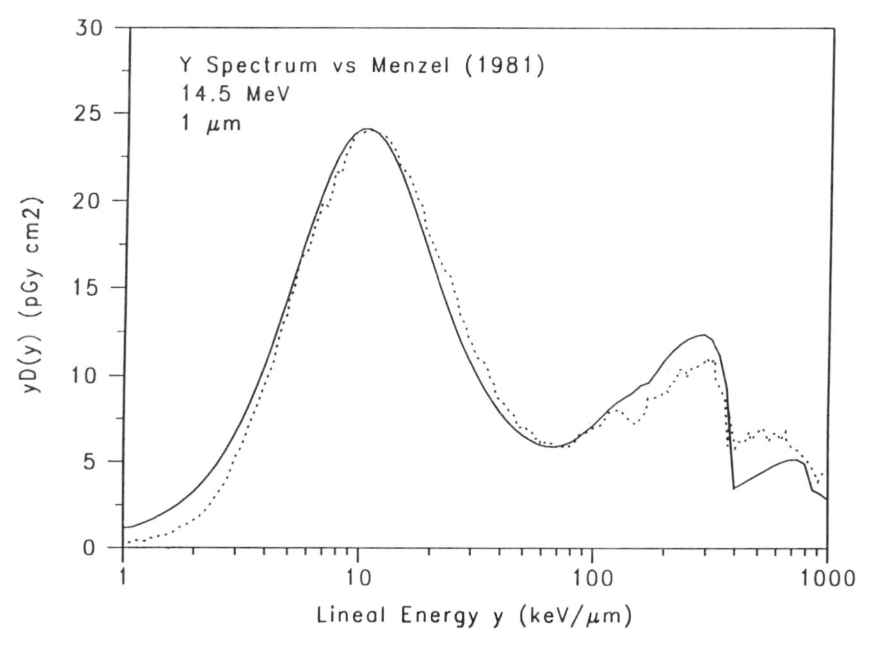

Figure 9. Calculated y spectrum including "crossers" and "passers" 14.5-MeV neutron energy for 1 μm site diameter compared to the experimental measurements of Menzel at 15.1 MeV.[53] The calculation is shown by a solid line, the experimental data by a dashed line.

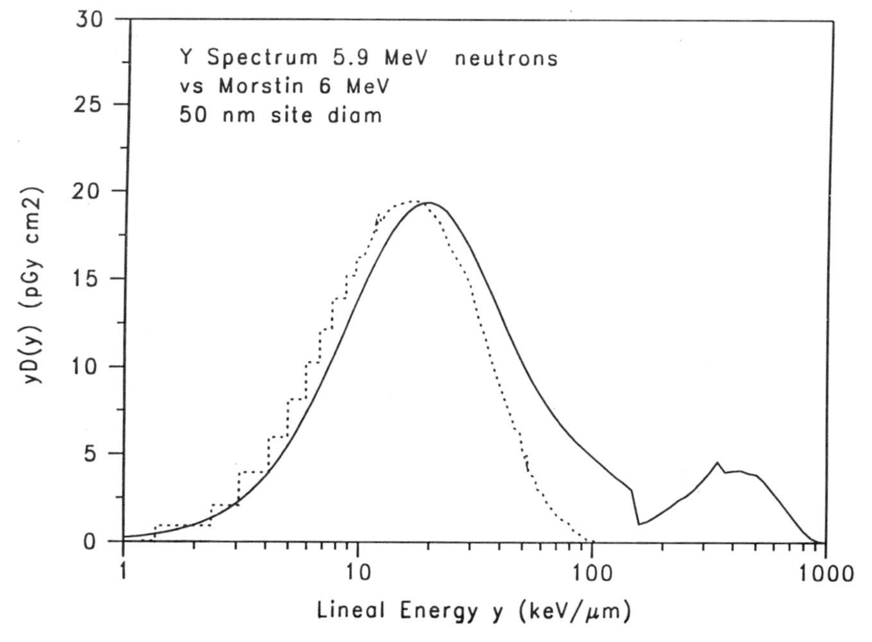

Figure 10. Comparison of y spectrum calculation for 5.9-MeV neutrons using the analytic-Monte Carlo method versus the Monte Carlo calculation of Morstin and Olko[58] for 6-MeV neutrons, both for 50 nm site diameter. The analytic calculation at 5.9-MeV neutron energy is indicated by a solid curve, and the Monte Carlo calculation at 6-MeV neutron energy is indicated by a dashed curve.

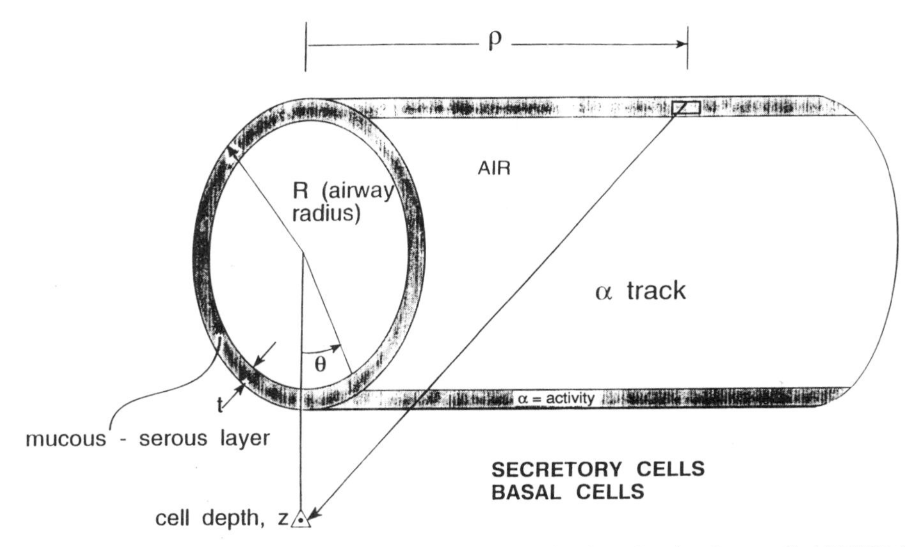

Figure 11. Geometry of cylindrical airway source calculation for radon daughters, called MODE 4.

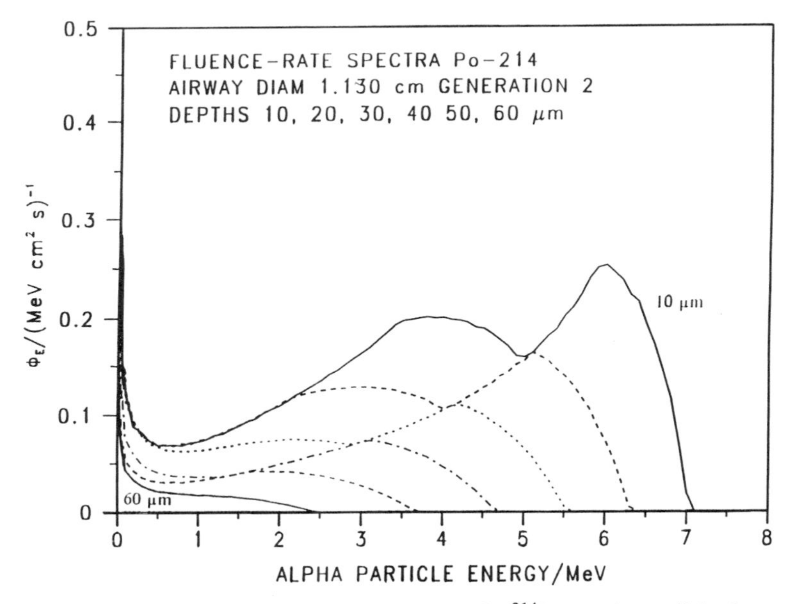

Figure 12. Alpha-particle fluence-rate (slowing-down) spectra for ^{214}Po at various cell depths, generation 2. The activity per unit area of the deposited source is 1 Bq cm^{-2}.

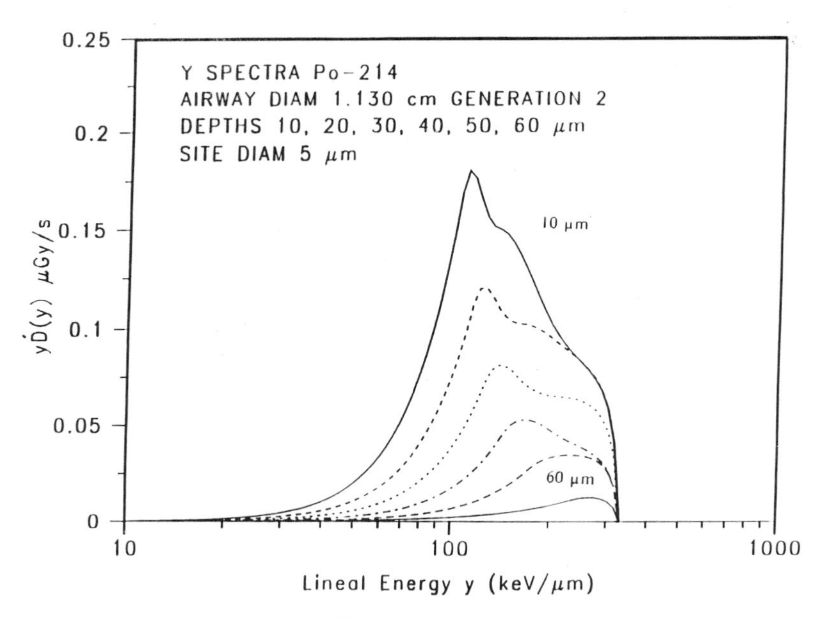

Figure 13. Lineal energy (y) spectra for ^{214}Po at various cell depths, generation 2, site diameter 5 μm.

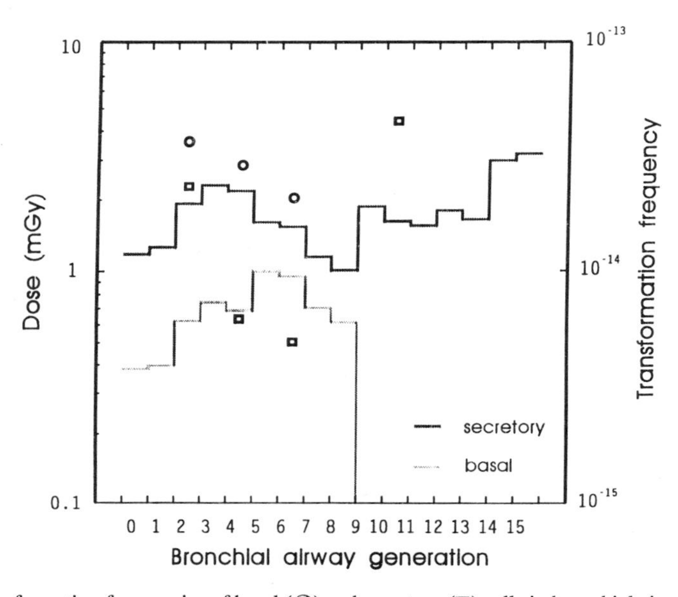

Figure 14. Transformation frequencies of basal (○) and secretory (□) cells in bronchial airway generations 2, 4, 6, 10 and comparison with cell-specific doses for defined radon progeny inhalation conditions (^{222}Rn concentration, 1 Bq l^{-1}; equilibrium factor F, 0.55; exposure time, 1 year).

REFERENCES

1. M.J. Berger. Spectrum of Energy Deposited by Electrons in Spherical Regions. *Proceedings of the Second Symposium on Microdosimetry*, pp. 541–559 (1969).
2. M.J. Berger. Energy Deposition by Low-Energy Electrons: Delta-Ray Effects in Track Structure, and Microdosimetric Event-Size Spectra. *Proceedings of the Third Symposium on Microdosimetry*, pp. 157–177 (1971).
3. M.J. Berger. Some New Transport Calculations of the Deposition of Energy in Biological Materials by Low-Energy Electrons. *Proceedings of the Fourth Symposium on Microdosimetry*, pp. 695–711 (1973).
4. H.G. Paretzke and M.J. Berger. Stopping Power and Energy Degradation for Electrons in Water Vapor. *Proceedings of the Sixth Symposium on Microdosimetry*, pp. 749-758 (1978).
5. M.J. Berger. On the Spatial Correlation of Ionization Events in Water. *Proceedings of the Seventh Symposium on Microdosimetry*, pp. 521–534 (1980).
6. M.J. Berger. Monte Carlo Calculation of the Penetration and Diffusion of Fast Charged Particles. *Methods in Computational Physics, Vol. 1*, B. Alder, S. Fernbach and M. Rotenberg, eds, pp. 135–215, Academic Press, New York (1963).
7. L. Landau. On the Energy Loss of Fast Particles by Ionization. *J. Phys. (USSR)* 8:201 (1944).
8. O. Blunck and S. Leisegang. Zum Energieverlust schneller Electronen in dünnen Schichten. *Z. Physik* 128:500 (1950).
9. ICRU Report 37. *Stopping Powers for Electrons and Positrons.* International Commission on Radiation Units and Measurements, Bethesda, MD (1984).
10. S. Goudsmit and J.L. Saunderson. Multiple Scattering of Electrons. *Phys. Rev.* 57:24 (1940).
11. L.V. Spencer. Theory of Electron Penetration. *Phys. Rev.* 98:1597 (1955).
12. M.E. Riley. *Relativistic Elastic Electron Scattering from Atoms at Energies Greater than 1 keV*. Sandia Laboratories report SLA-74-0107 (1974).
13. M.E. Riley, C.J. MacCallum and F. Biggs. Theoretical Electron-Atom Elastic Scattering Cross Sections. Selected Elements, 1 keV to 256 keV. *Atom. Data and Nucl. Data Tables* 15:443 (1975).
14. M.J. Berger and R. Wang. Multiple-Scattering Angular Deflections and Energy-Loss Straggling. *Monte Carlo Transport of Electrons and Photons*, T.M. Jenkins, W.R. Nelson and A. Rindi, eds., pp. 21–56, Plenum Press, New York (1988).
15. M.J. Berger, S.M. Seltzer, R. Wang and A. Schechter. *Elastic Scattering of Electrons and Positrons by Atoms: Database ELAST*. National Institute of Standards and Technology report NISTAR 5118 (1993).
16. S.M. Seltzer and M.J. Berger. Bremsstrahlung Spectra from Electron Interactions with Screened Atomic Nuclei and Orbital Electrons. *Nucl. Instr. Meth.* B12:95 (1985).
17. S.M. Seltzer and M.J. Berger. Bremsstrahlung Energy Spectra from Electrons with Kinetic Energy 1 keV – 10 GeV Incident on Screened Nuclei and Orbital Electrons of Neutral Atoms with Z = 1–100. *Atom. Data and Nucl. Data Tables* 35:345 (1986).
18. M. Gryzinski. Classical Theory of Atomic Collisions. I. Theory of Inelastic Collisions. *Phys. Rev.* 138:A336 (1965).
19. H. Kolbenstvedt. Simple Theory for K-Ionization by Relativistic Electrons. *J. Appl. Phys.* 38:4785 (1967).
20. J.H. Scofield. K- and L-Shell Ionization of Atoms by Relativistic Electrons. *Phys. Rev.* A18:963 (1978).
21. S.M. Seltzer. Cross Sections for Bremsstrahlung Production and Electron-Impact Ionization. *Monte Carlo Transport of Electrons and Photons*, T.M. Jenkins, W.R. Nelson and A. Rindi, eds., pp. 81-114, Plenum Press, New York (1988).
22. S.T. Perkins, D.E. Cullen and S.M. Seltzer. *Tables and Graphs of Electron-Interaction Cross Sections from 10 eV to 100 GeV Derived from the LLNL Evaluated Electron Data Library (EEDL), Z = 1–100.* Lawrence Livermore National Laboratory report UCRL-50400, Vol. 31 (1991).
23. S.M. Seltzer. Calculation of Photon Mass Energy-Transfer and Mass-Energy Absorption Coefficients. *Rad. Res.* 136:147(1993).
24. M.J. Berger and S.M. Seltzer. *Electron and Photon Transport Programs, I. Introduction and Notes on Program DATAPAC 4*. National Bureau of Standards Report 9836 (1968); *Electron and Photon Transport Programs, II. Notes on Program ETRAN 15*. National Bureau of Standards Report 9837 (1968).
25. M.J. Berger and S.M. Seltzer. Bremsstrahlung and Photoneutrons from Thick Tungsten and Tantalum Targets. *Phys. Rev.* C2:621 (1970).
26. S.M. Seltzer. An Overview of ETRAN Monte Carlo Methods. *Monte Carlo Transport of Electrons and Photons*, T.M. Jenkins, W.R. Nelson and A. Rindi, eds., pp. 153–181, Plenum Press, New York (1988).
27. M.J. Berger. Differences in the Multiple Scattering of Positrons and Electrons. *Appl. Radiat. Isot.* 42:905 (1991).

28. S.M. Seltzer. Electron-Photon Monte Carlo Calculations: The ETRAN Code. *Appl. Radiat. Isot.* 42:917 (1991).
29. J.A. Halbleib and T.A. Mehlhorn. *ITS: The Integrated TIGER Series of Coupled Electron/Photon Monte Carlo Transport Codes*. Sandia National Laboratories report SAND84-0573 (1984); also, *Nucl. Sci. Engr.* 92:338 (1986).
30. J.A. Halbleib. Structure and Operation of the ITS Code System. *Monte Carlo Transport of Electrons and Photons*, T.M. Jenkins, W.R. Nelson and A. Rindi, eds., pp. 249–262, Plenum Press, New York (1988).
31. J.A. Halbleib, R.P. Kensek, T.A. Mehlhorn, G.D. Valdez, S.M. Seltzer and M.J. Berger. *ITS Version 3.0: The Integrated TIGER Series of Coupled Electron/Photon Monte Carlo Transport Codes.* Sandia National Laboratories report SAND91-1634 (1992).
32. J.A. Halbleib, R.P. Kensek, T.A. Mehlhorn, G.D. Valdez, S.M. Seltzer and M.J. Berger. ITS: The Integrated TIGER Series of Coupled Electron/Photon Monte Carlo Transport Codes — Version 3.0. *IEEE Trans. Nucl. Sci.* 39:1025 (1992).
33. M.J. Berger. Energy Loss and Range of Electrons. *Nuclear and Atomic Data for Radiotherapy and Related Radiobiology*, pp. 323-345, International Atomic Energy Agency, Vienna (1987).
34. M.J. Berger and S.M. Seltzer. Calculation of Energy and Charge Deposition and of the Electron Flux in a Water Medium Bombarded by 20-MeV Electrons. *Ann. New York Acad. Sci.* 161:8 (1969).
35. L.V. Spencer and U. Fano. Energy Spectrum Resulting from Electron Slowing Down. *Phys. Rev.* 93:1172 (1954).
36. S.M. Seltzer, J. H. Hubbell and M.J. Berger. Some Theoretical Aspects of Electron and Photon Dosimetry. *National and International Standardization of Radiation Dosimetry, Vol.II*, pp. 3-43, International Atomic Energy Agency, Vienna (1978).
37. M.J. Berger, S.M. Seltzer and K. Maeda. Energy Deposition by Auroral Electrons in the Atmosphere. *J. Atmos. Terr. Phys.* 32:1015 (1970).
38. M.J. Berger. Microdosimetric Event Size Distributions in Small Sites in Water Vapour Irradiated by Protons. *Phys. Med. Biol.* 33:583 (1988).
39. J.E. Turner, R.N. Hamm, M.L. Souleyrette, D.E. Martz, T.A. Rhea and D.W. Schmidt. Calculations for ß Dosimetry Using Monte Carlo Code OREC for Electron Transport in Water. *Health Phys.* 55:741 (1988).
40. S.M. Seltzer. *Dose in Water from External Irradiation by Electrons: Radiation Protection Data.* National Institute of Standards and Technology report NISTIR 5136 (1993).
41. O.H. Crawford, J.E. Turner, R.N. Hamm and J.C. Ashley. Effects of the Tissue-Air Interface in Calculations of ß-Particle Skin Dose at a Depth of 70 μm. *Health Phys.* 61:641 (1991).
42. R.S. Caswell. Deposition of Energy by Neutrons in Spherical Cavities. *Radiat. Res.* 27:92-107 (1966).
43. R.S. Caswell and J.J. Coyne. Interaction of Neutrons and Secondary Charged Particles with Tissue: Secondary Particle Spectra. *Radiat. Res.* 52:448–470 (1972).
44. R.S. Caswell and J.J. Coyne. Microdosimetric Spectra and Parameters of Fast Neutrons. *Proceedings Fifth Symposium on Microdosimetry*, Verbania-Pallanza, Italy, Euratom Document EUR 5452 d-e-f, pp. 97–123 (1976).
45. R. Katz and S.C. Sharma (1973). Response of Cells to Fast Neutrons, Stopped Pions, and Heavy Ion Beams. *Nucl. Instrum. Meth.* 111:93-116 (1973).
46. R. Katz and S.C. Sharma. Cellular Survival in a Mixed Radiation Environment. *Int. J. Radiat. Biol.* 26:143–146 (1974).
47. G.W. Barendsen and J.J. Broerse. Dependence of the Oxygen Effect on the Energy of Fast Neutrons. *Nature* 212:722–724 (1966).
48. G.W. Barendsen and J.J. Broerse. Measurements of the Relative Biological Effectiveness and Oxygen Enhancement Ratio of Fast Neutrons of Different Energies. in *Biophysical Aspects of Radiation Quality*, International Atomic Energy Agency, Vienna (1968).
49. R. Railton, D. Porter, R.C. Lawson, and W.J. Hannan. The Oxygen Enhancement Ratio and Relative Biological Effectiveness for Combined Irradiations of Chinese Hamster Cells by Neutrons and Gamma Rays. *Int. J. Radiat. Biol*. 25:121–127 (1974).
50. International Commission on Radiation Units and Measurements (ICRU). *Microdosimetry*. ICRU Report 36, pp. 86-87, Bethesda, MD, USA (1983).
51. R.S. Caswell, J.J. Coyne, and L.J. Goodman. Comparison of Experimental and Theoretical Ionization Yield Spectra for Neutrons. *Proceedings of the Fourth Symposium on Neutron Dosimetry*, Edited by G. Burger and H. G. Ebert, Commission of the European Communities, Luxembourg, EUR 7448 en, pp.201–211 (1981).
52. A. Ito. Private Communication. University of Tokyo, Tokyo, Japan (1981).
53. J. J. Coyne, J. C. McDonald, H.-G. Menzel and H. Schuhmacher. Detailed Intercomparison of Calculated and Measured Ionization Yield Spectra for 20 MeV Neutrons and the Implications for High Energy Neutron Dosimetry. *Proc. Fourth Symposium on Neutron Dosimetry*, EUR 7448, pp. 213–223 (1981).

54. W.E. Wilson and H.G. Paretzke. Calculations of Distributions for Energy Imparted and Ionization by Fast Protons in Nanometer Sites. *Radiat. Res.* 87:521-537 (1981).
55. W.E. Wilson, N.F. Metting, and H.G. Paretzke. Microdosimetric Aspects of 0.3- to 20-MeV Proton Tracks: I. Crossers. *Radiat. Res.* 115:389–402 (1988).
56. W.E. Wilson and H.G. Paretzke. A Stochastic Model of Ion Track Structure. *Proc. 11th Symposium on Microdosimetry*, Gatlinburg, TN (1992).
57. J.J. Coyne and R.S. Caswell. Neutron Energy Deposition on the Nanometer Scale. *Radiat. Prot. Dosim.* 44:49–52 (1992).
58. K. Morstin and P. Olko. Calculation of Neutron Energy Deposition in Nanometric Sites. *Radiat. Prot. Dosim.* 52:89–92 (1994).
59. P. Kliauga. Nanodosimetry of Heavy Ions Using a Miniature Cylindrical Counter of Wall-Less Design. *Radiat. Prot. Dosim.* 52:317-321 (1994)..
60. M.N. Varma and V.P. Bond. Empirical Evidence of Cell Critical Volume Dose vs. Cell Response Function for Pink Mutations in Tradescantia. *Proceedings Eighth Symposium on Microdosimetry*, Edited by J. Booz and H. G. Ebert, Commission of the European Communities, pp. 439–450 (1982).
61. M.N. Varma, C.S. Wuu, and M. Zaider. Hit Size Effectiveness in Relation to the Microdosimetric Site Size. *Radiat. Prot. Dosim.*, 52:339–346 (1994).
62. R.S. Caswell and J.J. Coyne. Microdosimetry of Radon and Radon Daughters. *Radiat. Prot. Dosim.* 31:395–398 (1990).
63. R.S. Caswell and J.J. Coyne. Alpha Particle Spectra and Microdosimetry of Radon Daughters. *Indoor Radon and Lung Cancer: Reality or Myth?*, Edited by Frederick T. Cross, pp. 279–289, Battelle Press, Richland, WA (1992).
64. R.S. Caswell, L.R. Karam, and J.J. Coyne. Systematics of Alpha-Particle Energy Spectra and Lineal Energy (y) Spectra for Radon Daughters. *Radiat. Prot. Dosim.*, 52:377–380 (1994).
65. D.J. Crawford-Brown and W. Hofmann. An Effect-Specific Track-Length Model for Radiations of Intermediate and High LET. *Radiat. Res.* 126:162–170 (1991).
66. W. Hofmann, M. Nösterer, D.J. Crawford-Brown, and A. Hutticher. Spatial Distribution Patterns of Energy Deposition and Cellular Radiation Effects in Lung Tissue Following Simulated Exposure to Alpha Particles. *Radiat. Prot. Dosim.* 31:413–420 (1990).
67. W. Hofmann, M. Nösterer, M.G. Ménache, D.J. Crawford-Brown, R.S. Caswell, and J.J. Coyne. Microdosimetry and Cellular Radiation Effects of Radon Progeny in Human Bronchial Airways. *Radiat. Prot. Dosim.*, 52:381–385 (1994).
68. H.C. Yeh and G.M. Schum. Models of Human Lung Airways and Their Application to Inhaled Particle Deposition. *Bull. Math. Biol.* 42:461 (1980).
69. National Research Council. *Comparative Dosimetry of Radon in Mines and Homes.* National Academy Press, Washington, DC (1991).

DISCUSSION

Turner: Where do the alphas come from in tissue?

Caswell: Alpha reactions on oxygen, on carbon. If you are at high enough energy, one of the interesting ones is $C(n, n')3\alpha$. You get three alphas all at once.

Zaider: How did you calibrate the calculation?

Caswell: The calculation is absolute. It's absolute for 14 MeV.

Zaider: You said something about calibration which I did not understand.

Caswell: The calibration is in the experiment and in the experiment they do not, in fact, know the number of ionizations in their counter. They use typically an alpha particle source shooting across the counter with a known energy loss. And they see how big that pulse height is. And they say this is so much energy and then everything is scaled to that. So they don't absolutely know the number of ionizations.

Zaider: They know the number of ionizations; they may not know the energy per ionization. That's what they measure, ionizations.

Caswell: What they actually measure is pulse height.

Paretzke: The proportional effect is unknown. You have to either make a very complicated calculation, which nobody can do till today, or just use a calibration.

Zaider: Why do you use the Monte Carlo technique?

Caswell: It was partly just to see what the result is. Partly at the time we were interested in putting some effects in that you can't do very well in the analytic calculation. Monte Carlo has the advantage now, compared to what it used to have, of computers getting faster every generation and cheaper.

Zaider: Was this done analytically for 2 nanometers?

Caswell: We use the Wilson/Paretzke Code. I think there's nothing wrong with that.

Varma: He's using the analytical function that Wilson provided him. He's extending it to low site size. It might be interesting to compare this with the Monte Carlo and see if it breaks down at 2 nanometers and 10 nanometers.

Caswell: I think this is a very helpful thing. I think that it means you can calculate many results easily. However, there are some problems.

Miller: You mean they had the same production cross sections that you used?

Caswell: Morstin had made Monte Carlo calculations independently, so whether he used the same cross sections as we did, I don't know.

Miller: He could have used different neutron production cross sections.

Caswell: The proton curves should be the same. But why their proton curve stops at 100 keV/μ, I don't understand. So, I think this is important. I'm interested very much in comparing different codes.

Nikjoo: Are these attached particles in the lung?

Caswell: The particles are deposited in the mucous-serous layer; when you get to the biological modeling it depends only on position in the layer.

Paretzke: He was asking about attached and unattached fraction, but this determines only the efficiency of the position in the mucous layer different generations, and the rest is the same.

Caswell: This calculation is absolute for a given period of time. This is what Hofmann has actually used.

Varma: What was the dose?

Caswell: I think it's actually for a one-year exposure under certain conditions.

Curtis: Has this been published?

Caswell: Yes. It's in a paper at the Gatlinburg Microdosimetry Meeting.

Zaider: Is it a data point or calculation?

Caswell: Calculation. He claims it's the first one where we've had an explanation of why the cancers in radon appear at the low generations rather than further down in the bronchial tree.

Varma: This is a little bit controversial because if you look at this curve for the transformation, you will say that the basal cells are more sensitive than the secretory cells. And the biologists, like Sweet, have shown that the secretory cells are much more prone to transformation than basal cells.

Caswell: I should say something about the way his model works. Hofmann has actually used two cell lines to try to get, for particles of a given LET, a probability per track length of transformation or mutation or of cell killing. That is basically where he gets these numbers. I think this has to be looked at as a first shot; but on the other hand, I think it's an interesting approach.

Varma: Why do they call it Tiger series?

Caswell: I don't know why the name Tiger Series is given to the codes. It's a series of codes that were basically based on ETRAN and developed at Sandia to try to put in more complex geometries.

Hamm: You are using a water code also?

Caswell: Yes.

Varma: What is the difference in the cross sections in the two cases?

Hamm: I'm not sure we've resolved it.

Zaider: You also said event-by-event.

Hamm: Ours is an event-by-event calculation. That's a difference in the code.

Turner: I think we might have too large a scattering angle distribution for elastic scattering of an energetic electron.

Caswell: This difference came out in preparation for this meeting.

Turner: We have become suspicious of it.

Varma: I'm trying to understand whether it is in the methodology or the code, or is it in the basic input, cross section data.

Hamm: We looked at some differences in elastic scattering, but that does not appear to be the cause of the difference. There's some evidence that it's not in the energy loss cross sections per se, so I think that it may be in our algorithms for doing scattering of our secondary electrons, but we're not sure at this point.

Wilson: Do you know how old those ETRAN calculations were in your last graph? The reason I ask is it has been mentioned in the literature. That prior to about 1986 there was erroneous sampling of the Landau distribution in ETRAN. So maybe it's the ETRAN that's off.

Caswell: I think it is recent. What they do in ETRAN is they take a Landau distribution and they truncate it so that it gives the right average energy loss. There is also a Blunk-Leisegang binding correction.

Zaider: Surely they have to use other things than the Landau distribution.

Caswell: I can't defend that personally except I know that they have published Vavilov's calculation. If they thought there's any advantage to it, they could do it.

Paretzke: That's published 30 years ago. This curve didn't change the high energy curves.

Varma: ETRAN is modified -- what it was in '86, I think, it must be the latest one that he's using.

Caswell: I'm sure it is.

Zaider: What is the primary application of ETRAN?

Caswell: ETRAN is used for all kinds of electron transport and dosimetry calculations. Medical therapy, used for space, used for detector response functions. It's a general purpose electron transport code.

Paretzke: ETRAN was one of the first proton and electron transport codes 40 years ago.

Hamm: What's the energy range for this code?

Caswell: GeV's to 1 kilovolt.

Nikjoo: Earlier you mentioned that you would like to change the model of your track by using ionizations representing the track. What is the reason for this?

Caswell: Wilson and Paretzke provide both information on ionizations or I guess it's basically done in terms of ionizations and then energy deposition is calculated from that. The interest in that would be to try to relate ionization clusters to phenomenon in DNA, for example.

Nikjoo: Ionization is a poor representation of the track.

Zaider: Why is that?

Nikjoo: Because track deposits its energy not just by ionizations but by excitations, and that could be of relevance to biological damage.

Paretzke: But DNA needs an ionization to show a significant effect.

Nikjoo: As you know a large part of energy of a track is deposited by low energy electrons, nearly 40%. Is the energy biologically insignificant? But there is no experimental evidence that a single ionization is biologically significant.

Paretzke: This has been done over 50 years. Just change the energy of a proton which you shine on DNA and see what happens on DNA. There you can prove that it takes an ionization to do serious damage to DNA.

Miller: Just a point on this question, you know, the non-ionizing impact on DNA could be sort of like a collateral factor that influences this minimum distance to make a double strand break. I don't think you can ignore all the non-ionization damage to DNA as being unaffected, but it probably isn't a direct mover in strand break cause.

Caswell: You do score the excitations.

Nikjoo: Well, the track contains ionization and excitations, and the excitations could lead to strand breaks via production of OH radicals.

Varma: The question I want to ask you is that in terms of this analytical function, you show some spectra for the low nanometer dimension. And Marco is doing something similar in terms of calculating the spectra in the nanometer region. There is no experimental data to test which one is the correct distribution. Has a comparison been made, for example, with Marco's work with yours?

Caswell: Marco, have you compared with Morstin?

Zaider: No, I'm afraid not. Morstin doesn't have nanodosimetry, as far as I know.

Caswell: Yes, he does.

Varma: I think it's important that we make a comparison of the nanometer spectra of three codes. It seems to me it is the right thing to do to have this nanometer distribution compared within these three codes.

Paretzke: No. There are not three codes. Morstin uses Moca 16.

Varma: I'm talking about yours, Marco's, and Randy.

Paretzke: Randy uses our work.

Wilson: By ours in the comparison, do you mean our Monte Carlo or our analytic?

Paretzke: We did the Monte Carlo calculation and analytical function and then we used the analytical function.

Zaider: That should be easy to show.

Paretzke: We chose this because we compared with Jim Turner's code and the Oak Ridge code, which is for liquid, and we showed the energy deposition number of ionizations and small sizes and we showed that we should have differences because of the delocalization. It's what we expect.

Zaider: Unfortunately delocalization is what we know least about.

Paretzke: It might be real.

Zaider: Depending on what you assume for how delocalization occurs.

Varma: It seems to me that it's an obvious comparison that can be made and determine whether it agrees or doesn't agree.

Caswell: You are also checking whether we have put their analysis correctly into our analytic code, for example. Hopefully we have.

Turner: I have a view graph this afternoon comparing some results obtained with our code and Paretzke's.

Varma: Good. This is part of the objective of this workshop.

Terrisol: Also this afternoon Hooshang and Stan will show some comparison.

Varma: Randy, where are other areas where ETRAN Code is being used.

Caswell: I think that they're continually looking at things. One question is how to get below 1 kilovolt, which is very close to what kind of calculations are being done here. The other thing they're doing, I guess they have done, is putting in stopping powers from the ICRU. That's been done some time ago. In the PC version of the code, I'm putting in Berger and Inokuti stopping powers. We debated for a long time about whether we should try to make a user friendly version of the code other people could take, and we never could talk ourselves into it because it seemed like when you get done, the code gives the same answers as you had before. This is, in fact, a good time to do it because we want to put in as much as we can of the Wilson/Paretzke analysis. We want to put in the new proton and alpha stopping powers. And there are a lot of little changes and simplifications, then we have a code that we can mail to people on a diskette with a few pages of instructions.

Miller: Has there been an attempt made to make a marriage between the ETRAN condensed history formulation and the event by event that has to take over at around a kilovolt? There's no program that marries those two concepts?

Inokuti: I just wanted to make a general remark. That is, a slowing down calculation above several keV and below are really very different business. And it's not just a good idea to propose a marriage. They are entirely different businesses. Others develop ETRAN in the condensed history way just because the number of collisions is so big if you start from 10 MeV, and so that's the only sensible way to go. On the other hand, at high energies, the systemization of cross sections is in good shape. So there is a point in making a general program. By the time electron gets down to several keV and below, then the cross sections are complicated. And it's impossible to make a code for a general purpose. So these two are really completely different businesses.And why should you just plug them together? ETRAN results serve as input into lower energy calculations. I might even add that once the electrons slows down to 10 eV or so, then another business begins. I mean this electron slowing down has many different faces.

Miller: But when you use the output from ETRAN as the source term for event by event, isn't that implying some assumption about radiation equilibrium?

Inokuti: That's an additional circumstance you may consider. If it's in radiation equilibrium, of course, you know that the spectrum is stationary. It's time dependent.

Turner: Another application I might mention is our use of OREC in applied health physics for beta dosimetry, particularly for skin dosimetry. These codes offer a method of doing some checks on the Varskin code, for example, recognized by the NRC.

Caswell: Steve is also doing beta ray dosimetry, skin dosimetry calculations.

Turner: That's one of our interests, too.

PITS: A CODE SET FOR POSITIVE ION TRACK STRUCTURE

W.E. Wilson and J.H. Miller

Pacific Northwest Laboratory
Richland, WA 99352, USA

H. Nikjoo

MRC Radiobiology Unit
Harwell, Didcot, Oxon OX11 0RD, UK

ABSTRACT

A code set is described for simulating positive ion tracks in a variety of media and having the capability of interfacing with several secondary electron transport codes. The set derives out of the MOCA series of simulations and is motivated primarily by a perceived need for easier code maintenance (updating), documentation, and verification.

Cluster analysis of stochastic track structures is introduced as a quantifiable method to distinguish effects of different physical models and assumptions. The concept of distance distributions is generalized by extending the idea beyond pairs of individual interactions to pairs of clusters of interactions.

INTRODUCTION

PITS is a code set for simulating positive ion tracks in a variety of media, with and without secondary electron transport. There is a dearth of experimental results suitable for aiding in the development of computer simulations of charged particle track structure and also for verifying the written code itself. One reason for this is the fact that most of the focus has been on water as the absorber and water, in any phase, is not a tractable experimental target. To rectify this situation is the primary reason for introducing PITS; in one sense it does not represent "new code" because it draws heavily from the MOCA series of detailed-histories charged particle transport codes.

There has been considerable evolution in computer hardware and software since the introduction of MOCA two decades ago (Paretzke, 1973). Modernization of the code, to bring it into conformance with current standard coding practices is an additional motivation. This will make it considerably easier to maintain (update) and document. Measured against

Computational Approaches in Molecular Radiation Biology
Edited by M.N. Varma and A. Chatterjee, Plenum Press, New York, 1994

current professional software engineering standards, the extant codes that are the subject of this workshop are poorly documented, many are not documented at all.

The basic concept is schematically described in Figure 1; segmentation is emphasized with each code module written symbolically to the maximum extent possible. Model parameter values, i.e., the physical data specifying the projectile and the absorber, are gathered together and entered via data file(s) at the beginning. This makes it possible to see the effect of specific additions and changes in the physical data and model. It is not sufficient to know that codes A and B give different results if we do not know why. It is more important to know how the physical model assumptions in a given code effect the predictions made by that code. To be able to do this in an efficient manner requires carefully designed code and good documentation.

Modularization of the code provides a hidden benefit in that it is then possible to use different computer languages for different modules. For example, the C computer language offers features that most current FORTRAN implementations do not provide and which can be used to advantage, especially in scoring tracks.

By carefully defining the interface requirements, transport of the secondary electrons (delta-rays) can be accomplished by any of several existing codes. We have successfully used MOCA8, KURBUC and CPT(100) with a preliminary version of PITS (Nikjoo, this workshop).

This discussion presents the current status of PITS and also results of sensitivity tests of recent additions to the code; cluster analysis of naturally occurring track structures is introduced and used in the tests. Distance distributions for clusters of ionization are introduced for the first time as a generalization of the standard distance distributions between single ionizations or energy depositions.

MODEL BASIS

PITS is an implementation of a semiempirical model for inelastic interactions of charged particles with matter based on Bethe-Born theory and developed by J.H. Miller et al.

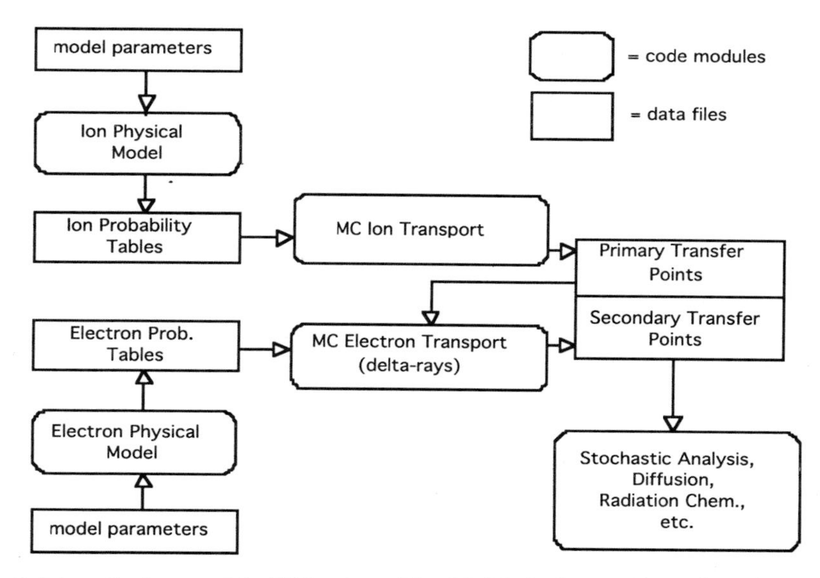

Figure 1. Schematic diagram of the PITS code modules. Modularization provides easier maintenance and documentation and also the opportunity to use any of several existing codes for delta ray transport.

(1985). The model is described well in the literature, so only a brief synopsis is included here.

The single differential cross section (SDCS) for ejection of electrons with kinetic energy between w and w+dw by a bare ion or electron with velocity v and charge z is approximated by

$$\frac{d\sigma}{dw} = \frac{4\pi a_0^2 z^2}{T} \left[A(w) \ln\left(\frac{4T}{R}\right) + B(w,T) \right] \tag{1}$$

where a_0 is the Bohr radius (0.529Å), R is the Rydberg energy (13.6 eV), and $T = mv^2/2$ with m being the electron mass. The A(w) coefficient is related to the optical properties of the target by the equation

$$A(w) = \sum_{k=1}^{N} \frac{R^2}{w + I_k} \frac{df_k}{dw} \tag{2}$$

where I_k is the ionization potential of subshell k and df_k/dw is the optical oscillator strength that is proportional to the cross section for ionization of subshell k by photons with energy I_k + w. The B(w,T) term in Equation 1 has two useful limits:

$$\lim_{w \to 0} B(w,T) = B(w) + O\left(\frac{w}{T}\right) \tag{3}$$

and

$$\lim_{w \to \infty} B(w,T) = \frac{T}{4\pi a_0^2 z^2} \left(\frac{d\sigma}{dw}\right)_{BE} \tag{4}$$

where O(w/T) denotes terms of order w/T and $d\sigma/dw|_{BE}$ denotes binary encounter SDCS. For small secondary electron energies, B(w,T) reduces to a function B(w) that is independent of projectile properties. This provides a basis for experimental determination of B(w) since substitution of Equation 3 into Equation 1 and neglecting O(w/T) gives the result

$$B(w) = \frac{T}{4\pi a_0^2 z^2} \left(\frac{d\sigma}{dw}\right)_{exp} - A(w) \ln\left(\frac{4T}{R}\right) \tag{5}$$

Given the optical data necessary to determine A(w), accurate SDCS at a single ion energy can be used to determine B(w). In practice results at several ion energies were averaged to reduce experimental error.

The relationship between the generalized oscillator strength df(Q,E)/dE of isolated molecules and the dielectric response function ε(Q,E) for a condensed medium,

$$\frac{R^2}{E} \frac{df(Q,E)}{dE} = - \frac{\mathrm{Im}[1/\varepsilon(Q,E)]}{8\pi^2 a_0^3 n} \tag{6}$$

where Q and E are momentum and energy transfer respectively, n is the density of scattering centers and Im denotes the imaginary part, provides the theoretical basis for including phase effects in our model of SDCS.

Our current data base of source model parameters includes, amorphous graphite (Arakawa et al., 1985), DNA (Inagaki et al., 1974), liquid water (Heller et al., 1974), hexagonal and amorphous ice (Seki et al., 1981; Kobayashi, 1983), and water vapor (Paretzke, 1987).

The angle of emission of an ionization electron ejected by the ion is obtained from the doubly differential cross sections (DDCS). The algorithm for the DDCS used in PITS is also based on Bethe-Born theory and developed similarly to that above for the SDCS. The DDCS for electron emission can be expressed as (Kim, 1972),

$$\sigma(\theta,w) = \frac{4\pi a_0^2}{T/R} \left[A(\theta,w) \ln\left(\frac{T}{R}\right) + B(\theta,w) + O(T^{-1}) \right] \quad (7)$$

The first term in Equation 7 describes the interactions involving small momentum transfers that arise from distant collisions and is often referred to as the "glancing collisions" term,

$$\sigma_{gc}(\theta,w) = \frac{4\pi a_0^2}{T/R} \ln\left(\frac{T}{R}\right) A(\theta,w) \quad (8)$$

The second term is the "hard collisions" term, which represents close collisions that often transfer significant momentum to the target particle,

$$\sigma_{hc}(\theta,w) = \frac{4\pi a_0^2}{T/R} B(\theta,w) \quad (9)$$

In Equations 7 and 8, T is the kinetic energy of an electron having a velocity equal to that of the ion. In the present PITS version, we ignore the higher order terms $O(1/T)$.

In the optical approximation, the angular dependent part, $A(\theta,w)$, is given by,

$$A(\theta,w) = \frac{1}{4\pi} \sum_j \left\{ \frac{R}{w + I_j} \frac{df_j}{dw} \left[1 + \frac{\beta_j}{2} P_2(\cos\theta) \right] \right\} \quad (10)$$

where

$$h_j(w) = \frac{R}{w + I_j} \frac{df_j}{dw} \quad (11)$$

is the dipole optical oscillator strength for shell j. In Equation 10, β_j is the photoelectron asymmetry parameter, P_2 is the Legendre polynomial of second order and I_j is the ionization threshold for shell j. For the $h_j(w)$, we use a piece-wise polynomial representation of the total oscillator strength.

The asymmetry parameter, β_j, is theoretically equal to two for S states, (the $2a_1$ orbital of water); for other states, β_j is energy dependent and we use a phenomenological fit based on to experimental DDCS data for water vapor (Toburen and Wilson, 1977). Empirically, β_j is taken to be,

$$\beta_j = \max \begin{cases} 1 - \exp(0.168 - 0.024\, w) \\ 0 \end{cases}, \qquad j = 1, 2, 3,$$

$$\beta_4 = 2. \quad (12)$$

Finally,

$$P_2 = \frac{3}{2}\cos^2\theta - \frac{1}{2}. \quad (13)$$

The hard collisions component of the experimental angular distributions are well represented by a (Gaussian) normal distribution (Wilson, 1978). Therefore, we represent this term, Equation 9, by a product of three functions,

$$\sigma_{hc} = g_w(\theta)\ S_{bea}(w)\ \ f_{bea}(w) \tag{14}$$

The angular dependence is described by a normalized Gaussian function $g_w(\theta)$,

$$g_w(\theta) = \frac{1}{\sqrt{2\pi}\,\Gamma} \exp\left[-\left(\frac{\cos\theta - \overline{\cos\theta}}{\Gamma}\right)^2\right] \tag{15}$$

For free electrons, the centroid of the binary encounter peak is related to the energy transferred, E_R, by

$$\overline{\cos\theta} = \sqrt{E_R}\ , \tag{16}$$

see for example Evans (1955). We find that better agreement is achieved with the experimental data for the bound electrons of water vapor if a small energy dependent correction term, δ, is added. Then

$$\overline{\cos\theta} = (\ 1 + \delta\ )\ \sqrt{E_R}, \tag{17}$$

where δ is given by a phenomenological power law relation which we do not detail here, (Wilson, 1992).

The Gaussian width parameter, Γ, is given by,

$$\Gamma = \frac{(E_2\ W)^{1/2}}{E_2 + W}\ , \tag{18}$$

with E_2 = 20 eV from least-squares fitting of the experimental data.

The amplitude of the hard collisions term is determined primarily by the second function, $S_{bea}(w)$, in Equation 14. For it, we use a simple differential binary-encounter treatment (Rudd and Macek, 1972).

Finally, the function f_{bea} is introduced to force the hard collisions component, σ_{hc}, to tend to zero as w goes to zero,

$$f_{bea}(w) = \max \begin{cases} 1 - \exp\left(\frac{9 - w}{16.5}\right) \\ 0 \end{cases} \tag{19}$$

For use in track structure simulations, the cross sections as probability density functions are integrated to give probability distributions and then inverted numerically to provide the cosine variate of the angle of emission.

MODEL EXTENSIONS

In anticipation of introducing an impact parameter representation of the inelastic cross sections (Miller et al., this workshop), a phenomenological model to disperse energy transfer points dispersed away from the trajectory of the particle has been added for both ion and secondary electron interactions. We use a functional form described by Paretzke et al. (1991) for the probability density P(b) for interaction at a lateral distance b from the path of the charged particle:

$$P(b) = C \frac{b}{b_0^2 + b^2} e^{-\omega b / v\gamma} \tag{20}$$

where v is the electron speed, $h\omega/2\pi$ is the energy transferred to the absorber, γ and b_0 are constants and C is an arbitrary normalization factor. b_0 is very nearly the most probable impact parameter value, i.e., P(b) is a maximum at essentially b_0. For preliminary sensitivity tests a constant average energy loss was assumed such that the coefficients in the exponential factor were held constant at $v\gamma/\omega = 20$ and localization of transfer points off the track axis was applied only to energy loss events of less than 50 eV. Three cases were tested, one representative of a metal (conductor) such as carbon, a second more representative of tissue and the last case for comparison is without dispersion of transfer points.

The probability density P(b) for the first two cases are shown in Figure 2. For the carbon case, $b_0 = 0.1$ nm and since the functional form of P(b) goes to infinite values of impact parameter b, localization was restricted to impact parameters of $b \leq b_{max} = 10$ nm. This case is depicted in the upper right panel of Figure 2. For the tissue-like case, $b_0 = 0.025$ nm and $b_{max} = 2$ nm; P(b) is plotted in the lower left of Figure 2 for this case.

CLUSTERING IN TRACK STRUCTURE

To provide an objective and quantitative measure of output sensitivity to variations of input parameters and model assumptions, we have introduced cluster analysis methods into the evaluation of stochastic features of charged particle tracks. The specific algorithm and

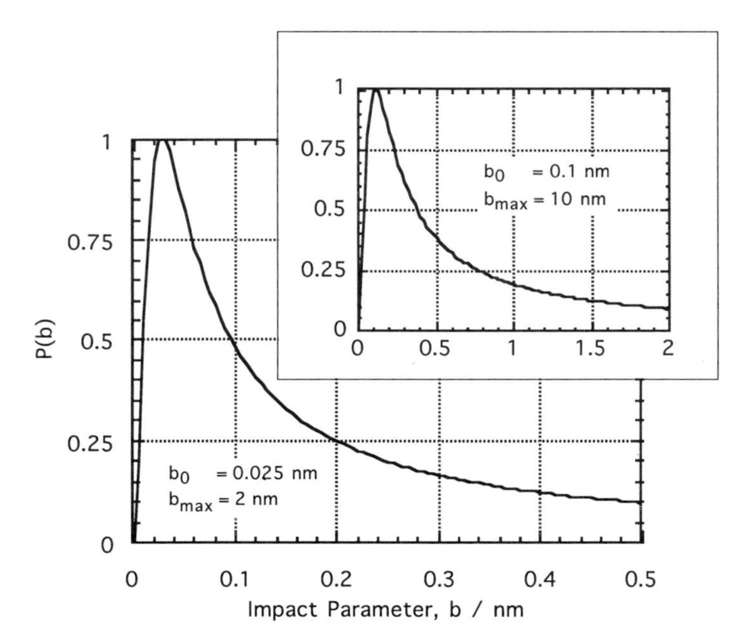

Figure 2. Probability density P(b) for impact parameter b. Truncating the function at b_{max} restricts the allowable impact parameters to $0 \leq b \leq b_{max}$. The two cases shown are representative of a conductor, upper panel, and tissue-like matter, lower panel.

method that we discuss here is the so called K-MEANS algorithm (Hartigan, 1975; Späth, 1980). The algorithm is a general computational tool for looking for correlation among diverse data in a systematic fashion, where correlated means "close" by some measure. Application to track structure features turns out to be simple because our usage is to discover items that are literally spatially close.

Input requirements of the algorithm include an initial partition of the ensemble of data into K groupings or clusters. Each datum must be in one and only one cluster; no cluster may be empty. In the simple algorithm discussed here, the cluster number K is conserved, that is, the algorithm neither adds nor deletes clusters. Within the algorithm, the centroid of each initial cluster is evaluated along with a measure of the discordance of the data within each cluster. The algorithm seeks to minimize the global discordance by attempting systematically to reassign a datum to every other cluster. The algorithm stops when no acceptable reassignment is found for one complete pass through the data ensemble. In this way the algorithm achieves a local minimum as apposed to a global minimum that would require testing all possible permutations of the data, a problem much too large even for the most powerful of hardware.

As the measure of discordance of the data within a cluster, we use the sum over the transfer points in the cluster of their squared distances from the centroid of the cluster, this is a quantity akin to χ^2. By taking the square root of this quantity finally returned by the algorithm we have an rms measure of the physical radius of each cluster. Also summing the energy associated with each transfer point in the final clusters defines the total energy deposited within a cluster and dividing the energy by the rms radius expresses the cluster energy density in terms of a radiation-quality quantity akin to LET.

For the examples that follow, we analyzed 100 nm segments of 1 MeV proton tracks simulated with PITS using a modified version of MOCA8 for transporting the delta rays. We scored only the ion track core in the sense of energy restricted LET by ignoring delta rays produced with more than 200 eV of energy. For the initial input partition to the K-MEANS algorithm, we first sorted each track into ascending order along the path of the ion, then placed the first N transfer points into the first cluster, the next N into the second, etc., until all transfer points were assigned and adjusting K, the number of clusters, accordingly for each track as generated. Values for N tried were 4, 5, 6, 8, and 10. No striking variation in results stood out, so we focussed on a mean cluster size for N of 6; all results discussed here were obtained using this value. In order to test only the sensitivity to the localization phenomenology, the other input physical model parameters were held fixed.

The probability distribution in cluster size, as measured by the number of transfer points in the cluster, is shown in Figure 3 for the three levels of dispersion described specified above (i.e., metal, tissue and none). Straight lines connect the data points to aid in identifying trends even though the variable on the abscissa can take on only integer values. One sees that introducing a dispersion feature restricted to short ranges as might be typical of tissue-like matter, gives rise to only minor differences in the cluster distribution compared with 'none.' The numbers of pairs and triples are only slightly reduced, the numbers of "fours, fives and sixes" are enhanced. However, for the conductor case, the number of medium sized clusters is significantly reduced and the number of large clusters, number of transfer points ≥ 11, is enhanced.

The sensitivity of the physical size of the clusters to the range of the dispersion function is illustrated by the frequency density distributions for the rms radius shown in Figure 4. The change in the clustering radii found by the algorithm as one varies the localization model is just what one would expect; for no dispersion the radius is the smallest, the most probable radius is about 0.5 nm. As one increases the range of the dispersion function the observed cluster radii are systematically larger, with the most probable radius increasing to about 0.8

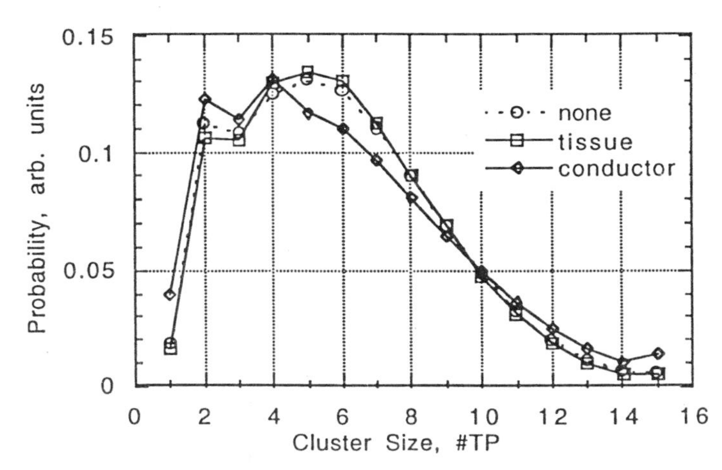

Figure 3. Cluster size distributions for all clusters. Three cases of dispersion are compared; one appropriate to a tissue-like absorber, one for a metal (conductor) and for comparison, no dispersion function employed. (#TP = number of transfer points).

and 1.5 nm for the tissue-like and metal cases respectively. What is, perhaps at first thought, surprising is that the radii for a given case do not depend on the number of transfer points in the cluster, all cluster sizes exhibit the same most probable value. However there is most likely no physical significance to this result because this is exactly the quantity that the clustering algorithm seeks to minimize for each cluster and therefore it tends to make them equal without regard to how many transfer points are in any given cluster.

Given the final assignment of each transfer point to a designated cluster, we can now ask about the total energy deposition associated with each cluster. For this purpose, we use the LET-like quantity defined above. Figure 5 presents the frequency density distributions for the ratio of energy deposited to rms radius of the cluster. The total energy quite obviously will depend on the number of transfer points in the cluster and since the rms radius of the cluster is independent of the number of transfer points one sees a monotonic increase in the energy density. For the three cases of dispersion, the energy density is on average least for the metal-like case because the mean rms radius is greatest for it. There is not much difference between the short range tissue-like case and none, the most probable energy density is approximately 200 eV/nm for a cluster size of 14 transfer points for the none case and only slightly less for the short range tissue-like case.

The preceding tests that addressed the question of the sensitivity of track-structure clustering to assumptions about the localization of energy transferred to the medium used water vapor as input model parameters. We can ask the same question with regard to absorber phase. To gain some insight into how condensed phase input parameters effect track structure clustering, we have generated and analyzed track segments simulated with PITS using liquid water parameters to compare with water vapor results. The comparison uses the short-range tissue-like dispersion for both liquid and vapor tracks throughout. Figure 6 shows the distribution of all cluster sizes as measured by the number of transfer points along with the appropriate water vapor case from Figure 3.

The obvious striking feature is the large increase in "triples." The harder primary delta-ray spectrum afforded by the shift in oscillator strength toward larger energy losses (Heller et al., 1974; Berkowitz, 1979) undoubtedly gives rise to this significant difference.

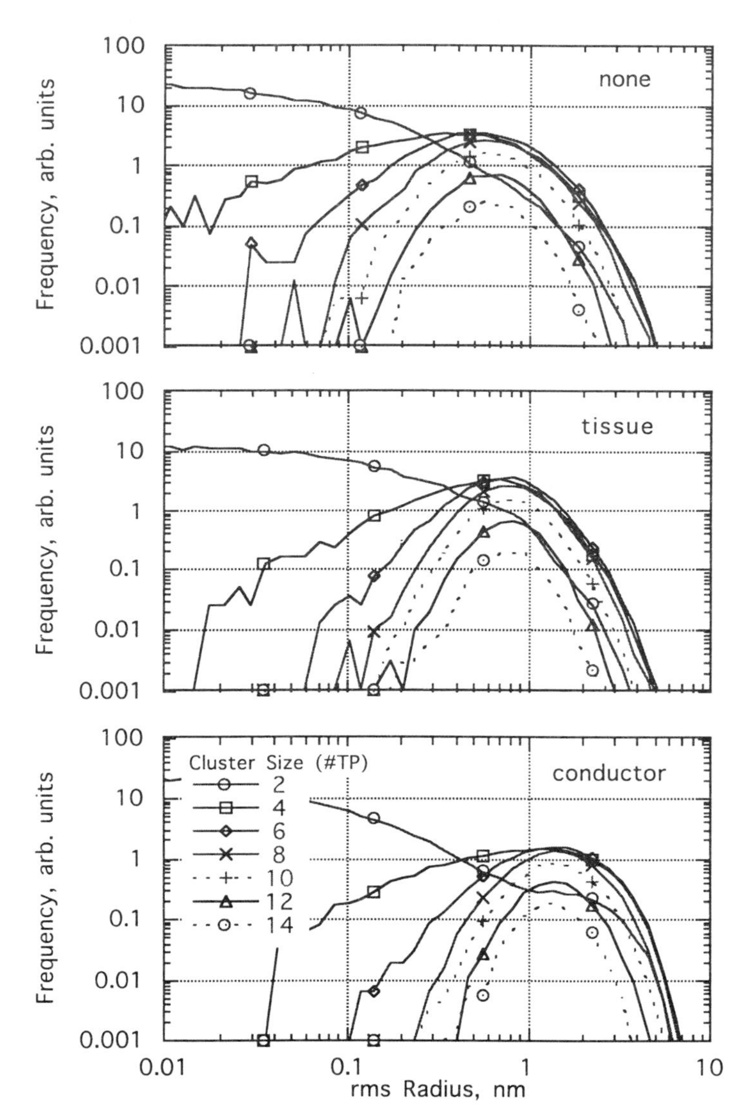

Figure 4. Probability density distributions for rms radius of the clusters returned by the K-MEANS algorithm. Three cases of dispersion are compared.

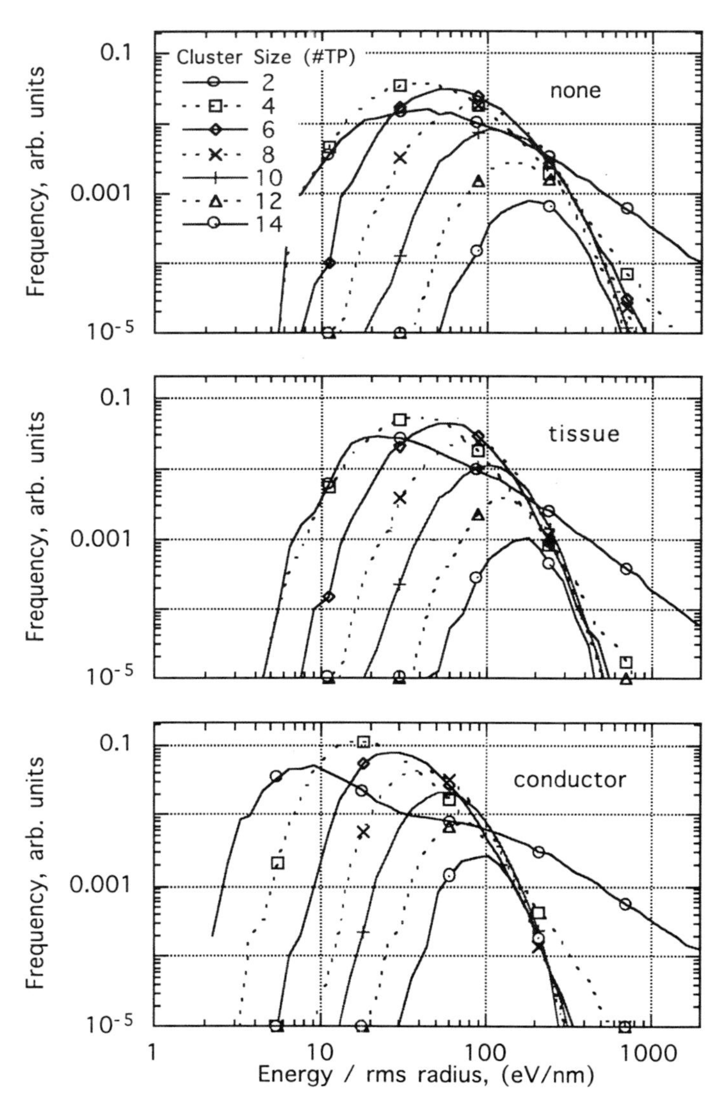

Figure 5. Probability density distributions for the ratio of total cluster energy to rms radius. Three cases of dispersion are compared.

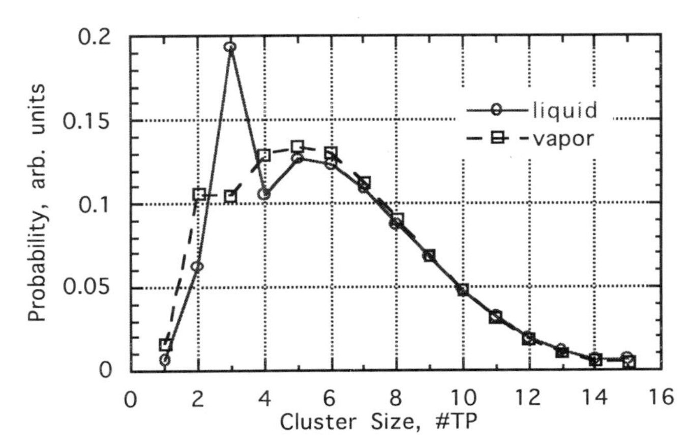

Figure 6. Cluster size distributions for all clusters. Input physical parameters for liquid water (Heller et al., 1974) and for water vapor (Paretzke, 1987) were used for the two cases.

The comparison of the frequency distributions for the rms radius and for the energy density in water vapor and liquid water are shown in Figures 7 and 8 respectively. The curves for water vapor are repeated from Figures 4 and 5 but with the data for cluster size equal to 3 added since this feature is a prominent difference between the two cases.The most probable rms radius is smaller for liquid water, being approximately 0.6 nm compared to about 0.8 nm for the vapor simulation. This presumably has to do with the larger number of ionizations produced in the liquid example; a larger number of transfer points in a given length of track segment simply implies their mean distance apart will be less.

The most probable energy density for large clusters, for example those with greater than ten transfer points, does not appear to change significantly by going to liquid phase.

CLUSTER DISTANCE DISTRIBUTIONS

The K-MEANS algorithm returns the coordinates of the centroids of the final clusters as well as the number of transfer points in each. It is a simple matter to calculate and score the distance between clusters, say between cluster i and j; furthermore, since the number of transfer points in each is known, one can score the distance between them as a function of their respective size (number of transfer points in each). We therefore define $f(d)_{m,n}$ as the differential distance distribution for cluster pairs of size m and n. Owing to the symmetry of the problem, the resulting data structure is an N by N symmetric matrix of distributions f(d), (actually, arrays or vectors, because discrete bins are used in scoring) where N is the maximum cluster size allowed. The distributions on the diagonal are the differential distance distributions for clusters of equal size, those close to but off the diagonal are for clusters of similar size, and distributions for dissimilar clusters are far from the diagonal. Visualization of such a complex of data on the standard printed page is limited; one possibility is to select a row of the matrix and display the *conditional* probability or distance distribution for cluster pairs of which one is of that selected size (row).

For illustration here, we selected row six, and present in Figure 9 the conditional probability densities for the distance between a cluster containing six transfer points and clusters of all other sizes. This result is for 1 MeV proton track segments of 100 nm length

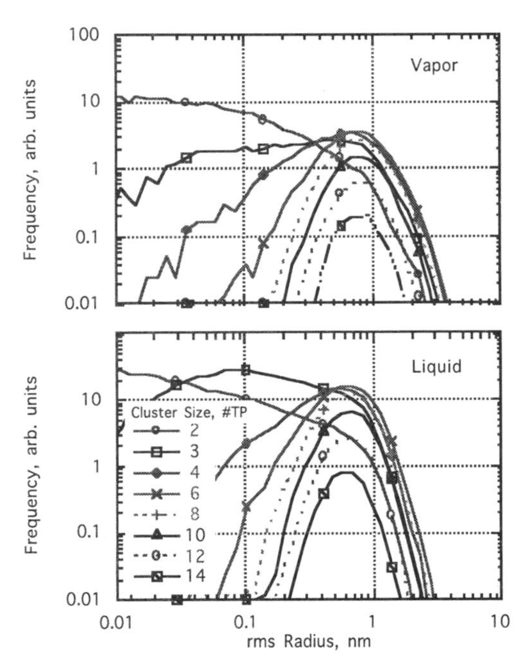

Figure 7. Probability density distributions for rms radius of the clusters returned by the K-MEANS algorithm. PITS output for liquid and vapor water are compared.

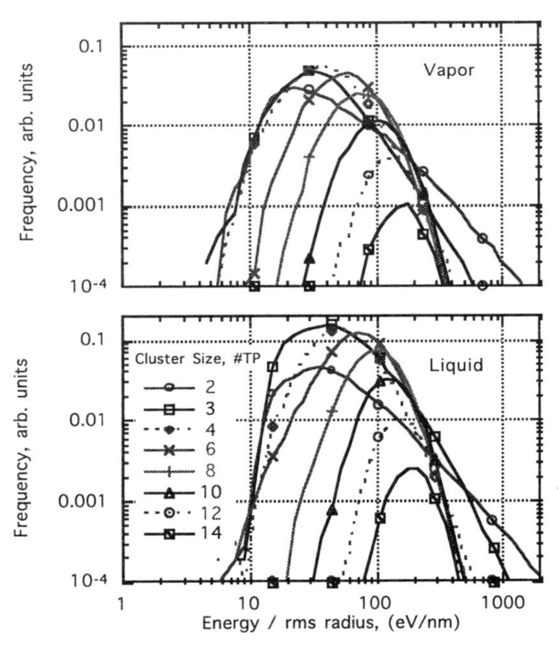

Figure 8. Probability density distributions for the ratio of total cluster energy to rms radius. Results for liquid and vapor water are compared.

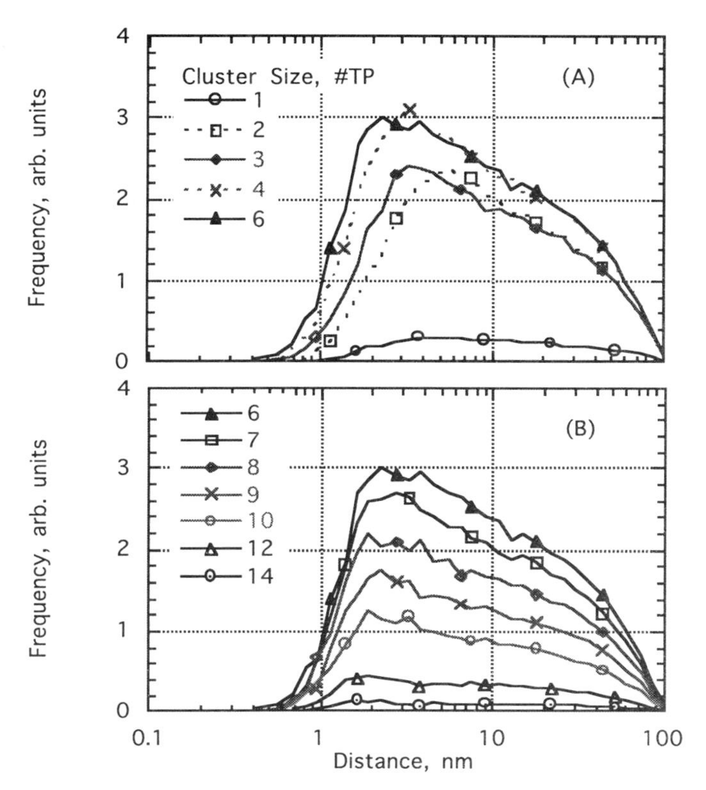

Figure 9. Differential distance distributions for cluster pairs, *conditional* on one of the pair containing 6 transfer points. The size of the other cluster of the pair is designated in the legend. This example is for 100 nm segments of 1 MeV proton tracks simulated in unit density water vapor and using tissue-like dispersion. For clarity some distributions are omitted.

using water vapor input parameters and tissue-like dispersion. A logarithmic scale is employed for the distance axis to expand the onset region of the distributions so the differences between curves is more evident. Panel A shows the distributions for the second cluster of the pair being of size 6 or smaller and panel B shows them for size 6 or larger.

Since the ion path segments were 100 nm, inter-cluster distances greater than 100 nm are naturally infrequent. The initial rise of the distributions beginning below 1 nm to their most probable value around 2 to 3 nm is related to the finite physical size of the clusters and the associated distributions in their rms radius (Figures 4 and 7). Distance to the nearest cluster neighbor of a given size (transfer points) is indicated by the most probable value. The distributions for cluster sizes of 1, 2, and 3 transfer points have a lesser initial slope and reach their respective maximum at a slightly greater distance than do the distributions for larger clusters. A possible explanation for this is that the larger clusters will preferentially be near the ion path where the ionization density is on average highest; conversely, small clusters will tend to be found in the periphery of the ion track and hence nearest small-cluster neighbors will be farther away.

ACKNOWLEDGEMENTS

Many years of collaboration with Dr. H.G. Paretzke in track structure simulations are gratefully acknowledged. This work is supported by the Office of Health and Environmental Research (OHER) of the U.S. Department of Energy under Contract DE-AC06-76RLO 1830.

REFERENCES

Arakawa, E.T., Dolfini, S.M., Ashley, J.C., and Williams, M.W., 1985, Arc-evaporated carbon films: optical properties and electron mean free paths, *Phys Rev B* 31:8097–8101.

Berkowitz, J., 1979, "Photoabsorption, Photoionization, and Photoelectron Spectroscopy," Academic Press, New York.

Blake, A.J., and Carver, J.H., 1967, Determination of partial photoionization cross sections by photoelectron spectroscopy, *J. Chem. Phys.* 47:1038–1044.

Evans, R.D., 1955, "The Atomic Nucleus," p. 835. McGraw-Hill, New York.

Hartigan, J.A., 1975, "Clustering Algorithms," John Wiley & Sons, New York, ISBN 0-471-35645-X.

Heller, J.M., Jr., Hamm, R.N., Birkhoff, R.D., and Painter, L.R., 1974, Collective oscillation in liquid water, *J. Chem. Phys.* 60:3483–3486.

Inagaki, T., Hamm, R.N., Arakawa, E.T., and Painter, L.R., 1974, Optical and dielectric properties of DNA in the extreme ultraviolet, *J. Chem. Phys.* 61:4246–4250.

Inokuti, M., 1971, Inelastic collisions of fast charged particles with atoms and molecules— the Bethe theory revisited, *Rev. Mod. Phys*. 43:297–347.

Kim, Y., 1972, Angular distributions of secondary electrons in the dipole approximation, *Phys. Rev. A* 6:666–670.

Kobayashi, K., 1983, Optical spectra and electronic structure of ice, *J. Phys. Chem.* 87:4317–4321.

Miller, J.H., Wilson, W.E., and Manson, S.T., 1985, Secondary electron spectra: a semiempirical model, *Radiat. Prot. Dosimetry* 13:27–30.

Miller, J.H., Wilson, W.E., and Ritchie, R.H., this workshop.

Nikjoo, H., this workshop.

Paretzke, H.G., 1973, Comparison of track structure calculations with experimental results, *in:* "Fourth Symposium on Microdosimetry," J. Booz, H.G. Ebert, R. Eickel, and A. Wacker, eds., pub. Commission of European Communities, Luxembourg, EUR 5122 d-e-f.

Paretzke, H.G., 1987, Radiation track structure theory, *in:* "Kinetics of Nonhomogeneous Processes," G.R. Freeman, ed., John Wiley & Sons, New York, ISBN 0-471-81324-9.

Paretzke, H.G., Turner, J.E., Hamm, R.N., Ritchie, R.H., and Wright, H.A., 1991, Spatial distributions of inelastic events produced by electrons in gaseous and liquid water, *Radiat. Res.* 127:121–129.

Rudd, M.E., and Macek, J.H., 1972, Mechanisms of electron production in ion—atom collisions, *in:* "Case Studies in Atomic Physics," 3:47–136.

Seki, M., Kobayashi, K., Nakahara, J., 1981, Optical spectra of hexagonal ice, *J. Phys. Soc. Jpn.* 50:2643-2648.

Späth, H., 1980, "Clustering Analysis Algorithms for Data Reduction and Classification of Objects," Halsted Press, div. of John Wiley & Sons, New York, ISBN 0-470-26946-4.

Tan, K.H., Brion, C.E., van der Leeuw, Ph.E., and van der Wiel, M.J., 1978, Absolute oscillator strengths (10—60 eV) for the photoabsorption, photoionisation and fragmentation of H_2O, *Chem. Phys*. 29:299–309.

Toburen, L.H., and Wilson, W.E., 1977, Energy and angular distributions of electrons ejected from water vapor by 0.3-1.5 MeV protons, *J. Chem. Phys.* 66:5202–5213.

Wilson, W.E., 1978, Analytical expression for cross section data, *in:* "Physical Sciences, Part 4 of Pacific Northwest Laboratory Annual Report for 1977 to the DOE Office of Energy Research," PNL-2500 PT4, p. 2.7. Pacific Northwest Laboratory, Richland, Washington.

Wilson, W.E., Metting, N.F., and Paretzke, H.G., 1988, Microdosimetric aspects of 0.3 to 20 MeV proton tracks, *Radiat. Res.* 115:389–402.

Wilson, W.E., 1992, A model of the DDCS for ejection of delta-rays by fast ions, *in:* "Physical Sciences, Part 4 of Pacific Northwest Laboratory Annual Report for 1991 to the DOE Office of Energy Research," PNL-8000 Pt. 4, p. 31, Pacific Northwest Laboratory, Richland, Washington.

DISCUSSION

Steele: If I understand this presentation correctly, the general thrust is mainly to gain some understanding of the stochastic geometry of a track record; also one wants to find features of merit that would allow aggregation of information about track records and to make comparisons?

Wilson: Exactly.

Steele: There are many different technologies that engage these goals. Here we have just picked one particular method and pursued it. We might want to go back to basics, before making such a choice. What are the real physical and biological principles that you might be able to add that could guide a systematic exploration of these track geometrise? There are lots of choices for metrics, features of merit and other things that we could look at. The reason I ask is that I think we get people interested in this topic if we knew how to point them toward what the real science was.

Wilson: The biologists among us are in a better position to answer that than I am, those who are close to the double strand break business and locally multiply damaged sites and so on. That is where this sort of information, I think, will contribute. We should be able to obtain from this sort of data, the mean free path for producing nearby clusters of a given size. This should be useful to the multiple damaged site people. It is also very interesting that the average size of these clusters matches well with the size of the DNA double helix.

Steele: A quick follow-up question. The data that you are viewing in this track really was just the most elementary binary event—either an event occurred or an event did not occur at this special location. Does your data also allow you to tell what the differing natures of some of these events are?

Wilson: Yes, of course. Initially I chose to partition into six transfer points per cluster, and do the initial study analyzing all transfer points. That is all ionizations, excitations, and subionization dead electrons together. You can do the same thing looking at subcategories by analyzing just the clustering of ionizations alone. For example, I chose six initial transfer points per cluster because with PITS and MOCA 8, there are about equal numbers of ionizations, excitations, and sub-ionizations. So six gives about two or three ionizations per cluster.

Zaider: If you started with four, would you obtain the same distributions?

Wilson: I did a few comparisons; when I went from six to five, I saw nothing exciting to pursue initially. No, it doesn't seem to make much difference.

Nikjoo: I just wanted to make further comments to what Walt is saying. The primary interest is to go down further in the chemical analysis of the OH radicals deposition, how they cluster, and then further from that into laboratory application of this track.

Curtis: Let me give you one answer to your question. I have developed a theory that uses as its hypothesis that there are two different types of lesions formed in DNA. One is a repairable lesion; one is an irreparable lesion. I have hypothesized that the formation of such lesions depends upon how much energy is deposited locally. Ionization cluster sizes determine how much energy is deposited locally. If one can then relate the frequency of clusters to, let's say, a parameter in the theory, and we get a resultant agreement with experimental data of some biological end point, we might have more confidence that this type of hypothesis is correct.

Ritchie: Walt, I'd like to quibble a little bit about some of your terminology. You call this delocalization, but I think of this as follows: the ion generates a coherent excitation along its path which is unlocalized. It is the process of localization that gives rise to single or multiple electron ejection. And another point about the quantity gamma in the expression, we originally set this up as a fitting parameter because we don't know as much as we'd like to about the way this energy goes into single particle excitations. For example, the excitation might migrate along the DNA chain, before becoming localized. The value that we originally gave to gamma has sort of stuck, but it certainly can be treated, and really should be treated, as an adjustable parameter to match biological data as Stan was saying.

Sachs: Two things. First, did I hear you say that if you are sitting on a cluster of six, then either the nearby guys have many members in the cluster; or if you look at many member clusters, they're more likely to be nearby? Either one of those seems completely anti-intuitive to me.

Wilson: And I must admit I'm not real sure myself on how to interpret this distribution.

Varma: With the type of work that we have seen here, Michael is asking the question: Are there some better statistical methods and mathematical approaches to this problem. Maybe I should ask Michael, to suggest a couple of examples where this type of thing can be done with other methodologies.

Steele: There are a couple of issues. Number one, I think most probabilists or statisticians, if given data with that picture, would probably model it as a marked Poisson process. So it is a Poisson process along the line, perturbations off the line determined by random variables which are independent. Once you posit a stochastic model for a track, then you have a different set of technologies for analysis than just simple cluster analysis, which is kind of like opening doors with a sledgehammer. It's indelicate.

Chatterjee: Could some of the techniques in the imaging processes people use, for example, pattern recognition, be used here?

Steele: I don't think pattern recognition would have any immediate impact. There are lots of tools in the tool kit of pattern recognition, but there is not likely to be any "off the shelf" tool of pattern recognition would have any direct bearing. Pattern recognition, still the ideas that sit behind could be important here. One example — the technology of empirical processes as you would study the empirical process along the time line corresponding to this track. That's why it would be interesting to have the data that speaks to the specific genre of these points. In the simplest part of point process theory all of the points are indistinguishable points, but you've got a whole other set of technologies if you have red, blue, and green points, or better yet, if you can assign energy values to each of the points themselves. The key issue remains the ability to connect the models to using of biological significance. The better this connection, the more sense it makes to take the models seriously.

Varma: One area where the technology developed here can be used is, for example, in the study of neuron firings.

Steele: The problems of back analysis seems easier. You've got a nice geometry. You've got a meaningful line. It's a one-dimensional object with vectors and noise associated with it. That's an easier thing to understand. It is somewhat squishily located in space, but this is a tractable novelty that just makes the problem interesting. This problem is more amenable to analysis because it is close to things that people have already studied in great detail.

Sachs: Suppose we try answering one of Steele's questions as follows, and I hope to be shot down: suppose we say that the only relevant transfer points are ionizations and all ionizations are equal. Is that a good approximation or is that a bad approximation?

Inokuti: Very bad.

Sachs: What's a good substitute?

Chatterjee: But I think the importance of this comes in the next step after ionization, that those ionizations lead to what kind of spatial distribution of radicals. That's where the importance comes in.

Steele: There are many tools that are relevant to that data. Yet even to choose a tool out of the sack, you need to evolve a clear consensus about the features of the model that are needed to address issues of biological significance — and I think we've heard several useful suggestions of places to begin.

MONTE CARLO TRACK-STRUCTURE CALCULATIONS FOR AQUEOUS SOLUTIONS CONTAINING BIOMOLECULES

J.E. Turner, R.N. Hamm, and R.H. Ritchie

Health Sciences Research Division
Oak Ridge National Laboratory
Oak Ridge, TN 37831

W.E. Bolch

Department of Nuclear Engineering
Texas A&M University
College Station, TX 77843

ABSTRACT

Detailed Monte Carlo calculations provide a powerful tool for understanding mechanisms of radiation damage to biological molecules irradiated in aqueous solution. This paper describes the computer codes, OREC and RADLYS, which have been developed for this purpose over a number of years. Some results are given for calculations of the irradiation of pure water. Comparisons are presented between computations for liquid water and water vapor. Detailed calculations of the chemical yields of several products from X-irradiated, oxygen-free glycylglycine solutions have been performed as a function of solute concentration. Excellent agreement is obtained between calculated and measured yields. The Monte Carlo analysis provides a complete mechanistic picture of pathways to observed radiolytic products. This approach, successful with glycylglycine, will be extended to study the irradiation of oligonucleotides in aqueous solution.

MONTE CARLO APPROACH

The objective of this work is to develop and apply a computer model of events that take place from the time radiation interacts in an aqueous solution until measurable radiolytic products are formed. For the highly transient conditions produced within the tracks of charged particles, Monte Carlo procedures are ideally suited to simulate what actually happens. The task can be carried out to the extent that relevant cross sections and interaction

Computational Approaches in Molecular Radiation Biology
Edited by M.N. Varma and A. Chatterjee, Plenum Press, New York, 1994

mechanisms are known. Since there is only incomplete knowledge of these factors, the Monte Carlo simulation can be used to test various hypotheses. In this way it becomes an important and effective tool for studying radiation-damage mechanisms to biological molecules. Predicted results can be compared with measurements, and the simulation can suggest new experiments. Agreement with measured data is a necessary, but not sufficient, condition for establishing the validity of the model and its inherent assumptions. However, successful testing against a wide variety of measurements under different conditions adds confidence in its essential correctness and usefulness.

CALCULATIONS FOR LIQUID WATER WITH OREC

Calculations with OREC start with a specific type of primary charged particle, having a given initial position and velocity in a liquid-water medium. A flight distance is selected for the location of the first collision, based on the total inverse mean free path for inelastic and elastic scattering. Numerical values for the inverse mean free paths, as well as all other numerical data, are based on a combination of measurements, where available, and theory. These extensive input data to the code are under continuing review and revision as further studies of details are conducted. Given the site of collision, the code then selects a type of interaction, elastic or inelastic. If the latter, then a choice is made between excitation and ionization; and a transition between specific quantum states is assigned. Algorithms have been developed to treat all possibilities and provide explicitly the residual state of the water molecule and the momentum eigenstate of the scattered primary particle as well as that of the secondary electron, when ionization occurs. All secondary electrons are transported until their energies drop below the assumed threshold of 7.4 eV for electronic transitions in liquid water. To simulate the transport of the primary particle, this process can be repeated until its energy is degraded below threshold; and the entire track is thus generated.

In OREC the passage of a primary charged particle thus leaves in its wake an array of ionized and excited water molecules, H_2O^+ and H_2O^*, in explicit quantum states and subexcitation electrons, e^-_{sub}, with specified energies < 7.4 eV. This array is formed very rapidly ($\leq 10^{-15}$ s) over several hundred angstroms in the track of the primary particle. Starting at 10^{-15} s, the computer code RADLYS, also using Monte Carlo techniques, then simulates the development of the track through the prechemical (10^{-15} s to 10^{-12} s) and chemical stages (> 10^{-12} s). The initial products from OREC give rise to the following chemical species: OH, H, H_3O^+, e^-_{aq}, and H_2, where e^-_{aq} denotes the hydrated electron. The passage of the primary particle and all secondary electrons thus results in a spatial array of these species at the time 10^{-12} s. The identity and position of each species, as determined by Monte Carlo procedures, is known explicitly. Starting at 10^{-12} s, RADLYS then simulates the further passage of time by random walk (diffusion) of the individual reactants and their possible reactions with one another (e.g., $OH + H \rightarrow H_2O$, $OH + OH \rightarrow H_2O_2$, etc.) or with solute molecules, if present. Details of the code are given in several publications.[1–5]

Figure 1 shows an example of a complete track calculated with OREC and RADLYS for a 4–keV electron, which starts moving upward from the origin of the coordinate axes. Each dot represents the position of one of the active radiolytic species. The results of diffusion and intratrack reactions can be seen at the four times shown. The number of reactants decreases from 924 at 10^{-12} s to 403 at 10^{-7} s, after which relatively few additional reactions occur. Such calculations provide an explicit representation of the chemical development within a charged–particle track in water. One obtains some understanding of the distances over which different parts of a track can affect one another and the times at which details of track structure tend to become unimportant.

COMPARISON WITH MEASUREMENTS

Calculations such as those represented by Fig. 1 also provide quantitative chemical yields of the radiolytic species as functions of time. As shown in Fig. 2, the time-dependent yields of e^-_{aq} and OH have been determined experimentally for water from pulse radiolysis with high-energy electrons.[6–8] These measured values played a key role in the development of OREC and RADLYS, particularly in formulating the hydration/ thermalization algorithms for subexcitation electrons and in the channeling of excited states into different species from 10^{-15} s to 10^{-12} s. As pointed out above, agreement of calculated results with such data is a necessary condition for the simulation to be valid, but does not constitute verification of the model.

Calculations have also been performed of the Fricke dosimeter yield for tritium beta particles. The computed and measured yields agree.

As described in more detail below, we have carried out extensive calculations and measurements for glycylglycine irradiated in aqueous solution. In these studies, the calculated yields of a number of radiolytic products as functions of solute concentration are found to be in excellent agreement with the measured values.

We are presently conducting a detailed study of the extensive experimental data collected by LaVerne and Pimblott.[9] These data provide time-dependent yields for radicals and molecular products from water irradiated by electrons. Demonstrating agreement

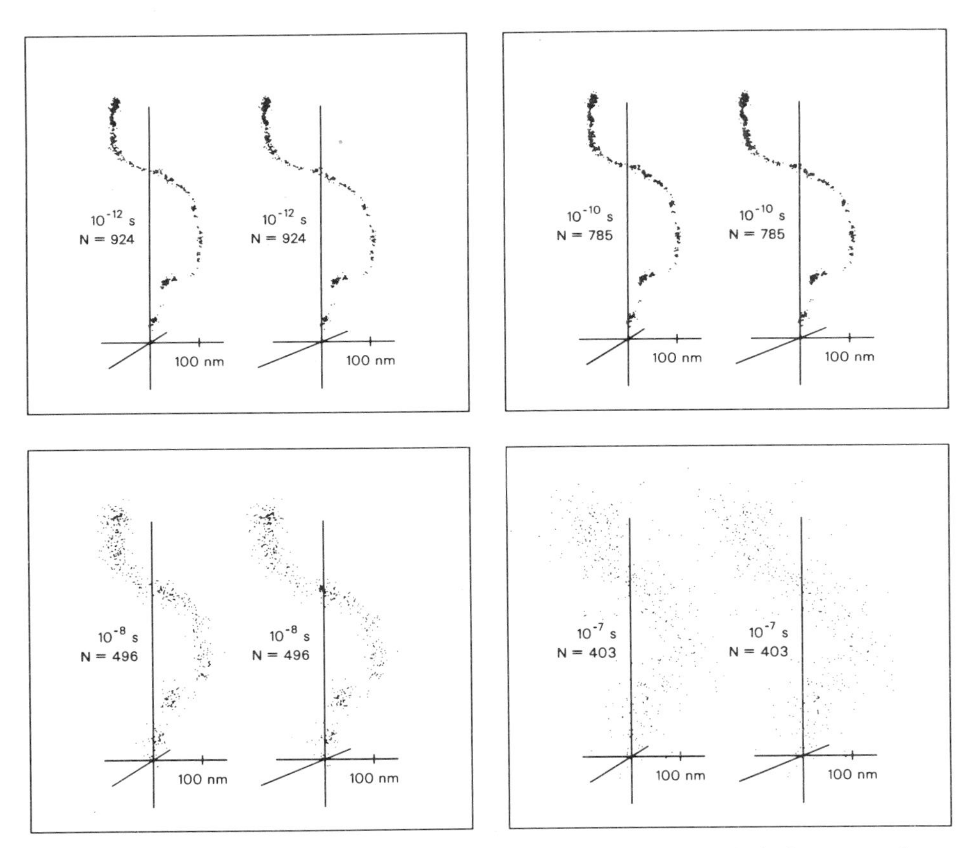

Figure 1. Complete track of a 4-keV electron calculated with OREC. Each dot in these stereo views represents the position of a radiolytic species (OH, H, e^-_{aq}, or H_3O^+) at the times shown.

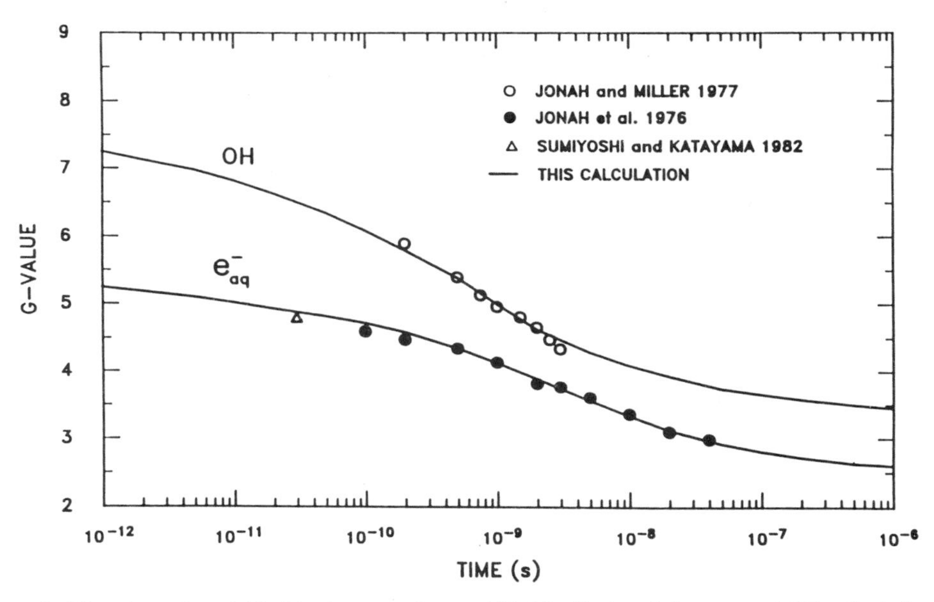

Figure 2. Time-dependent yields (G-value = number per 100 eV) of hydrated electrons and OH radicals from pulse radiolysis with high-energy electrons. Points show measurements[6–8] and lines give the yields calculated with RADLYS.

between the results of calculations and this considerable body of data would add significantly to confidence in the code.

In summary, results from OREC and RADLYS have been and are being compared with available experimental data. In addition, the basic assumptions and numerical values used in the code are under continuing study and are subject to revision in the light of new information.

COMPARISON OF CALCULATIONS FOR LIQUID WATER AND WATER VAPOR

Several Monte Carlo computer codes have been developed for charged-particle transport in water vapor, for which much more extensive experimental data exist than for the liquid. Comparisons of detailed calculations for the two states of condensation offer a valuable opportunity to gain further understanding of the physics simulated by the codes. We have conducted a number of studies with H.G. Paretzke to compare his vapor code, MOCA, with OREC.[10–12]

The total inelastic and ionization inverse mean free paths (IMFP) in the two codes are shown in Fig. 3 for electrons with energies up to 10 keV. The partitioning of the total inelastic IMFP into ionization and excitation cross sections, σ_{ion} and σ_{exc}, is shown in Fig. 4. At energies above a few tens of eV, ionization contributes over 90% to the inelastic IMFP in the liquid, compared with 60–70% in the vapor. This large difference contributes to the substantially smaller value of the average energy needed to produce an ion pair: W ~ 22 eV/ip estimated for the liquid as compared with W = 33 eV/ip measured and calculated for the vapor. An additional contribution (in the same direction) to the difference in W values is made by the relatively harder single-collision spectrum for energy loss in the liquid. A comparison of these spectra for 5-keV electrons in the two phases is provided in Fig. 5. The trend of the differences is the same except at very low energies.

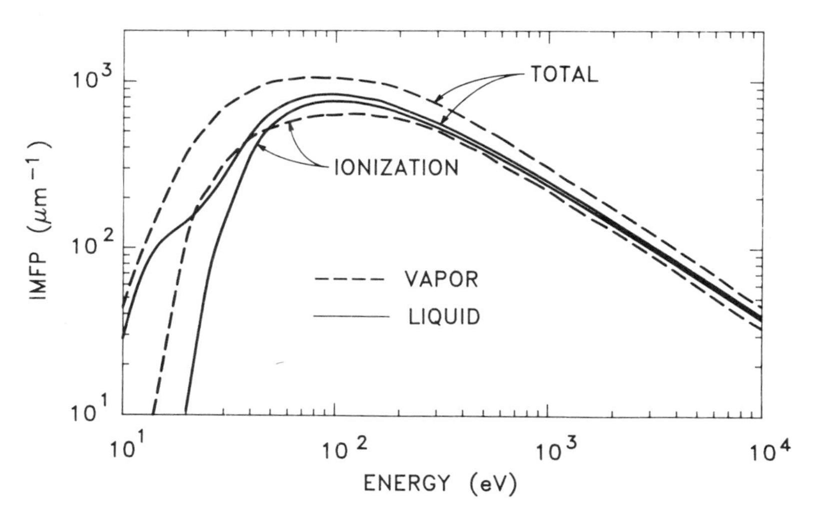

Figure 3. Total inelastic and ionization inverse mean free paths (IMFP) as functions of electron energy used in the computer code MOCA for water vapor (dashed lines) and OREC for liquid water (solid lines). (Values apply to unit density in both phases.) (Ref. 12.)

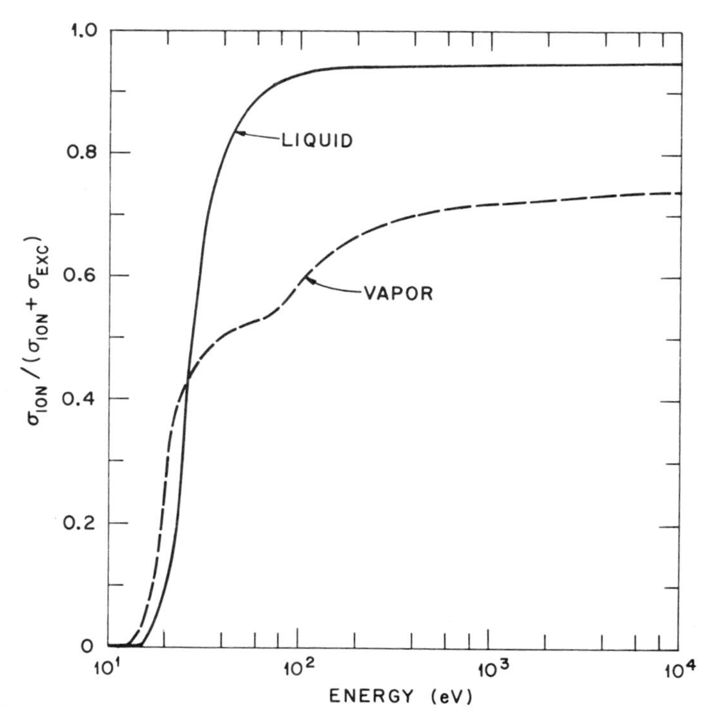

Figure 4. Ratios of ionization cross section to total inelastic cross section as functions of energy for electrons in liquid water (solid curve) and water vapor (dashed curve). (Ref. 10.)

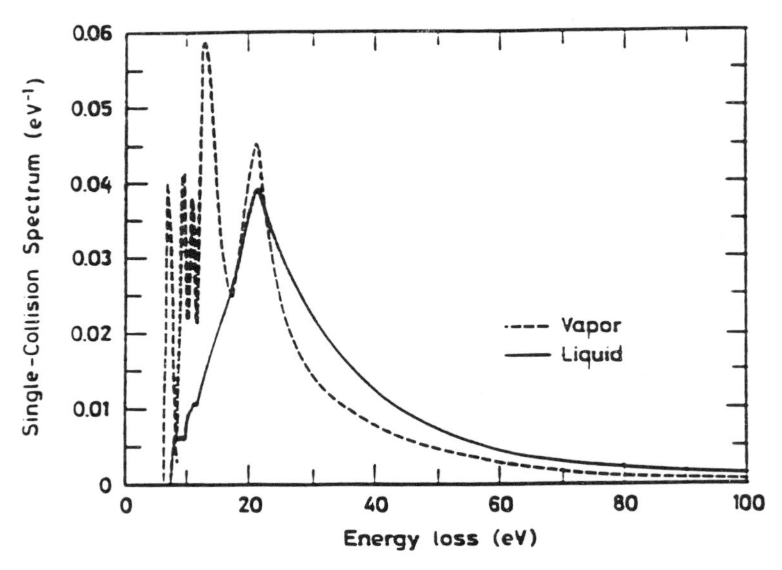

Figure 5. Normalized single-collision energy-loss spectra for 5-keV electrons in water vapor (dashed curve) and liquid water (solid curve). (Ref. 11.)

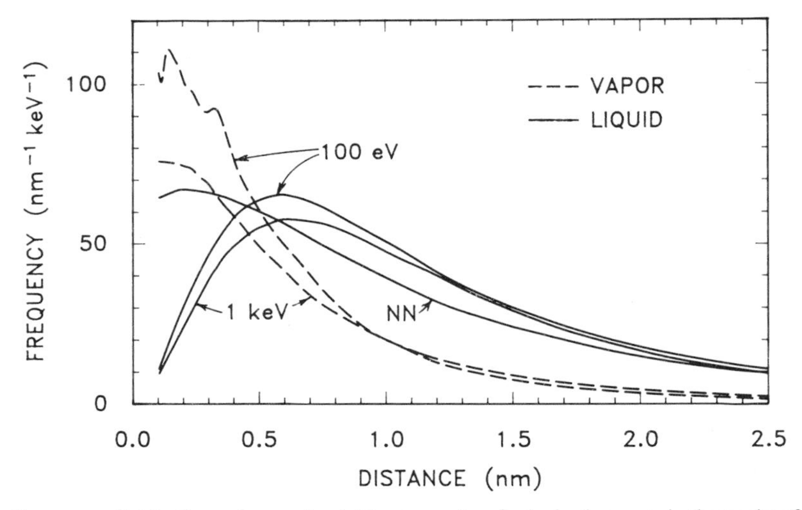

Figure 6. Frequency distributions of nearest-neighbor separation for inelastic events in the tracks of 100-eV and 1-keV electrons in water in the vapor (dashed curves) and liquid (solid curves) phases. The curve labeled NN is for the liquid without collective effects. See text. (Ref. 12.)

The liquid/vapor differences discussed here pertain to the initial, physical stage of the electron-water interaction, at 10^{-15} s. Such differences could affect the subsequent chemical development of a track (e.g., by using RADLYS). Implied differences in the later chemistry can be illustrated by comparing the frequency distribution for nearest neighbors of all inelastic events in tracks, as shown in Fig. 6 for 100-eV and 1-keV electrons. (The curves for 10-keV electrons are almost the same as for 1-keV electrons). At both energies, the most probable nearest-neighbor separation is about 0.6 nm in the liquid, compared with only about 0.1 – 0.2 nm in the vapor. This difference is due principally to the presence of collective effects in the liquid. Collective action, which is absent in the vapor, results in a non-localization of events when energy losses are ≤50 eV,[12] resulting in a greatly diminished probability for finding two neighboring events as close together in the liquid as in the vapor. The curve marked "NN" in Fig. 6 was calculated with the collective effects "turned off" in OREC. It shows that other substantial differences, like those already mentioned, exist in the codes for the liquid and vapor. The different distributions, such as those shown in the figure, would lead to different chemical development within the electron tracks and to different product yields.

TRACK STRUCTURES ON A BIOMOLECULAR SCALE

Figure 7 shows a track segment of a 1-MeV proton in liquid water. Each dot shows the location of an OH, H, e^-_{aq}, or H_3O^+ radical at 10^{-12} s after passage of the proton, the time at which diffusion-controlled intratrack reactions begin in RADLYS. The lower segment shows the middle portion of the track on an enlarged scale. The distribution of the reactants in groups, or spurs, is clearly evident.

In Fig. 8, the middle portion of the proton track is displayed on a still larger scale superimposed on a double-helical array of dots. The array corresponds approximately to the positions of successive bases and sugars of DNA along a cylinder with a diameter of 2 nm.

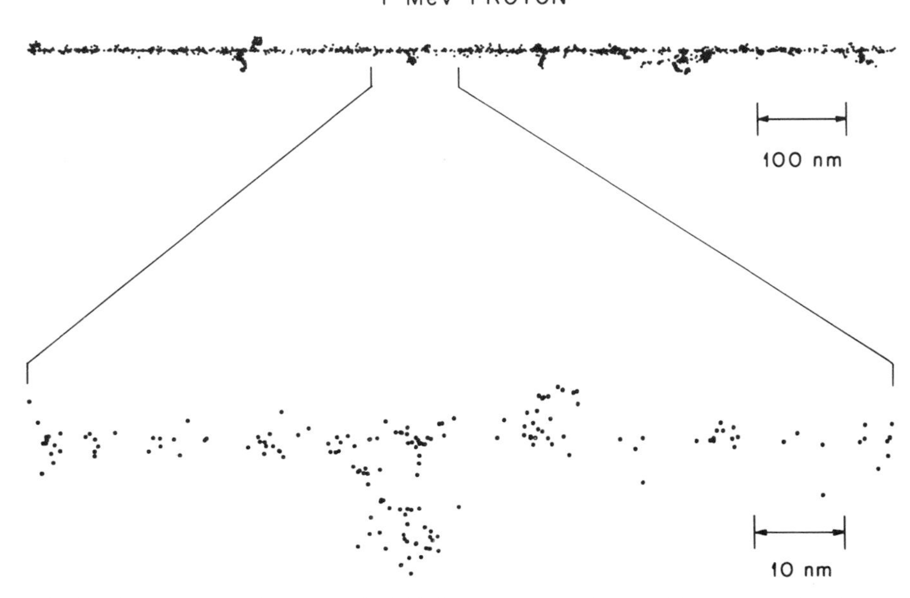

Figure 7. Calculated segment of the track of a 1-MeV proton in liquid water with central portion displayed on a large scale in lower portion. Each dot represents the location of one of the reactive species OH, H, e^-_{aq}, or H_3O^+ at 10^{-12} s.

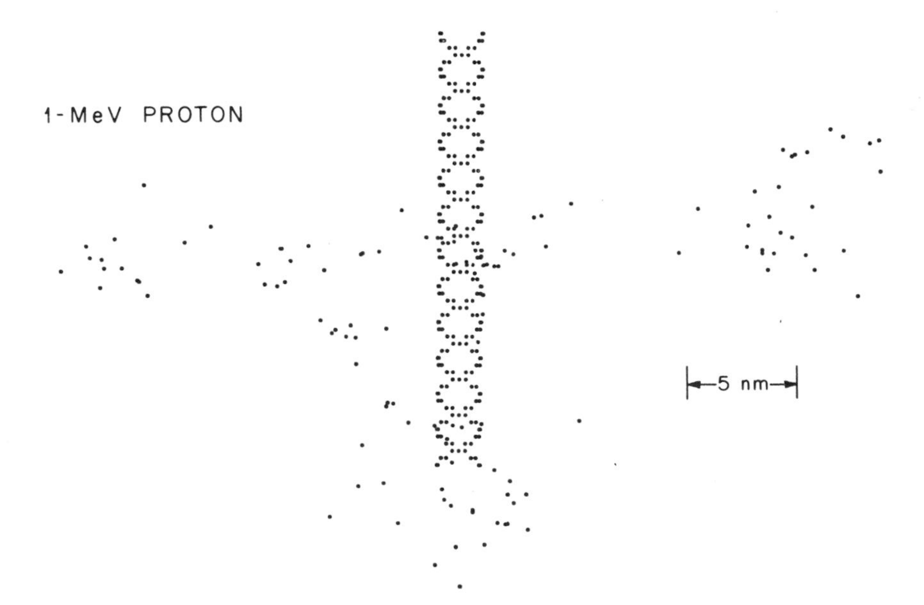

Figure 8. Central portion of track from Fig. 7 superimposed on double-helical pattern of DNA structure. See text.

The densities of the reactants along the track and the reaction sites on the DNA are comparable in this example, in which the proton LET = 28 keV/μm. By contrast, LET = 4.8 keV/μm for a 10-MeV proton, 240 keV/μm for a 1-MeV alpha particle, and 4.3 keV/μm for a 100-keV electron. Thus, the density of reactants in the water can be comparable to, or much less than or much greater than the density of reaction sites in DNA, depending on the type of particle and its energy. Also, a charged particle can traverse the DNA structure itself with a high or low probability of producing a quantum transition there (direct effects), depending on these factors.

STUDIES TO LINK PHYSICAL INTERACTIONS TO MECHANISMS OF RADIATION DAMAGE TO BIOMOLECULES

The RADLYS computer code has been extended to study the action of biological molecules in aqueous solution, with and without the presence of dissolved oxygen. As with pure water, the Monte Carlo technique is aptly suited to the simulation of the transient, non-equilibrium conditions and intratrack chemical changes that occur. As a test case of the feasibility of such a study, we investigated the photon irradiation of the dipeptide, glycylglycine, in oxygen-free aqueous solution. This molecule was chosen because of its relative simplicity and because the chemical yields of a number of radiolytic products had been measured and interpreted by Garrison and coworkers.[13,14] At the same time as we began the theoretical studies, we started a laboratory program to provide experimental measurements of yields. Calculations were performed for the same conditions of solute concentration, photon energy spectrum, dose rate, and other factors under which the measurements were made.[15]

The structure of the glycylglycine molecule is shown by the formula

$$(GLY)_2 = NH_3^+CH_2CONHCH_2COO^-.$$

When irradiated in solution at concentrations not exceeding ~ 1 M, the direct action of radiation on the molecule itself is negligible. The radiolytic products of water attack the molecule with the following initiating reactions:

$$e^-_{aq} + (GLY)_2 \rightarrow NH_3^+ + \dot{C}H_2CONHCH_2COO^-$$

$$OH + (GLY)_2 \rightarrow NH_3^+CH_2CONH\,\dot{C}HCOO^- + H_2O$$

$$H + (GLY)_2 \rightarrow NH_3^+CH_2CONH\,\dot{C}HCOO^- + H_2.$$

The resulting peptide radicals react with one another, and the deamination radical also reacts with the solute. A number of detailed reaction pathways, postulated by Garrison and coworkers,[14] were simulated in the RADLYS code. Explicit computations were carried out for the yields of free ammonia, total ammonia (following hydrolysis of some products), acetylglycine, diaminosuccinic acid, aspartic acid, and succinic acid as functions of glycylglycine concentration between 0.05 M and 1.0 M. The yields were both measured and calculated for 250-kV X rays, delivered at a dose rate of 2.80 Gy/min in oxygen-free solutions. Results of the measurements and calculations are shown in Fig. 9, where the agreement is very good. As found in the calculations and observed experimentally, the yields of aspartic and succinic acids decrease with increasing solute concentration, while the reverse is true for the other products. This is attributed to the fact that the deamination radical is a precursor to the former two, and it is scavenged by the solute.

As mentioned above, the Monte Carlo simulation furnishes a detailed picture of events as they are thought to occur in the experiment. We cite two examples to illustrate how informative the simulation can be.

First, Fig. 10 shows the concentrations of several of the molecular products as functions of the time after the X-ray machine is turned on. The glycylglycine concentration is 0.05 M. At early times, the product concentrations build up at a constant rate, all of the products

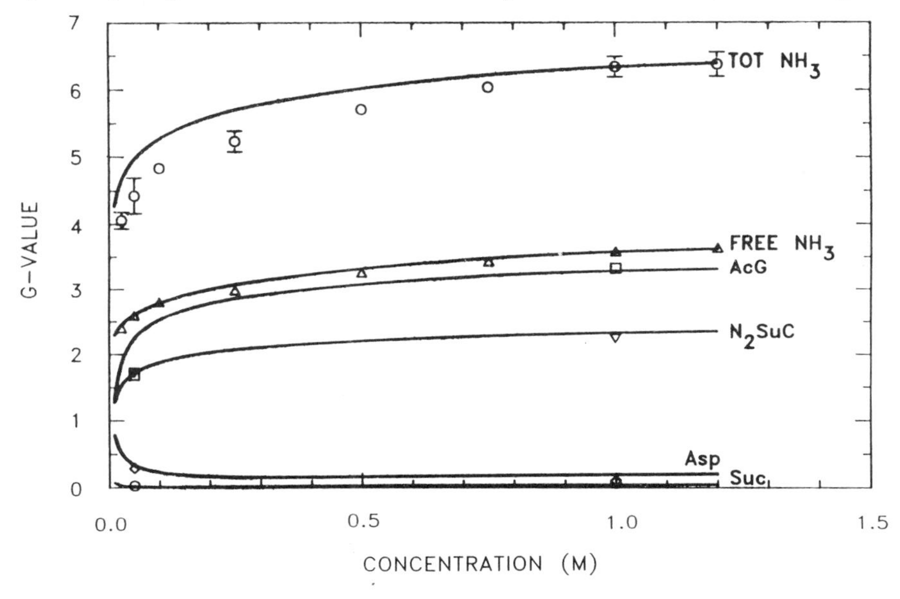

Figure 9. Comparison of measured (points) and calculated (lines) yields of total ammonia, free ammonia, acetylglycine (AcG), diaminosuccinic acid (N_2Suc), aspartic acid (Asp), and succinic acid (Suc) as functions of glycylglycine concentration. The yields were obtained for irradiation of oxygen-free aqueous solutions by 200 kV X rays at a dose rate of 2.80 Gy/min.

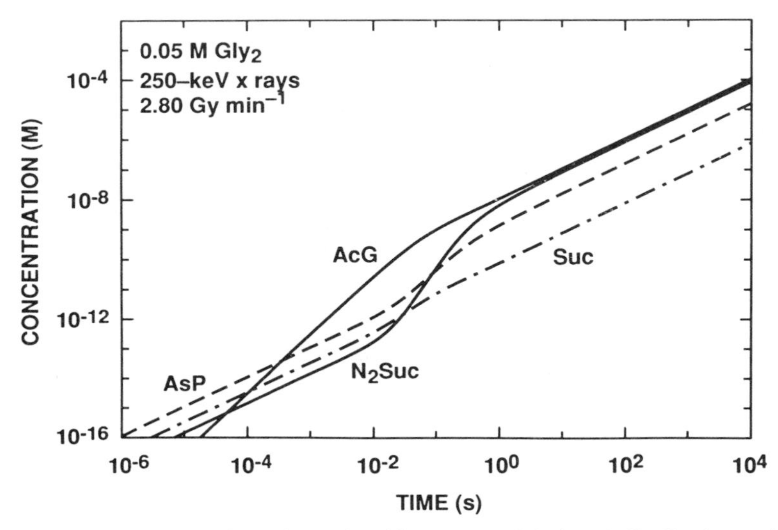

Figure 10. Calculated concentrations of aspartic acid (Asp), acetylglycine (AcG), diaminosuccinic acid (N_2Suc), and succinic acid (Suc) as functions of time for the irradiation of 0.05 M glycylglycine. Slopes change at times when the overlap of radicals produced in the tracks of different electrons becomes important.

being formed in the intratrack reactions of the individual electrons produced by the X rays. Depending on the rate constants and diffusion constants of the precursors, track overlap begins to occur at about $10^{-2} - 10^{-1}$ s; and intertrack reactions take place. The slopes in Fig. 10 then rise and become parallel to the original slopes for the remainder of the time, ≥ 1 s.

Second, Table 1 shows the fractions of some of the products that are made outside the tracks of individual electrons at 0.05 M and 1.0 M. The rate constant for the reaction of e^-_{aq} with $(GLY)_2$ to form ammonia is very large. None of the hydrated electrons escape from the vicinity of the electron tracks where they are originally produced. (At 0.01 M, however, some are able to escape.) Formation of succinic acid requires the dimerization of two deamination radicals. Since these radicals react with glycylglycine, succinic acid formation is depressed by increasing solute concentration (Fig. 9). Table 1 reflects quantitatively the way in which the deamination radical is scavenged by the solute at 1.0 M before it can participate in intertrack reactions. As a result, its overall yield decreases with increasing $(GLY)_2$ concentration.

Table 1. Fractions of products made outside individual tracks.

Product	0.05 M	1.00 M
NH_3	0.000	0.000
N_2Suc	0.999	0.994
Suc	0.584	0.002
Asp	0.940	0.115
AcG	1.000	0.998

SUMMARY

We have reviewed our research over the past two decades on the development of Monte Carlo track-structure codes for calculations of radiation transport and energy deposition in aqueous solutions. After the investigations of the physical interactions with pure water, the studies were extended to treat the formation of radicals and molecular species and their chemical reactions in water. In the late 1980s the work was further extended to analyze damage to biological molecules irradiated in aqueous solution. These investigations carried through for the first time a simulation of all events that occur from the initial physical interactions of radiation to the appearance of new radiolytic species with measurable chemical yields.

The "test-case" studies of measurements and calculations for glycylglycine led to a successful, detailed simulation and quantitative accounting of the radiation chemistry of this molecule. The Monte Carlo simulation, coupled with experimental measurements, offers a powerful approach to understanding mechanisms and pathways of radiation damage to biological molecules. Our next objective is the study of oligonucleotides and double-stranded segments of DNA in aqueous solution.

ACKNOWLEDGMENT

Research sponsored by the Office of Health and Environmental Research, U.S. Department of Energy under contract DE-AC05-84OR21400 with the Martin Marietta Energy Systems, Inc., in the Health Sciences Research Division at the Oak Ridge National Laboratory, P.O. Box 2008, Oak Ridge, TN 37831-6123.

REFERENCES

1. R.H. Ritchie, R.N. Hamm, J.E. Turner, H.A. Wright, and W.E. Bolch, Radiation interactions and energy transport in the condensed phase, *in* Physical and Chemical Mechanisms in Molecular Radiation Biology, W.A Glass and M.N. Varma, eds., pp. 99–135, Plenum Press, New York (1991).
2. C.E. Klots, R.N. Hamm, H.A. Wright, and J.E. Turner, Non-homogeneous kinetics in irradiated matter: An appraisal of computational strategies, *Radiat. Prot. Dosimetry* 31:29–31 (1990).
3. J.E. Turner, R.N. Hamm, H.A. Wright, R.H. Ritchie, J.L. Magee, A. Chatterjee, and W.E. Bolch, Studies to link the basic radiation physics and chemistry of liquid water, *Radiat. Phys. Chem.* 32:503–510 (1988).
4. R.N. Hamm, J.E. Turner, R.H. Ritchie, and H.A. Wright, Calculation of heavy-ion tracks in liquid water, *Radiat. Res.* 104:S-20–S-26 (1985).
5. J.E. Turner, J.L. Magee, H.A. Wright, A. Chatterjee, R.N. Hamm, and R.H. Ritchie, Physical and chemical development of electron tracks in liquid water, *Radiat. Res.* 96:437–449 (1983).
6. C.D. Jonah, M.S. Matheson, J.R. Miller, and E.J. Hart, Yield and decay of the hydrated electron from 100 ps to 3 ns, *J. Phys. Chem.* 80:1267–1270 (1976).
7. C.D. Jonah and J.R. Miller, Yield and decay of the OH radical from 200 ps to 3 ns, *J. Phys. Chem.* 81:1974–1976 (1977).
8. T. Sumiyoshi and M. Katayama, The yield of hydrated electrons at 30 picoseconds, *Chem. Letters* 1887–1890 (1982).
9. J.A. LaVerne and S.M. Pimblott, Scavenger and time dependences of radicals and molecular products in the electron radiolysis of water: Examination of experiments and models, *J. Phys. Chem.* 95:3196–3206 (1991).
10. J.E. Turner, H.G. Paretzke, R.N. Hamm, H.A. Wright, and R.H. Ritchie, Comparative study of electron energy deposition and yields in water in the liquid and vapor phases, *Radiat. Res.* 92:47–60 (1982).
11. H.G. Paretzke, J.E. Turner, R.N. Hamm, H.A. Wright, and R.H. Ritchie, Calculated yields and fluctuations for electron degradation in liquid water and water vapor, *J. Chem. Phys.* 84:3182–3188 (1986).
12. H.G. Paretzke, J.E. Turner, R.N. Hamm, R.H. Ritchie, and H.A. Wright, Spatial distributions of inelastic events produced by electrons in gaseous and liquid water, *Radiat. Res.* 127:121–129 (1991).

13. W.M. Garrison, Reaction mechanisms in the radiolysis of peptides, polypeptides, and proteins, *Chem. Rev.* 87:381–398 (1987).
14. W.M. Garrison, H.A. Sokol, and W. Bennett-Corniea, Radiation chemistry of glycylglycine in oxygen-free systems, *Radiat. Res.* 53:376–384 (1973).
15. W.E. Bolch, I.E. Turner, H. Yoshida, K.B. Jacobson, R.N. Hamm, and H.A. Wright, Monte Carlo simulation of free radical attack to biomolecules irradiated in aqueous solution, *Radiat. Prot. Dos.* 31:43-46 (1990).

DISCUSSION

Zaider: How do you decide which of the products occur?

Turner: There are certain sites where the radicals can attack, and we have to try to decide what happens, and make a hypothesis. The graphs that I've shown indicate that many complicated things are going on.

Miller: This morning we talked about the approximation of separating the spur chemistry from the subsequent attack on the macromolecule. I don't think that your glycylglycine experiment is capable of testing this approximation. Are you going to design DNA experiments in a way that could address that point?

Turner: Well, that's a good question. I think the glycylglycine, to a certain extent, was a proof of principle. For example, can you do the calculation in reasonable computer time? Can you keep track of everything? Should you use rate constants? Should you use a model for all of the atomic positions in detail along a molecule?

Miller: So it looks as if we need to design experiments in a way that will answer critical questions about spurs overlapping the macromolecule and sampling early time aspects of the track.

Turner: I agree with that.

DuBois: In the last viewgraph, was that gas phase data?

Turner: No, those were all in aqueous solution.

DuBois: I was going to ask how did that differ? Where are your weaknesses in the code? Where do you feel information needs to be supplied?

Turner: I'll give you my evaluation of it. Well, I think the doubly differential cross-sections that I mentioned are largely unknown. Also, we have some indication from Randy Caswell that perhaps our elastic scattering angles might be too large at high electron energies. This can be inferred from depth-dose curves we calculated for 800 kilovolt electrons. Apparently, that doesn't affect the chemistry. The chemistry takes place in the last 100 or 300 eV of electron energies in the track and spurs. There are also uncertainties, I believe, in the hydration distance of the electrons. I think possibly the partitioning in the liquid of some of the states that are identified as excitation in the vapor is uncertain also. How do thresholds change in the liquid?

Hamm: If we're going into larger molecules we need to look at direct effects on these molecules; then we are going to need cross-sections for these molecules and not the values in water.

Turner: For the molecular products at high dose rates, the curves for product yields would be steeper and the track overlap effect would kick in at earlier times.

COMPARISON OF VARIOUS MONTE CARLO TRACK STRUCTURE CODES FOR ENERGETIC ELECTRONS IN GASEOUS AND LIQUID WATER

Hooshang Nikjoo

MRC Radiobiology Unit
Chilton, Oxon OX11 0RD, U.K.

Shuzo Uehara

School of Health Sciences
Kyushu University
Higashi-Ku, Fukuka, Japan

ABSTRACT

Cross sections for *kurbuc*, a Monte Carlo track structure code simulating histories of electrons, molecular interaction by interaction, in the energy range of 10 eV to 10 MeV, have been presented. Comparisons have been made for four independent Monte Carlo track structure codes for energetic electrons in gaseous and liquid water. The comparisons have been made in terms of point kernels for interactions and energy absorbed, and frequencies of energy depositions in cylindrical volumes of sizes similar to biological macromolecules. Comparisons have been made for 100 eV, 300 eV, 500 eV, 1 keV, 10 keV and 100 keV monoenergetic electrons. The four electron codes used in this study are *moca8b* and *kurbuc* for water vapour and *orec* and *cpa100* for liquid water. A summary of cross sections used in each code has been presented. The comparisons show similarities and differences in clustering properties of the four codes.

INTRODUCTION

Monte Carlo track structure simulation has become an important tool in biophysical modelling of radiation effects in mammalian cells. The critical importance of clustered properties of radiation tracks in determining their biological properties cannot be measured over dimensions of less than tens of nm by current experimental methods. The availability of Monte Carlo track structure codes which simulate radiation tracks, molecular interaction by interaction, has in more recent years made it possible to make theoretical investigation of those clustered properties down to dimensions of less than 1 nm. Although the first article

Computational Approaches in Molecular Radiation Biology
Edited by M.N. Varma and A. Chatterjee, Plenum Press, New York, 1994

on Monte Carlo methods appeared in 1949,[1] it was not until 1963 that Martin Berger[2] showed us how it can be used to carry out electron Monte Carlo track structure calculations. Since then many review articles have been published on the subject dealing with the principles and methods of Monte Carlo track structure calculations.[3-7] Currently, there are a number of codes available for simulation of electron tracks in simple environments such as water.[1,8-16] The complexity of biological structures such as DNA make such simulations very difficult or impossible because of the lack of availability of primary input cross section data. Table 1 lists most of electron codes written by various authors to simulate passage of electron tracks in water to date. All these codes simulate histories of electrons from initial energy down to a cut-off energy of nominally about 10 eV. Some of the codes have extended the simulation by following the thermalised electrons into physicochemical stage down to ~ 10^{-12}s when stable chemical species are produced. Although all codes listed in Table 1 and others, broadly speaking, have incorporated the same primary input data in the form of experimental cross sections for water but methods by which subsequent processes involved have been treated produce tracks with different clustering properties. These differences could lead to significant differences in physical, chemical and biological interpretations. Therefore, it is pertinent to assess the differences between these codes and the confidence one might place in present theoretical understanding of the physics and chemistry of track structure.

Previously, comparisons have been made between Monte Carlo track structure codes for energetic electrons in water.[17-20] Most previous comparisons have been made in terms of physical properties pertaining to the physics of Monte Carlo track structure calculations. A major parameter of the present paper is comparison of clustering properties of tracks in terms of amount of energy deposited at 10^{-15}s in cylindrical volumes of sizes similar to DNA, nucleosome and chromatin fibre.

The data used in this study have been generated by computer codes *moca8b* and *kurbuc* for water vapour and *cpa100* and *orec* for the liquid water medium. All data for *moca8b*, *kurbuc* and *cpa100* were generated at the MRC Radiobiology Unit on Unix machine while those for *orec* were generated at Oak Ridge. Detailed description of codes *moca8*b, *orec* and *cpa100* have previously been published elsewhere.[8-10]

As detailed description of code *kurbuc* is not yet published a summary of the code is given here.

Table 1. Electron Track Structure Code for Energetic Electrons @ 10^{-15}s.

Code	Author	Date	Medium	Energy Range	Ref
ETRAN	Berger & Seltzer	1963	All medium	10 keV - GeV	2
MOCA8B*	Paretzke	1970	H_2O vapour	10 eV - 100 keV	8
OREC*	Turner *et al*	1976	liq. H_2O	10 eV - 1.0 MeV	9
CPA100*	Terrissol	1976-	liq. H_2O	10 eV - 100 keV	10
DELTA	Zaider & Brenner	1983	vapour	10 eV - 10 keV	11
ETRACK*	Ito. A	1985	vapour	10 eV - 1 MeV	12
KAPLAN*	Kaplan	1990	vap. + liq. H_2O	1-10 keV	13
TRION	Lappa *et al*	1992	H_2O vapour	10 eV - 1 MeV	14
ETS*	Hill & Smith	1993	vap. + liq. H_2O	10 eV - 10 keV	15
KURBUC	Uehara *et al*	1993	vapour	10 eV - 10 MeV	16

*These codes have extensions to physicochemical time domain

The Code *Kurbuc*

The code *kurbuc* simulates tracks of energetic electrons, molecular interaction by interaction, in water vapour. Detailed description of the code is given by Uehara *et al.*[16]

Total elastic, ionization and excitation cross-sections used in the code *kurbuc* are given in Fig (1). The elastic scattering cross-sections were obtained from Rutherford's formula. As Rutherford's formula is inadequate in describing elastic scattering at low energies various experimental data were used to obtain elastic cross-sections below 1 keV. The double differential cross-sections for elastic scattering below 1 keV have been shown in Fig (2) as a function of scattering angle (θ). The experimental data are from Nishimura *et al* at 15 eV,[21] Trajmar *et al* at 20 and 50 eV,[22] and Katase *et al* for energies greater than 100 eV.[23] The ionization cross-sections above 10 keV were calculated by Seltzer's[24] method. The ionization cross-sections below 10 keV were taken from published data of Paretzke.[8]

The total excitation cross-sections above 10 keV were derived by the formula given by Berger and Wang.[25] Similarly the total excitation cross-sections below 10 keV were taken from Paretzke.[8] The total ionization, excitation and elastic cross-sections were fitted in such a way to produce a smooth curve, between the high and low energy regions, details of which are given by Uehara *et al.*[16] In Fig (1) also shown are data of Hayashi[26] for elastic, ionisation and excitation cross sections. Angular distributions of secondary electrons were obtained by fitting the data of Opal *et al.*[27] The different ionization cross sections as a function of secondary electrons scattering angle for 500 eV primary electron is shown in Fig (3).

The inelastic cross sections and the collison stopping power for water vapour have been presented in Fig (4) and Fig (5) respectively. The input data for *kurbuc* have been compared with the data for *moca, eltran* and *orec*.

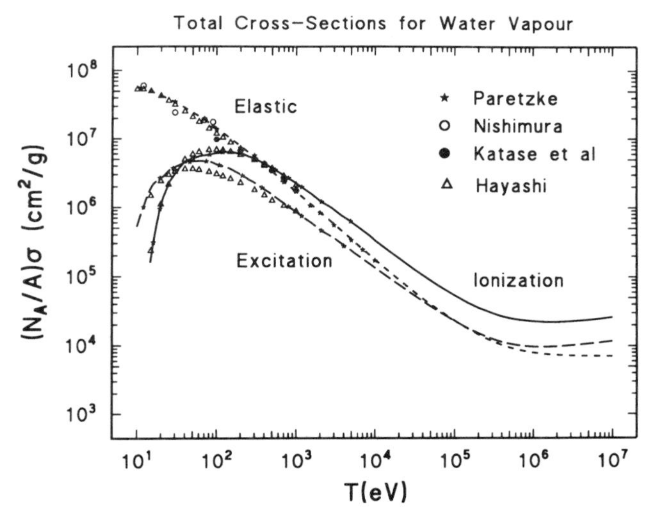

Figure 1. Total elastic, excitation and ionization cross-sections for water vapour. Experimental data of Nishimura[21] and Katase *et al*[23] and published data of Paretzke[8] were used to obtain cross-sections for code *kurbuc* by methods explained in the text. Data of Hayashi[26] have been included for comparisons. For electron energies greater than 10 keV calculated cross-sections were adapted as there are no experimental data available.

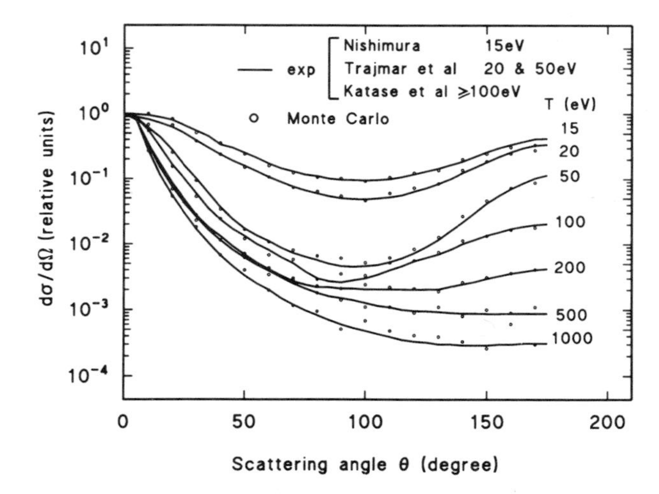

Figure 2. Differential cross-sections dσ/dΩ in relative units for elastic scattering in water as a function of scattering angle θ. Various experimental source were used to obtain the fitted data for the code *kurbuc*. The solid lines are the experimental data of Nishimura[21] for electron energy of 15 eV, at 20 eV and 50 eV experimental data of Trajmar *et al*[22] and for electron energies 100 eV, 500 eV and 1 keV experimental data for Katase *et al*[23] were used. The fitted data O were obtained by direct sampling of the experimental data for 10^4 trials.

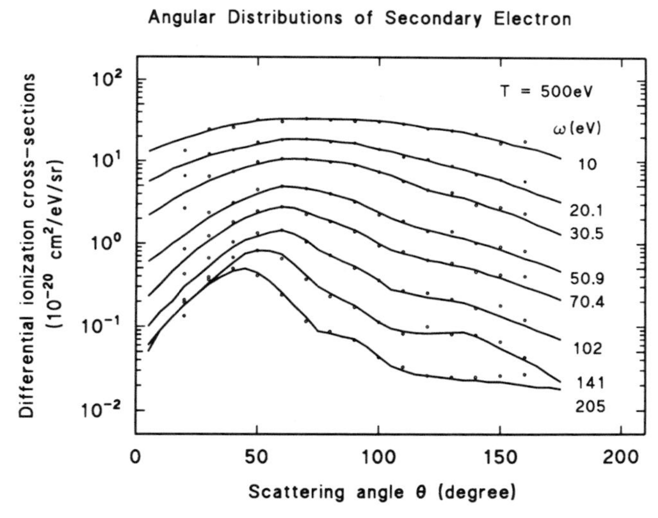

Figure 3. Electron production cross-sections differential as a function of differential ionization cross-sections in 10^{-2} cm^2/eV/sr. Experimental data of Opal *et al*[27] were used to obtain angular distributions of secondary electrons, for primary electron energy of 500 eV, by direct sampling. The sampled data O were obtained for 10^4 trials.

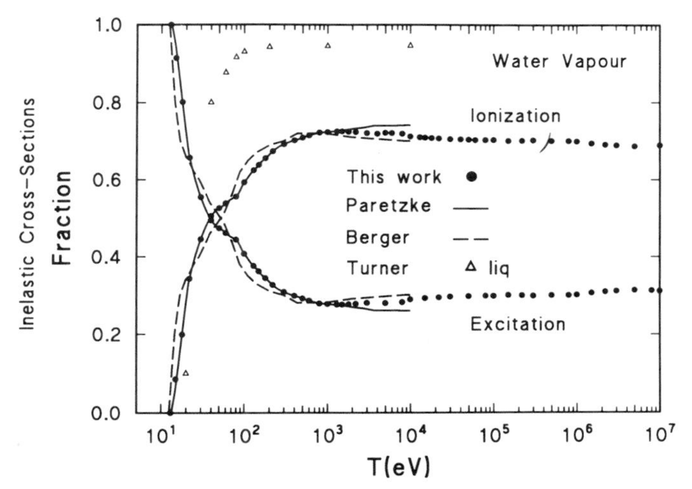

Figure 4. Inelastic cross-sections as a function of initial electron energies from 10 eV to 10 MeV for water vapour. Cross-sections calculated for *kurbuc* have compared with cross-sections for Monte Carlo track structure codes *moca8*, *etran* from Paretzke[8] and that of *orec* obtained from Turner *et al*[17] for liquid water.

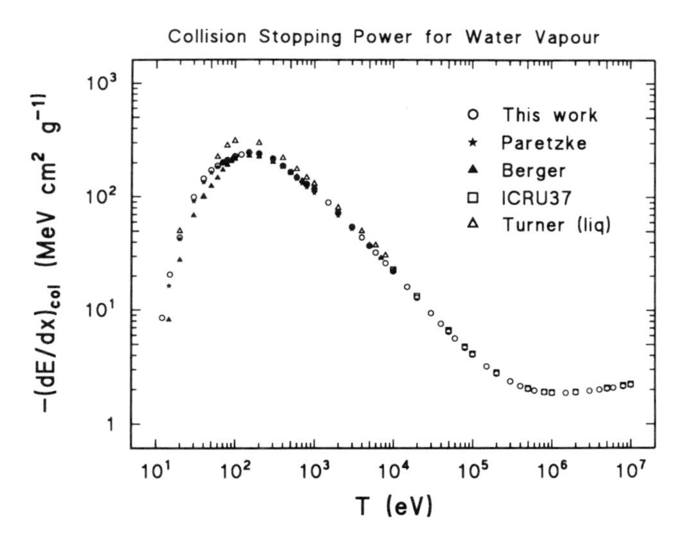

Figure 5. Calculated mass stopping powers for codes *kurbuc*, *moca8*, *etran* for water vapour and *orec* for liquid water. The ICRU37 data for the higher energy range has been compared with the calculated data for *kurbuc*.

Comparison of Monte Carlo Track Structure Codes

Comparison of four Monte Carlo track structure codes have been made. The codes are *moca8b*, *cpa100*, *orec* and *kurbuc*. All four codes simulate electron histories, molecular interactions by interactions, from a few eV to keV and MeV regions as listed in table (1).

A summary of cross-sections used as input data for each code and source of cross sections are given in table (2).

Table 2. Sources of Cross-Sections Used in Each Code.

Cross Sections	MOCA8 H_2O vap	KURBUC H_2O vap	CPA H_2O liq	OREC H_2O liq
Total inelastic	Derived from various sources	< 10 keV: Derived from published data of Paretzke. > 10 keV: Summation of calculated ionization and excitation cross sections.	Derived from various sources.	From the measured dielectric function, constrained by sum rules.
Ionisations	Inner shell: Drawing formula and fitted to the data of Glupe & Mehlhorn. Outer shell: Tan et al data and own analysis	< 10 keV: Derived from published data of Paretzke. > 10 keV: Seltzer formula based on Weizsacker-Williams method for both inner and outer shells.	Use of experimental data and dielectric response function for 4 outer shells.	Partitioning, based on various sources, of the imaginary part of the dielectric function between the inner and four outer shells.
Excitations	Derived from various sources and own analysis	< 10 keV: Adopted published data of Paretzke. > 10 keV: Fano-plot using Berger-Wang parameters. One excited state with mean excitation energy assumed. For excitation energies > 10 keV extrapolated.	Differential oscillator strengths with dielectric response function from experiments. liq=ice	Partitioning, based on various sources, of the imaginary part of the dielectric function between six excitation types.
Elastic	> 300 eV Berger and Moliére formula. < 300 eV Moliére theory and use of experimental data.	Moliére formula with modified screening parameter fitted to experimental data below 50 keV.	< 8.4 eV experimental data. > 8.4 eV Mott-Dirac formula.	< 10 eV liq=0.6* vapour. < 1 keV phase-shift calculations. > 1 keV Thomas-Fermi theory.
Energy differential sec. e^- spectrum	Calculated according to Platzman, Fano and Kim.	Seltzer formula based on Weizsacker-Williams method.	According to Platzman, Fano, Kim and adopted Osc. Strength.	Energy loss from Born approx. calculation less binding energies.
Auger e^- spectrum	Analysis of experimental data.	Auger yield of inner shell.	Auger yield of one.	Auger yield of one for inner shell.
For 1 keV e^- g(ion) W(eV) at 10^{-15} s	3.33 yield/100 eV 30.0 eV/ip	3.22 yield/100 eV 31.0 eV/ip	4.50 yield/100 eV 22.2 eV/ip	3.99 yield/100 eV 25.1 eV/ip

Previously codes *moca* and *orec* have been subject of extensive comparison.[17-20] Recently, the four codes, *moca*, *kurbuc*, *cpa* and *orec* have also been the subject of a preliminary comparison.[28] Apart from *orec*, electron tracks used in this study were generated on the same computer and the same random number generator at the MRC Radiobiology Unit. The data from *orec* were generated at Oak Ridge. Large numbers of 'physical tracks' (> 10^3), were generated for 100 eV, 300 eV, 500 eV, 1 keV, 10 keV and 100 keV and compared according to the following tests. 'Physical track' here means track at 10^{-15}s including all interactions generated by an electron with a given initial energy T_o (eV) followed down to a cut-off energy of T_c. Any electron below the cut off energy T_c is considered to be a 'dead electron'. All 'dead electrons' were diffused a short distance and its energy deposited at the point. In short, 'physical track' here is a photographic image of all ionization and excitations at 10^{-15}s.

1. Point Kernal for Interactions. Radial distributions of all interactions were obtained at 1 nm intervals. The origin of all tracks were positioned at the centre of a sphere with radius larger than the maximum lateral extensions of all tracks. The sphere was divided into spherical shells and the number of interactions in each shell were determined. The radial distributions for various electron energies from 100 eV to 100 keV are shown in Fig (6). These data show the variation of electron range at each initial electron energy and differences between the liquid and vapour data.

2. Point-Kernal for Absorbed Energy. The differential radial distribution of energy was determined as a fraction of energy deposited in each spherical shell. The distributions for 100 eV, 300 eV, 500 eV and 1 keV monoenergetic electrons are plotted in Fig (7). The distributions for larger initial energies present the same trends as seen in Fig (7).

3. Clustering Analysis of Interactions. An understanding of the mechanisms of damage in biological targets by ionizing radiation requires a knowledge of spatial distribution of energy loss in subcellular structures such as DNA and larger macromolecules. The method of analysis has extensively been reported elsewhere.[29-31] In principle the method seeks to calculate the absolute frequencies of energy depositions in small cylindrical volumes of dimensions 1 nm upward. The preliminary step is to simulate the radiation track, molecular interaction by interaction, in the given medium. Large numbers of electron tracks generated by each code were used to score absolute frequency of energy depositions in cylindrical targets. Each track was placed in a large volume to contain the entire track. Subsequently, small cylindrical targets were randomly generated in the spherical volume using method of μ-randomness. The energy deposited in the target cylinder was obtained by comparing the positions of all ionizations and excitations with the volume occupied by the target. The energy deposited in slices of 0.1 nm interval were grouped together to give the total energy in the target. Fig (8-10) show clustering of all events as integral frequency distributions for various electron energies for target sizes similar to chromatin fibre, nucleosomes and DNA. The left ordinates show comparison of distributions of absolute frequencies of energy depositions in cylindrical volumes of dimensions similar to a segment of chromatin fibre, a nucleosome and a segment of DNA when randomly positioned and oriented in water irradiated with 1 Gy of given radiation. The distributions in Fig (8-10) shows similarity and differences between the four codes.

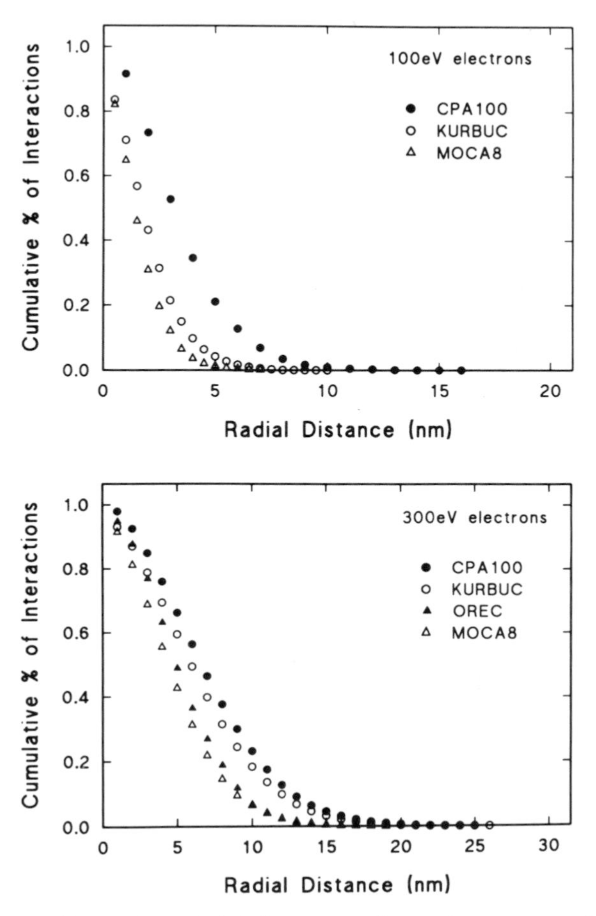

Figure 6a. Point kernel of interactions. Radial distribution of interactions as a normalized cumulative percentage of interactions at nm intervals from the center of a sphere placed at origin. Each point represents total number of interaction in a spherical shell from the origin. For 100 eV electrons 5000 tracks starting at origin were generated for each code. For 300 eV electrons the same as above was applied except for 'orec' only 100 electron tracks were used.

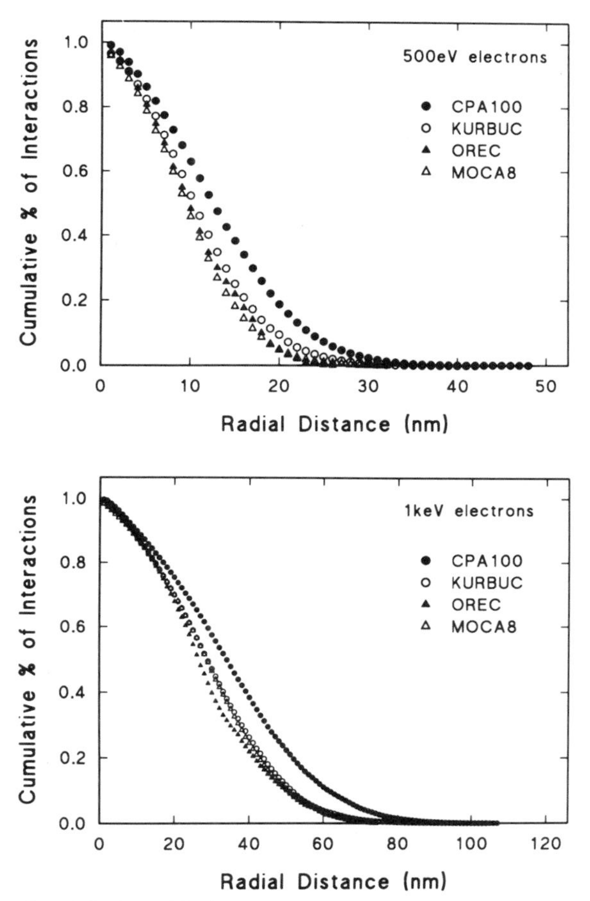

Figure 6b. As in Fig (6a). Point kernel for interactions. Radial distribution of interactions for 500 eV and 1 keV electrons for the four codes.

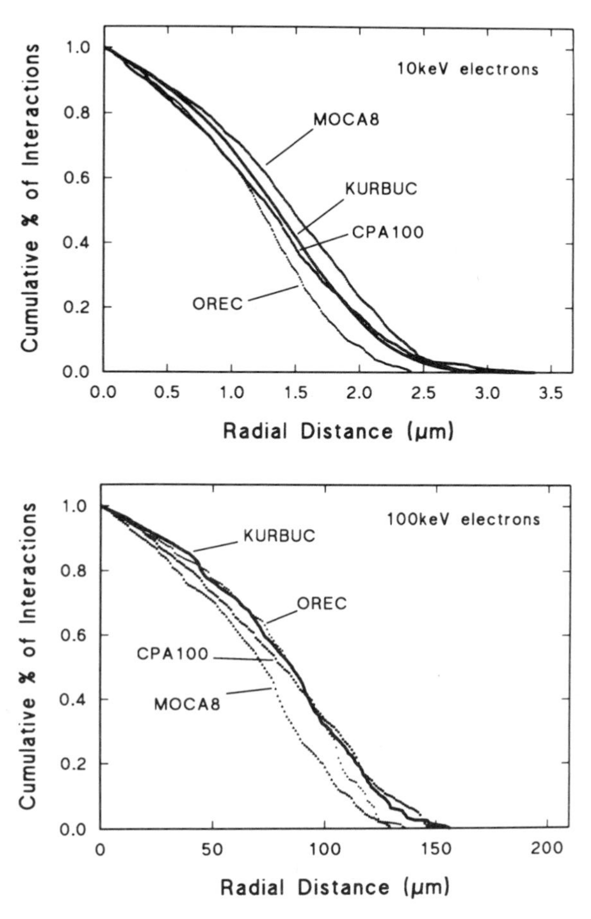

Figure 6c. As in Fig (6a). Point kernel for interactions. Radial distribution of interactions for 10 keV and 100 keV electrons for the four codes. The distributions for 100 keV electrons shown large scatter due to much lower number of electron tracks used.

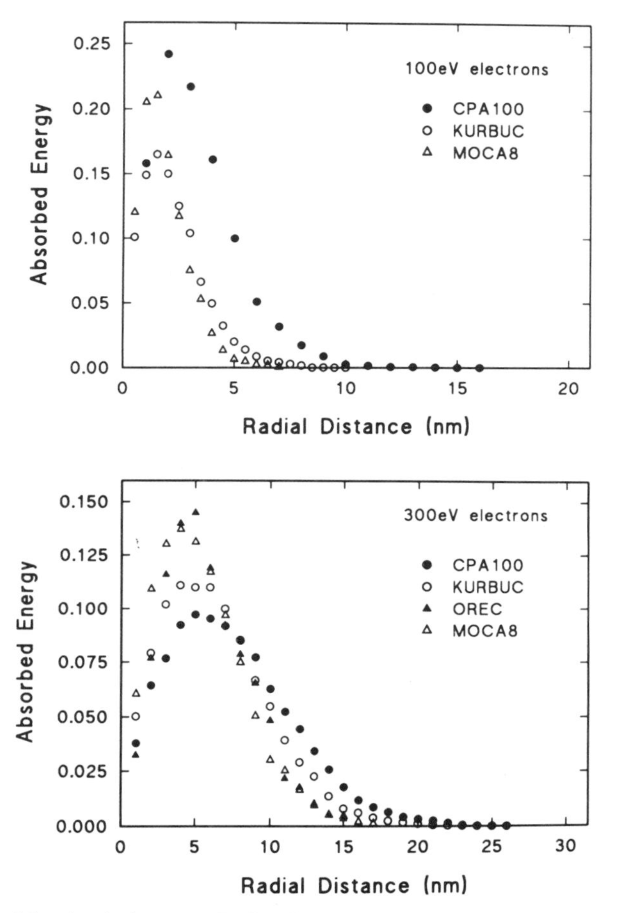

Figure 7a. Point kernel for absorbed energy. Radial distribution of energy deposited in a sphere positioned at origin and divided into spherical shells. The ordinate shows the differential energy deposited in sub-shells for 100 eV and 300 eV electrons.

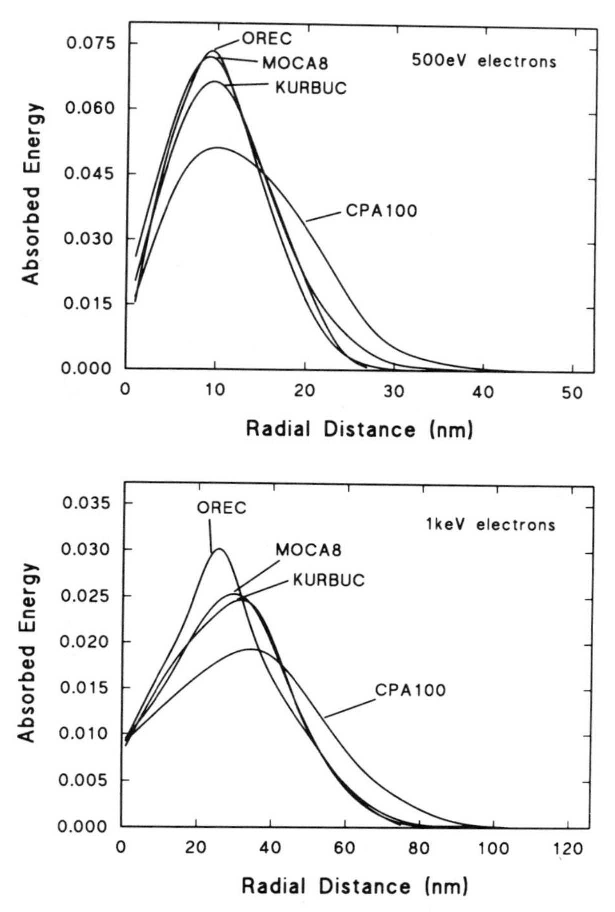

Figure 7b. As in Fig (7a) but for electron energies 500 eV and 1 keV. The solid lines show the fitted lines to the distributions for the four codes.

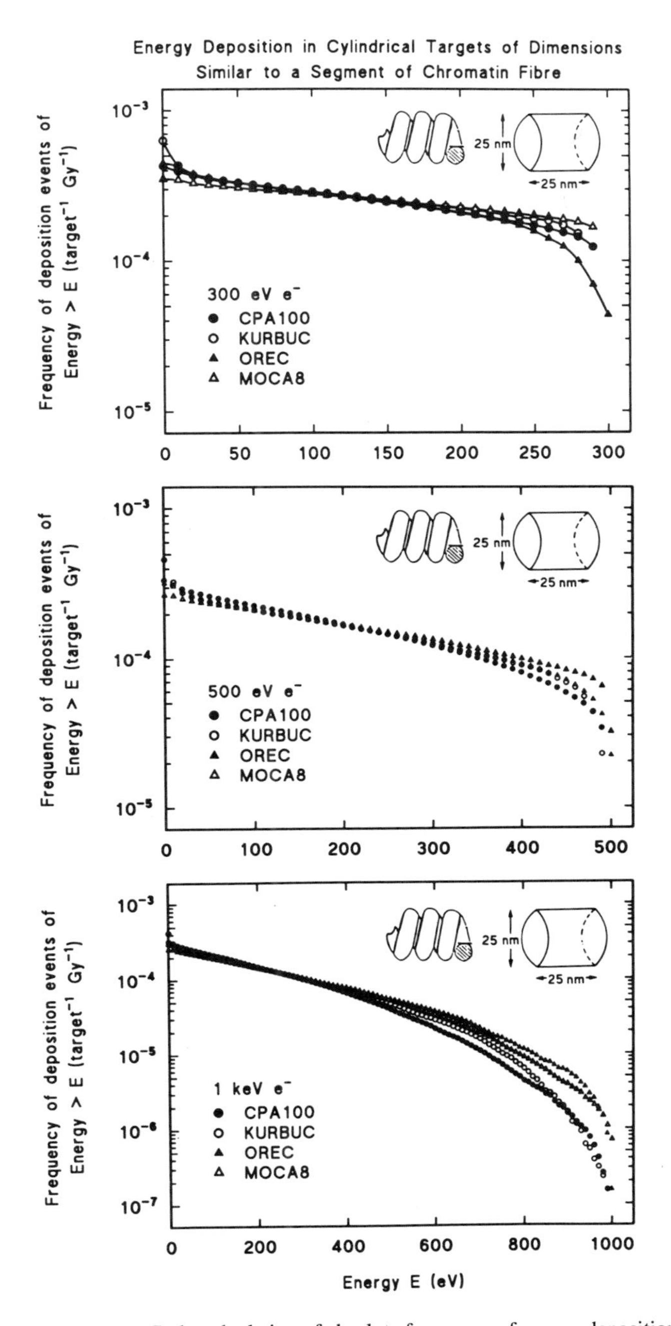

Figure 8. Monte Carlo calculation of absolute frequency of energy deposition greater than an amount, E(eV), in a cylindrical volume representing a short segment of chromatin fibre (25 nm by 25 nm) in water irradiated with various electron energies as obtained by each code. The left ordinate give the absolute frequencies of deposition of energy greater than E(eV) in a cylindrical volume randomly positioned in water irradiated with 1 Gy of the given radiation. The scatter at larger energy depositions are mainly due to lower statistics.

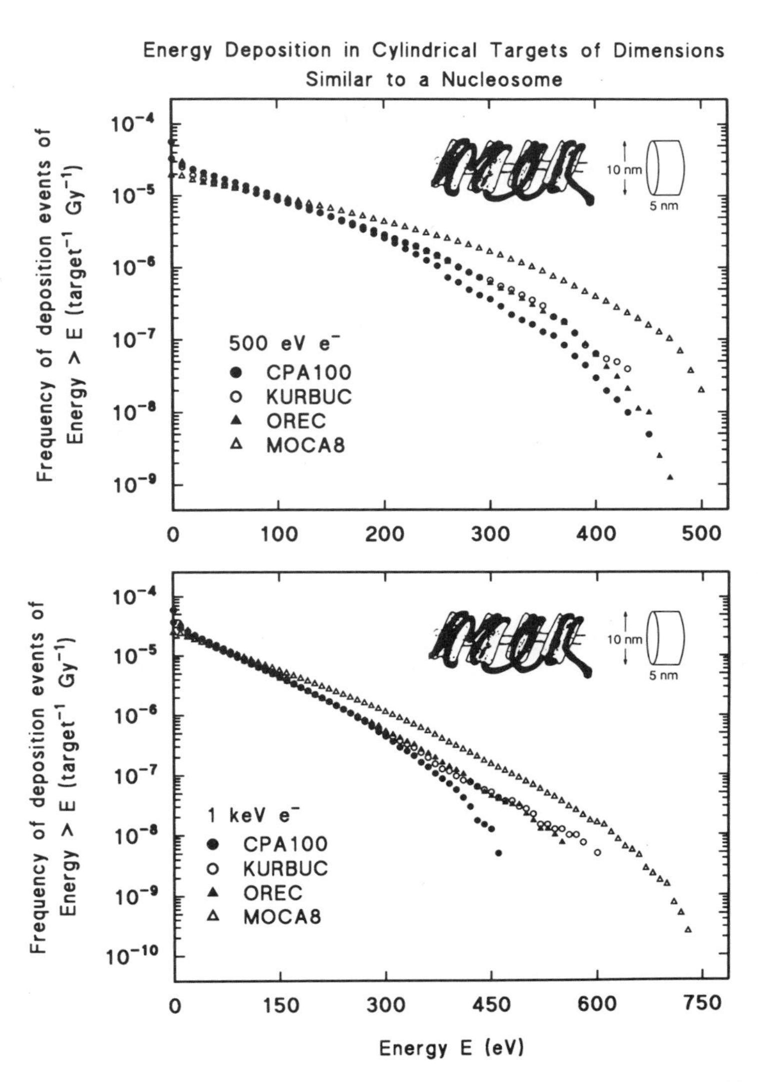

Figure 9. As in Fig (8) but showing the distributions of absolute frequencies of energy deposition per Gy in a cylindrical volume (diameter 10 nm by length 5 nm) approximately the size of a nucleosome irradiated with 300 eV and 1 keV electrons.

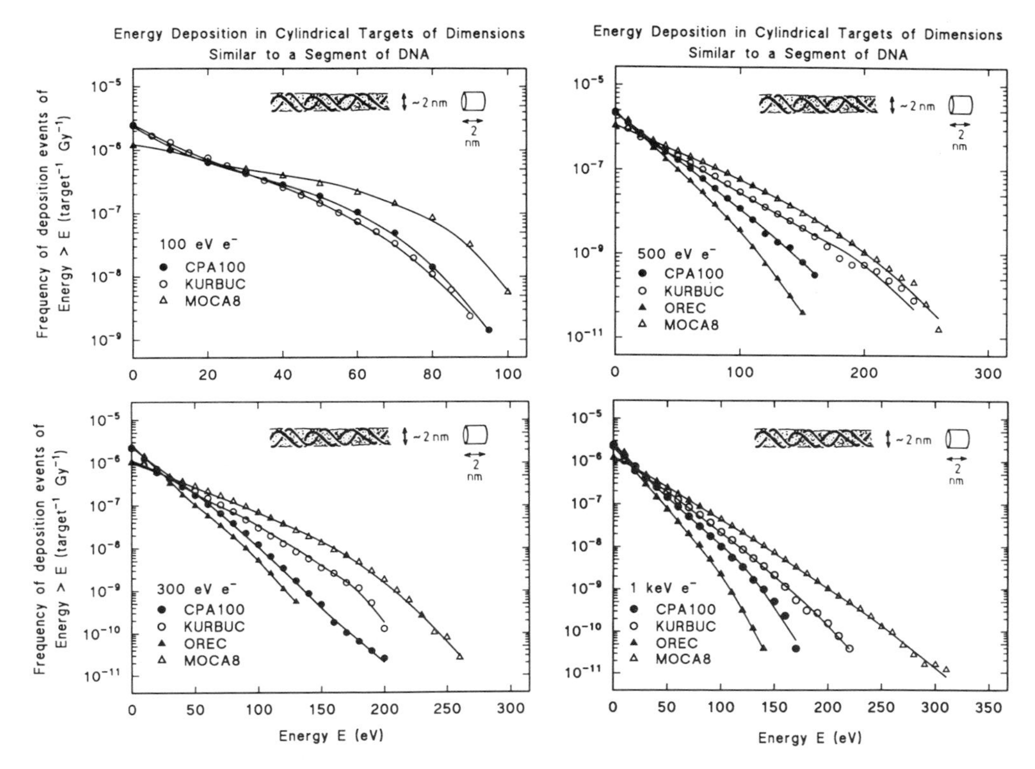

Figure 10. Frequencies of energy deposition in a short segment of an approximately DNA-sized cylinder in water irradiated with 100 eV, 300 eV, 500 eV and 1 keV electrons. The ordinate gives the absolute frequencies of deposition of energy E(eV) (or greater) in a single cylinder randomly positioned, and randomly oriented, in water irradiated with 1 Gy of the given radiation.

SUMMARY AND DISCUSSION

Although Monte Carlo track structure simulation is a very new tool in the field of radiobiology since its inception but in the past two decades it has made significant contribution to global understanding of the role of physical and chemical damage in cellular DNA.[32-37] These advances have come about in spite of severe short comings of fundamental input data in the form of differential cross sections in liquid water, DNA and cell environment. With availability of super computers it is now possible, broadly speaking, to simulate physical and chemical stages of interaction of ionizing radiation in aqueous and simple cellular environments.[38-42]

For the purpose of this study all codes were operated without the explicit treatment of inelastic or elastic collisions by electrons of kinetic energy less than the cut off limit of 10 eV (T_c = 10 eV). The residual energy of such a 'dead electron' is deposited in a single step at a single point. Approximately ~ 15% of total energy is deposited in this way. The energy of these 'dead electrons' may substantially affect the result for only very small target volumes and small energies. For example, if it is of interest to calculate, on a threshold assumption, the number of DNA strand breaks obtained from the energy deposited by these tracks, the extra energy deposits may suffice to raise the energy deposited above the assumed small threshold value for producing strand breaks (Charlton *et al* 1989).[38] In contrast, for the longer energies and volumes, the exclusion of these 'dead electrons' should have little effect other than to reduce the energy axis by ~ 15%.

This study has presented comparisons of four major independent Monte Carlo track structure codes for monoenergetic electrons for gaseous and liquid water. Cross sections have been presented for electrons in water vapour from 10 eV to 10 MeV. Comparison of the codes shown that although, broadly speaking, all codes have been developed from the same initial experimental information but distinct differences are observed between the codes for the tests presented here. The tests indicate differences in clustering characteristics between the codes due to conceptual difference in the process involved following the initial interaction of the primary electron in the medium. Some codes have adapted empirical or semi-imperial treatment of the input data. Each of the tests in the form of integral radical distributions of interactions, differential distributions of energy deposited and clustering characteristics of tracks in different size volumes, shows different characteristics of the tracks. In summary the following broad conclusion have been derived.

1. Comparisons have been made for four independent Monte Carlo track structure codes for monoenergetic electrons for water vapour and liquid.
2. Broadly speaking, all codes have been developed from the same initial experimental information
3. Distinct differences exist between the codes in subsequent processes involved which have produced tracks with different clustering characteristics
4. Results show similarities in many aspects, mainly in relation to large-size targets
5. Results show substantial differences in relation to small-size targets such as DNA
6. Differences observed could lead to differences in physical, chemical and biological interpretations.
7. The differences observed are not solely due to phase effects.

In view of the above conclusions the following proposals are made. First, setting up of a cross section database which would be of interest to track structure calculations in radiobiology. Second, to agree on a rigorous bench marking for Monte Carlo track structure codes for electrons and ions.

REFERENCES

1. N. Metropolis and S. Ulem. The Monte Carlo Method. *Journal of the American Statistical Association* 44:335-341 No. 247 (1949).
2. M.J. Berger. Monte Carlo Calculations of the Penetration and Diffusion of Fast Charged Particles. In Methods in Computational Physics, B. Alder, S. Fernback and M. Rotenberg (eds.), pp. 135, Vol 1. Academic Press, New York (1963).
3. D.E. Raeside. Monte Carlo Principles and Applications. *Phys. Med. Biol.* 21:181 (1976).
4. J.E. Turner, H.A. Wright and R.N. Hamm. Review Article: A Monte Carlo Primer for Health Physicists. *Health Phys.* 48:717 (1985).
5. I.M. Sobol. The Monte Carlo Method. The University of Chicargo Press, London (1974).
6. R.L. Morin. Monte Carlo Simulation in the Radiological Sciences. CRC Press (1988).
7. I. Lux, L. Koblinger. Monte Carlo Particle Transport Method: Neutron and Photon Calculations. CRC Press (1990).
8. H.G. Paretzke. Simulation von Elektronenspuren im Energiebereich 0,01-10 keV in Wasserdampf
9. R.N. Hamm, J.E. Turner, R.H. Ritchie, H.A. Wright. Calculation of heavy-ion tracks in liquid water. *Radiat Res.* 104:S-20--S-26 (1985).

10. M. Terrissol, J.P. Patau, T. Eudaldo. Application a la microdosimetrie et a la radiobiologie de la simulation du transport des electrons de basse energie dans l'eau a l'etat liquide. In Sixth Symposium on Microdosimetry J. Booz and H.G. Ebert (eds.), pp. 169-178, Harwood Academic Publishers Ltd (1988).
11. M. Zaider, D.J. Brenner and W.E. Wilson. The Applications of Track Calculations to Radiobiology 1. Monte Carlo Simulation of Proton Tracks. *Radiat. Res.* 95:231-247 (1983).
12. A. Ito. Calculation of Double Strand Break Probability of DNA for Low LET Radiations Based on Track Structure Analysis. In Nuclear and Atomic Data for Radiotherapy and Related Radiobiology, International Atomic Energy Agency, Vienna (1987).
13. I.G. Kaplan, A.M. Miterev and V.Ya. Sukhonosov. Simulation of the Primary Stage of Liquid Water Radiolysis. *Radiat. Phys. Chem.* 36:493-498 (1990).
14. A.V. Lappa, E.A. Bigildeev, D.S. Burmistrov and O.N. Vasilyev. "Trion" code for radiation action calculations and its application in microdosimetry and radiobiology. *Radiat. Environ. Biophys.* 32:1-19 (1993).
15. M.A. Hill and F.A. Smith. Calculation of initial and primary yields in the radiolysis of water. *Radiat. Phys. Chem.* (in press) (1993).
16. S. Uehara, H. Nikjoo and D.T. Goodhead. Cross-sections for water vapour for Monte Carlo electron track structure code from 10 eV to MeV region. *Phys. Med. Biol.* 38:1841-1858 (1993).
17. J.E. Turner, H.G. Paretzke, R.N. Hamm, H.A. Wright and R.H. Ritchie. Comparative study of electron energy deposition and yields in water in the liquid and vapour phases. *Radiat. Res.* 92:47-60 (1982).
18. J.E. Turner, H.G. Paretzke, R.N. Hamm, H.A. Wright and R.H. Ritchie. Comparision of electron transport calculations for water in the liquid and vapour phases. In *Proceedings of 8th Symposium on Microdosimetry,* J. Booz and H.G. Ebert (eds.), Commission of the European Communities, Luxembourg (1983).
19. H.G. Paretzke, J.E. Turner, R.N. Hamm, H.A. Wright and R.H. Ritchie. Calculated yields and fluctuations for electron degradation in liquid water and water vapour. *J. Chem. Phys.* 84:3182-3188 (1986).
20. H.G. Paretzke, J.E. Turner, R.N. Hamm, R.H. Ritchie and H.A. Wright. Spatial distributions of inelastic events produced by electrons in gaseous and liquid water. *Radiat. Res.*127:121-129 (1991).
21. H. Nishimura. Elastic Scattering Cross-Section of H_20 by Low Energy Electrons. *Electronic and Atomic Collisions*, Proc. XIth Int. Conf., eds N. Oda and K. Takayanagi, North Holland, Amsterdam, p.314, (1979).
22. S. Trajmar, D.F. Register and A. Chutjian. Electron Scattering by Molecules. II. Experimental Methods and Data. *Physics Reports 97*, 219-356 (1983).
23. A. Katase, K. Ishibashi, Y. Matsumoto, T. Sakae, S. Maezono, E. Murakami, K. Watanabe and H. Maki. Elastic Scattering of Electrons by Water Molecules over the range 100-1000 eV. *J. Phys. B: At. Mol. Phys.*, 19:2715-2734 (1986).
24. S.M. Seltzer. Cross-Sections for Bremsstrahlung Production and Electron-Impact Ionization. In: *Monte Carlo Transport of Electrons and Photons*. eds T.M. Jenkins, W.R. Nelson and A. Rindi, Plenum Press, New York, pp.81-114 (1988).
25. M.J. Berger and R. Wang. Multiple Scattering Angular Deflections and Energy-Loss Straggling. In: *Monte Carlo Transport of Electrons and Photons*. eds. T.M. Jenkins, W.R. Nelson and A. Rindi, Plenum Press, New York, pp.21-56, (1988).
26. M. Hayashi. Electron collision cross-sections for atoms and molecules determined from beam and swarm data. In Atomic and Molecular Data for Radiotherapy. IAEA, Vienna (1989), IAEA-TECDOC-SO6.
27. C.B. Opal, E.C. Beaty, W.K. Peterson. Tables of secondary-electron-production cross sections. *Atomic Data*, 4:209-253 (1972).
28. H. Nikjoo, M. Terrissol, R.N. Hamm, J.E. Turner, S. Uehara, H.G. Paretzke and D.T. Goodhead. Comparison of Energy Deposition in Small Cylindrical Volumes by Electrons Generated by Monte Carlo Track Structure Codes for Gaseous and Liquid Water. *Radiat. Protec. Dosimetry,* (in press) (1994).
29. H. Nikjoo, D.T. Goodhead, D.E. Charlton and H.G. Paretzke. Energy Deposition in Small Cylindrical Targets by Ultrasoft X-rays. *Phys. Med. Biol.* 34:691-705 (1989).
30. D.E. Charlton, H. Nikjoo and D.T. Goodhead. Energy deposition in sub-microscopic volumes in Radiation Research (A twentieth-century perspective) Vol II, W.C. Dewey, M. Edington, R.J.M. Fry, E.J. Hall, G.F. Whitmore (eds.), Academic Press, London (1992).
31. D.T. Goodhead and H. Nikjoo. Track Structure Analysis of Ultrasoft X-rays Compared to High- and Low-LET Radiations. *Int. J. Radiat. Biol.* 55:513-529 (1989).
32. E.M. Fielden and P. O'Neill. The Early Effects of Radiation on DNA. Springer-Verlag, London (1990).
33. G.R. Freeman. Kinetics of Nonhomogeneous Processes. John Wiley and Sons (1987).
34. K.F. Baverstock and J.W. Stather. Low Dose Radiation. Taylor and Francis (1989).
35. K.F. Baverstock and D.G. Charlton. DNA Damage by Auger Emitters. Taylor and Francis (1988).
36. R.W. Howell, V.R. Narra, K.S.R. Sastry and D.V. Rao. Biophysical Aspects of Auger Processes. AAPM (1992).

37. M. Terrissol and A. Beaudré. Simulation of Space and Time Evolution of Radiolytic Species Induced by Electrons in Water. *Rad. Prot. Dosim.* 31:175-177 (1990).
38. D.E. Charlton, H. Nikjoo and J. Humm. Calculation of Initial Yields of Single and Double Strand Breaks in Cell Nuclei from Electrons, Protons and Alpha-Particles. *Int. J. Radiat. Biol.* 55:1-19 (1989).
39. W.R. Holley and A. Chatterjee. The Application of Chemical Models to Cellular DNA. In: *The Early Effects of Radiations on DNA*. eds E.M. Fielden and P. O'Neill, Springer-Verlag, pp.195-209 (1990).
40. R.N. Hamm and J.E. Turner. Model Calculations of Radiation-Induced DNA Damage. In: *Biophysical Modelling of Radiation Effects*. eds K.H. Chadwich, G. Moschini and M.N. Varma, Adam Hilger, New York, pp.53-60 (1992).
41. S. Henβ and H.G. Paretzke. Biophysical Modelling of Radiation Induced Damage in Chromosomes. In: *Biophysical Modelling of Radiation Effects*. eds K.H. Chadwich, G. Moschini and M.N. Varma, Adam Hilger, New York, pp.69-76 (1992).

DISCUSSION

Zaider: First I want to thank you for doing a very useful job. Having said that, however, I think we have to make comparisons in a very careful way that is agreed upon ahead of time or else we may obtain results that will confuse us rather than help us. The first thing we ought to do is define energy deposited in a volume. The definition that we use is to simply count ionizations and excitations, take their total and add it up. Apparently that is equivalent with your definition. A 10 eV subexcitation delta ray stopped dead at a particular point, may actually distort the picture of a 100 eV electron track.

Nikjoo: Electrons with energy less than 10eV have been considered dead. Nearly 13% of energy is deposited in this manner, then electrons are usually allowed to diffuse a short distance and the energy deposited at that point is considered.

Zaider: They're still out there and they should be totally discarded.

Nikjoo: Yes, it can be done.

Paretzke: I might give you the justification why we do this. This is to stay in line with the definition of dose.

Nikjoo: I have described the method of how dead electrons have been treated here. This treatment could be changed if we could agree on how to deal with dead electrons.

Miller: Are the four codes the same or different in treating the dead electrons?

Nikjoo: All four codes treat 'dead electrons' in the same way.

Zaider: Do you consider that energy is deposited locally or do you take that energy and displace it one nanometer away?

Terrissol: At 10^{-15} sec, the subexcitation electron normally goes on. But if you stop it cold, then it is there and does not move.

Zaider: But this is wrong.

Varma: Are thermalized electrons treated the same way in all the codes you have compared?

Nikjoo: Thermalized electrons are not important here.

Hamm: I gave Hooshang a set of tracks in which I provided the coordinates of the dead electrons where they drop below the threshold for inelastic interactions. I do not know what he did with that.

Varma: That is a difficult thing. He's telling you the way he did the comparison. It may not be right. You can agree or disagree with that, but we need to define certain specific parameters as to how they should be measured. It's not a question of right or wrong. I think what Zaider is saying is an important point. What he's saying is that if you do neglect the thermalized electrons, then comparisons are not quite correct for different codes, since in some codes computation is stopped when electrons are thermalized

Zaider: In the codes that I've seen, the electron diffuses away.

Chatterjee: Marco, will that change the data?

Zaider: I have no idea.

Nikjoo: I think we need to set up a protocol for testing and bench marking these codes.

Varma: This is a process we need to discuss further. We should set up a group of scientists who do the Monte Carlo Code development and have them come up with recommendations on how the code results should be compared.

Caswell: It's slightly different. I want to say something about bench marking or comparing to the gold standards. It seems to me that where you have something like a stopping power, which you know is a well-known quantity, you can make a comparison of the codes using things that you know. And if you do your model right and you've done the calculation right, you should come out with the stopping power.

Zaider: But for the stopping powers you don't need Monte Carlo codes.

Caswell: You can check the codes by the stopping power. I'm not saying that you must.

Varma: I think we all agree that comparison and bench marks are a very important aspect of the development of those codes. To see where we are going it is important that we carefully determine what we compare and how we compare it. I will agree with Zaider that if you compare apples and oranges, then it will do more harm than good in further development of these codes.

Nikjoo: Well, the purpose of this presentation was to show you the differences that could exist between various Monte Carlo track codes for liquid and water vapor.

Varma: Let us hope that Marco Zaider, Bob Hamm, Hooshang Nikjoo and Michel Terrissol will come up with some kind of recommendation for the parameters that should be compared with the different codes and bench marks.

A COMPARISON BETWEEN TWO MONTE CARLO CODES ON DETERMINATION OF TRANSIENT CHEMICAL YIELDS

R.N. Hamm and J.E. Turner

Health Sciences Research Division
Oak Ridge National Laboratory
Oak Ridge, TN 37831-6123

A. Chatterjee

Life Sciences Division
Lawrence Berkeley Laboratory
University of California
Berkeley, CA 94720

ABSTRACT

Monte Carlo computer codes have been independently developed at several laboratories for performing calculations of the radiolysis of water. The different codes involve a wide variety of models and related assumptions in treating the many physical and chemical processes that occur. Because few detailed aspects of such computations can be directly checked by experiment, it is important to make comparisons of various predicted microscopic distributions. In this paper we compare results obtained with the codes developed at Lawrence Berkeley Laboratory and at Oak Ridge National Laboratory. Both codes were used to calculate the spatial distributions of various radical species in spurs along the tracks of energetic electrons. Similarities and differences in the results of this preliminary study are shown. Additional work is planned.

INTRODUCTION

Radiation-induced damage to a living cell begins with the physical stage of energy deposition on the various target molecules present in the system. Among the variety of target molecules with which radiation can interact, water molecules and DNA are considered to be critically important.[1] Both of these molecules can undergo the primary processes of ionization and excitation, which are generally complete in about 10^{-15} s. However, the damage to DNA (e.g., strand breaks, base alterations, etc.) from these fast processes can evolve over several orders of magnitude in time, but the kinetics associated with them have not been studied in great detail. On the contrary, a time-dependent analysis of the ensuing chemical stage that follows after the absorption of energy by water molecules has provided us with a better understanding of the time evolution of water-radical-induced DNA damage. Most of these studies are theoretical in nature, and it has been increasingly clear that experimental techniques are not likely to improve enough to allow exploration of the time interval between 10^{-15} s and 10^{-12} s. Following the absorption of energy by water molecules, new chemical species such as hydroxyl radicals, hydrogen radicals, solvated electrons, etc., are produced within the cellular complex. These species become thermalized in about 10^{-12} s. At this point in the time scale, diffusion of radicals begins and various chemical reactions follow, some of which result in DNA damage. Water radicals can react with each other and produce molecular products, such as H_2, H_2O_2, H_2O, etc. Radicals which escape such reaction can lead to DNA damage. Hence it is extremely important to evaluate quantitatively the yields of various radical species as functions of time including information on spatial distributions.

There are several theoretical approaches which deal with the calculation of yields of chemical species produced as a result of radiolysis of water. For example, Zaider et al.[2] have developed a Monte Carlo scheme using gas-phase cross sections for electronic energy loss. Then they extrapolate to unit density for applicability in the liquid phase. Their strategy is based on the fact that much more extensive experimental data are available for the vapor phase than for the liquid phase. It is not clear whether this procedure adequately describes the liquid-phase cross sections. However, their time-dependent decay curves for OH as well as e^-_{aq} have similar shapes to those observed experimentally by Jonah et al.[3,4] In a different approach, Brenner and Zaider[5] use Monte Carlo methods to obtain the spatial distributions of water radicals at $\sim 10^{-12}$ s, which become the input to another code, based on Smoluchowski's equation, that computes events to later times. In contrast to these approaches, Chatterjee and Magee[6] initiate the computation with an assumed Gaussian distribution following the passage of a charged particle. In their method, they utilize track entities called "spurs," "blobs," and "short tracks" for electron irradiation and "core" and "penumbra" for heavy-charged particles. This approach has now been extended to evaluate the yields of DNA strand breaks. However, in the Chatterjee and Magee model developed at the Lawrence Berkeley Laboratory (LBL), the stage between 10^{-15} s and 10^{-12} s is treated very approximately with several assumptions. A local energy deposit between 6 eV and 100 eV in water is called a spur; an energy deposit between 100 eV and 500 eV is called a blob; and for short tracks these values lie between 500 eV and 5000 eV.

Perhaps the most complete model available at present is the one developed at the Oak Ridge National Laboratory.[7,8] In this model, attempts are made to account for every ionization or excitation event in a statistical manner that follows through to the essential completion of intratrack radical reactions at about 10^{-6} s. In liquid water, the initial physical processes are represented by writing

$$H_2O \rightarrow H_2O^+ \quad , \quad H_2O^* \quad , \quad e^- \quad .$$

The Monte Carlo computer code, OREC, was developed at ORNL to perform the detailed transport and energy-loss calculation of the primary charged particle and all of its secondary electrons. For a primary or secondary particle of given energy, the code selects a flight distance to the next event, the type of event that occurs, the energy loss, the scattering angle, and the energy and angle of the secondary electron when an ionization occurs. The inelastic cross sections are derived from a complex dielectric response function developed specifically for liquid water.[9] OREC explicitly considers the physico-chemical and chemical stages. The theoretical decay curves obtained for OH and e^-_{aq} are in good agreement with the experimental data of Jonah et al.,[3,4] measured between 3×10^{-11} s and 4×10^{-8} s. In addition, the computed ferric yield for tritium beta rays agrees with experiment.

In view of the several theoretical techniques available, one of the objectives of this conference has been to make a comparison between various methods. This chapter provides a quantitative comparison between various results obtained from the ORNL and LBL codes.

In the LBL code, the prethermalization stage is approximated by assuming that the radicals are distributed according to a Gaussian function with a characteristic standard deviation σ for each radical. These σ values have been adjusted to obtain the best fit of experimental data on steady-state yields of radicals. Conversion of the energy absorption to production of chemical species is accounted for by assuming that 17 eV is required on an average to create a water-radical pair. Based on the σ values and the number of radical pairs produced in a spur, random numbers are used to select from a three-dimensional Gaussian distribution, and the various chemical species are positioned in space accordingly. Subsequent to this procedure, the next steps in the overall Monte Carlo scheme are very similar to those of the Oak Ridge code (OREC). Since OREC is capable of localizing the initial chemical species without making any assumptions such as prescribed Gaussian functions, it is extremely important to compare the various characteristics of these distributions by the two codes.

The two main characteristics chosen for the present comparison are the spur-size distribution and the various σ values. As mentioned earlier, spurs are created by the energy losses below 100 eV. Hence, the number of ion pairs (approximately same as radical pairs) will depend upon the energy in a spur. Mozumder and Magee[10] have computed the spur size and its corresponding probability. These can also be computed by the Oak Ridge code and hence form the basis for comparison.

COMPARISON OF RESULTS

Calculations were made using OREC for segments of 1-MeV electron tracks. In these calculations each energy-loss event by the primary electron was treated independently and categorized as a spur, blob or short track according to the criteria of Chatterjee and Magee. The average yields of the various radicals and their spatial distributions were then tabulated for the different categories from a large number of trials and compared with the corresponding distributions from the LBL model.

Figure 1 shows the probability density $P(x)$ for the location of an OH radical at a perpendicular distance x from the path of a primary charged particle when the number of radical pairs in a spur is $n = 1$. The open circles give the distribution computed from OREC, and the dashed curve (LBL) shows the Gaussian distribution used by Chatterjee and Magee.[11] The solid curve (ORNL) represents a Gaussian distribution, having the same standard deviation as that of the open circles. The dashed and solid curves virtually coincide. All distributions are normalized, both here and in the following figures.

Figure 2 presents the distribution of OH-radical positions when there are $n = 2$ radical pairs in the spur. The calculated ORNL distribution (open circles) is more widely spread than that of LBL (dashed curve). When a normal distribution is plotted with the same standard deviation as that of the open circles, the solid curve (ORNL) results. The OREC distribution is sharper than Gaussian near the trajectory of the primary particle and extends to larger values of x, where the probability density is small.

Figure 3 shows the distribution for the hydrated electron for spurs in which there is only a single radical pair. The Gaussian LBL (dashed curve) and ORNL (open circle) distributions are markedly different. When a normal curve is assumed with the same standard deviation as that of the open circles, the solid curve is obtained. Figure 4 presents the same information for the hydrated electron in spurs where two radical pairs are formed. It is apparent that the ORNL model gives much greater displacements in the initial

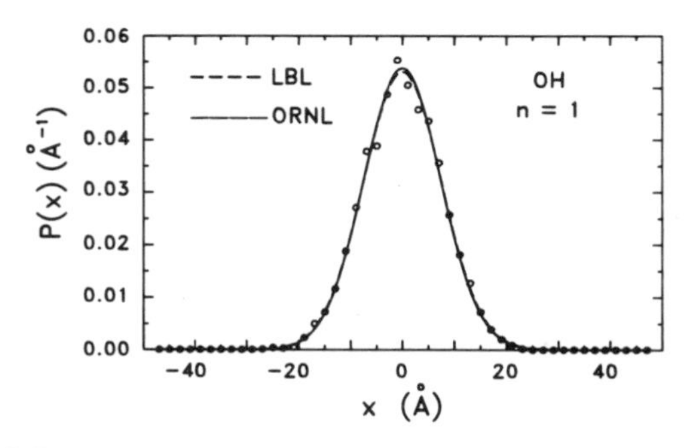

Figure 1. Distribution of displacements x of OH radicals from the path of a primary charged particle in spurs that contain exactly $n = 1$ radical pair. The dashed curve represents the LBL Gaussian distribution, the circles show the distribution calculated with OREC, and the solid curve gives a Gaussian curve having the same standard deviation as the OREC points.

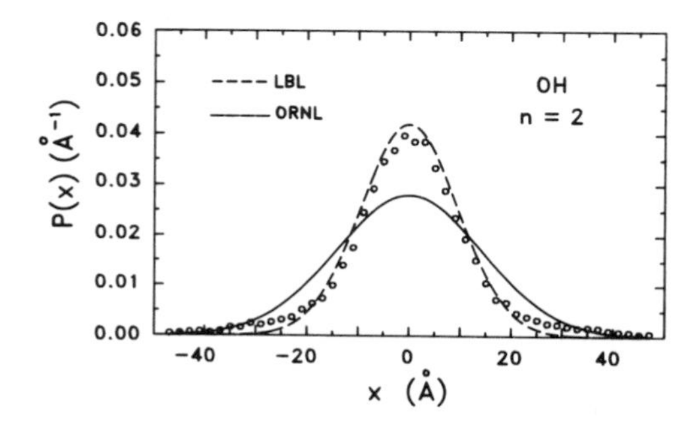

Figure 2. Same as Fig. 1 for spurs with exactly $n = 2$ radical pairs.

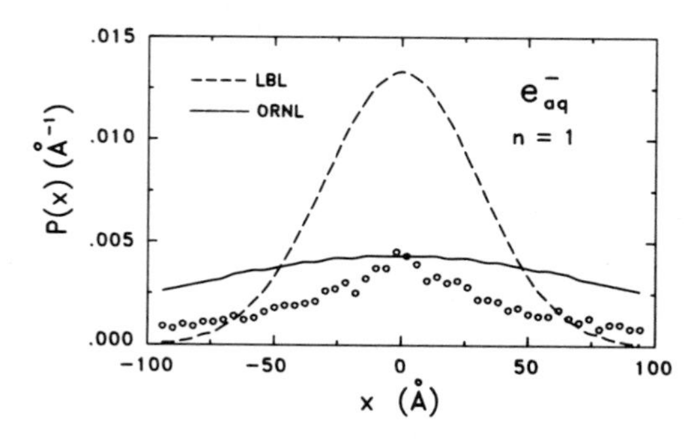

Figure 3. Distributions (same notation as in Figs. 1 and 2) for hydrated electrons with $n = 1$.

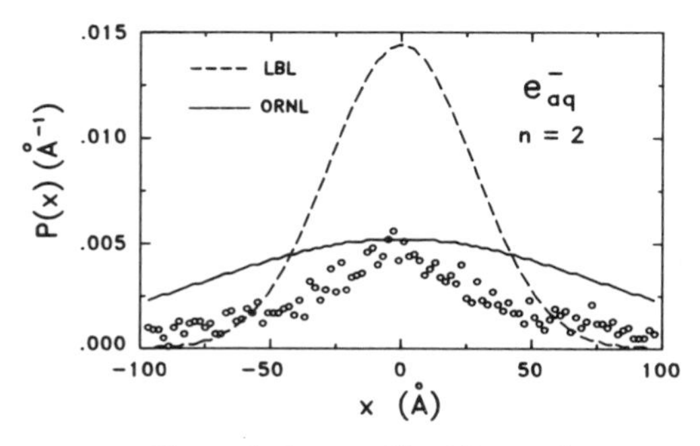

Figure 4. Same as Fig. 3 for *n* = 2.

(~10^{-12} s) positions of the electrons from the path of the primary particle than the LBL model. Furthermore, the ORNL distribution is only approximately Gaussian.

The three distributions for the hydronium ion for spurs with two radical pairs are plotted in Fig. 5. Here the ORNL distributions are tighter than that in the LBL model. The ORNL distribution (circles) is somewhat more peaked than the Gaussian curve with the same standard deviation.

Table 1 summarizes the standard deviations of the radical distributions computed with the LBL and ORNL models. *P(n)* gives the probability that a spur has exactly *n* radical pairs. The standard deviations for OH and H are shown for the two models. In the LBL calculations, the standard deviations for e^-_{aq} and H_3O^+ are taken to be the same as that of the H radical. The ORNL standard deviations are shown in the last two columns of the table. They are both different from that for H in the ORNL model.

Some of the distributions shown here are quite similar for the two models, but others differ greatly. Despite this, both models give good agreement with the measured decay

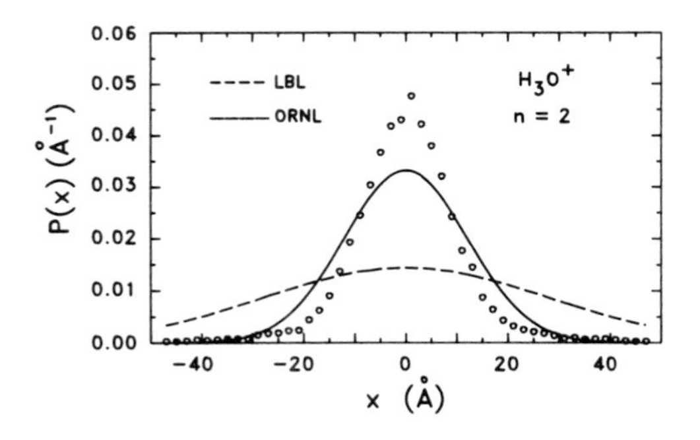

Figure 5. Distributions for H_3O^+ with *n* = 2 (same notation as before).

Table I. Spur size distributions and standard deviations for radicals.

n	P(n)		σ_{OH} (Å)		σ_H (Å)		σ_{eaq}	σ_{H3O+}
	LBL	ORNL	LBL	ORNL	LBL	ORNL	ORNL	ORNL
1	0.48	0.304	7.5	7.4	30.0	4.2	92.1	7.5
2	0.35	0.377	9.5	14.2	27.7	18.6	74.6	12.0
3	0.10	0.165	10.8	16.6	25.5	20.6	76.8	14.2
4	0.04	0.074	11.9	17.3	23.5	21.7	75.4	14.9
5	0.02	0.039	12.8	17.2	21.7	21.4	78.1	14.8
6	0.01	0.024	13.6	19.1	20.0	23.7	79.0	16.6
7		0.012		18.8		23.1	77.5	16.2
8		0.004		19.9		23.1	71.2	17.4
9		0.001		19.3		23.6	69.1	16.5

curves for OH radicals and for hydrated electrons.[3,4] Further work is underway to attempt to understand this and to delineate which differences in these distributions are most significant for the final calculated yields.

ACKNOWLEDGMENT

Research sponsored by the Director, Office of Energy Research, Office of Health and Environmental Research, U.S. Department of Energy, under contract DE-AC05-84OR21400 with Martin Marietta Energy Systems, Inc. and under contract DE-AC03-76SF00098 with the University of California.

REFERENCES

1. J.B. Little, J.N. Whaley and H.L. Liber, Role of energy distribution in DNA on the mutagenic effects of internal emitters, *in* "Mechanisms of Anti-Carcinogenesis and Radiation Protection," O.F. Nygaard and A.C. Upton, eds., p. 201, Plenum Press, New York (1991).
2. M. Zaider, D.J. Brenner, and W.E. Wilson, The application of track calculations to radiobiology, *Radiat. Res.* 95:231–247 (1983).
3. C.D. Jonah and J.R. Miller, Yield and decay of the OH radical from 200 ps to 3 ns, *J. Phys. Chem.* 81:1974–1976 (1977).
4. C.D. Jonah, M.S. Matheson, J.R. Miller, and E.J. Hart, Yield and decay of the hydrated electron from 100 ps to 3 ns, *J. Phys. Chem.* 80:1267–1270 (1976).
5. D.J. Brenner and M. Zaider, Stochastic and deterministic treatments of the time decay of species created by heavy-charged particle interactions, *Radiat. Prot. Dosimetry* 13:127-132 (1985).
6. A. Chatterjee and J.L. Magee, Radiation chemistry of heavy particle tracks. 2. Fricke dosimeter system, *J. Phys. Chem.* 84:3537–3543 (1980).
7. J.E. Turner, J.L. Magee, H.A. Wright, A. Chatterjee, R.N. Hamm, and R.H. Ritchie, Physical and chemical development of electron tracks in liquid water, *Radiat. Res.* 96:437–449 (1983).
8. J.E. Turner, R.N. Hamm, R.H. Ritchie, and W.E. Bolch, Monte Carlo track-structure calculations for aqueous solutions containing biomolecules; paper in this volume.
9. R.H. Ritchie, R.N. Hamm, J.E. Turner, H.A. Wright, and W.E. Bolch, Radiation interactions and energy transport in the condensed phase, *in* "Physical and Chemical Mechanisms in Molecular Radiation Biology," W.A. Glass and M.N. Varma, eds., p. 99, Plenum Press, New York (1991).
10. A. Mozumder and J.L. Magee, The early events of radiation chemistry, *Int. J. Radiat. Phys. Chem.* 7:83 (1975).
11. A. Chatterjee and J.L. Magee, Theoretical investigation of the production of strand breaks in DNA by water radicals, *Radiat. Prot. Dosimetry* 13:137–140 (1985).

DISCUSSION

Varma: What is the difference between your and Aloke's Gaussian calculations? What is the physical significance of this difference?

Hamm: I think we understand why they're different. You ask what the significance is. Part of that is to look at how these affect the subsequent chemistry.

Varma: In the subsequent chemistry studies you showed comparison with the G values. They were fairly close.

Hamm: Not the initial G values.

Chatterjee: The OH yield in the Oak Ridge code is 7.2 and we took 5.88. The respective slopes are much different. The rate of decay for the Oak Ridge code is much faster, because the initial yield is larger.

Zaider: Why would it be faster then?

Hamm: There's another point and that is that there may be more correlations. Presumably, since Aloke is putting in all of his species randomly through the Gaussian and we're putting these in pair-wise, we should have a higher correlation.

Holley: Your spurs are probably not going to be spherical?

Hamm: The best I could tell, they were. These are at very low energy, of course.

Inokuti: I have a number of suggestions. For example, when you say whether the spur is spherical, you ought to calculate moments. I would like to see what the moments are. A set of numbers would be very easy to get.

Hamm: I've only calculated the second moments. They are essentially the same.

Inokuti: Also, whether the distribution is Gaussian or not, again, you calculate all the moments, or the characteristic function, then I can see numbers. I can tell right away. How nearly spherical is completely determined by moments.

Miller: There's been one formulation of the radiation chemistry in the literature that was based on pair correlation functions. That's another way to go.

Inokuti: That's another way, yes.

Terrissol: In the Rome Microsdosimetry Symposium I published a curve of decay of OH, but not for 1 MeV.

AN ATTEMPT TO MODIFY THE *MOCA* WATER-VAPOR-ION CODE TO SIMULATE LIQUID PHASE

S.B. Curtis

Fred Hutchinson Cancer Research Center
1124 Columbia Street, MP-665
Seattle,WA 98104

J.B. Schmidt

Lawrence Berkeley Laboratory
University of California
Berkeley, CA 94720

W.E. Wilson

Pacific Northwest Laboratories
Richland, WA 99352

ABSTRACT

An attempt was made to apply to liquid water the Monte Carlo code, MOCA14, which calculates track-structure (i.e., three-dimensional distribution of ionizations) for heavy charged-particle tracks in water-vapor. The criterion used was that the stopping powers for four energies of protons (1, 2, 5 and 10 MeV) should correspond to the stopping powers in liquid water found in ICRU Report #49. To adjust phenomenologically the vapor source data, two modifications were made: (1) several vapor-phase excitation interactions were assumed to result instead in ionizations in liquid; and (2) the optical oscillator strength along the energy-loss scale was shifted by adding (or subtracting) a constant while simultaneously and independently subtracting a constant energy from the inelastic ionization energy thresholds. The total energy-loss was used to overcome the binding energy (which was decreased by an arbitrary amount) and to provide energy to the ejected secondary electron. To be consistent, a similar adjustment was made to the secondary electron energy-loss processes. It was found that no adjustment of the two constants brought the values of the average stopping powers into agreement with the ICRU stopping power values. It is concluded that no simple and physically meaningful manipulation of the water vapor code for protons in the energy range between 1 and 10 MeV could bring the stopping powers into agreement with currently accepted values.

Computational Approaches in Molecular Radiation Biology
Edited by M.N. Varma and A. Chatterjee, Plenum Press, New York, 1994

INTRODUCTION

The purpose of this study was to determine if a simple phenomenological, but physically meaningful, adjustment to the charged particle Monte Carlo code (MOCA14) could be found that would provide an adequate representation of ion track structure in the liquid phase of water, that is, that the modified code should at least predict a reasonable stopping power for protons in liquid water. This was done as a preliminary step in trying to use the modified code to study ionization clustering and its importance in producing lesions of interest in DNA.

Unfortunately, no unanimous agreement exists on what the stopping power for charged particles in liquid water should be (see, e.g., Turner, et al. (1982), LaVerne (1986), ICRU (1984), and references therein). During the course of this study, a new compilation for protons and alpha particles appeared in ICRU Report 49 (ICRU, 1993). For comparison purposes, we have chosen the numbers from that publication to be the most recent and accurate authoritative compilation of values of stopping power for liquid water. We note with some surprise, however, that the ICRU 49 proton values for water vapor are slightly larger than those for liquid water at the same energy.

METHODS

The primary data which reflect charged particle interactions with matter, and so are the core of the Monte Carlo simulations, are data on the differential optical oscillator strengths. The code uses oscillator strengths derived from the Berkowitz compilation (Berkowitz, 1979; see also, Zeiss, et al., 1975). Oscillator strengths for the liquid phase are often deduced from the reflectance measurements of Heller, et al. (1974).

To adjust phenomenologically the vapor source data for these basic differences, we tried two things, separately and in combination. First, like Turner, et al. (1982) and Paretzke, et al. (1991), we assumed several vapor-phase excitation interactions would result instead in ionizations in liquid; and second, we shifted the optical oscillator strength along the energy-loss scale by adding (or subtracting) a constant and simultaneously and independently subtracting a constant energy from the inelastic ionization energy thresholds. In selecting both constants, non-physical values, e.g., negative total energy losses, were avoided.

The assumption that some vapor phase excitation interactions would, in liquid phase, result in ionizations is physically reasonable because large energy transfers into high Rydberg states are likely to result in free electrons in liquid. Furthermore, it is known that W-values in liquid are lower (ca. 20 eV/ion pair), and therefore more ion pairs are produced per unit energy loss, so more ionizations must be phenomenologically simulated in liquid than in vapor phase. The vapor interactions that were assumed to produce ionizations were the dissociative continuum line at 9.7 eV, the diffuse bands + Rydberg $C_3 + D_3$, $A_3 + B_3$, $C_{>3} + D_{>3}$, $A_{>3} + B_{>3}$ the Lyman Alpha, Beta, and Gamma lines, and the Alpha and Beta Balmer lines (Paretzke, 1987).

Tests were made by generating 100 tracks of protons at each of four energies (1, 2, 5 and 10 MeV). The length of each track was one-half micrometer. The total energy- loss for each track was tabulated and the mean stopping power for a given energy was calculated from the average energy lost over the total 50 micrometers of track length.

RESULTS

Without any shift of the energy scale, the (extra) excitations converted to ionizations produced stopping powers that were between 17% and 31% too large, depending on energy, compared to the ICRU liquid water numbers (ICRU, 1993).

By shifting the vapor phase oscillator strength vs. energy curve upward along the energy scale, one should in effect increase the energy-loss in an event. This total energy loss is then used to overcome the binding energy (which was decreased by an arbitrary amount) and to provide energy to the ejected secondary electron. To be consistant, a similar adjustment was made to the secondary electron energy-loss processes. It was found that no adjustment of the two constants brought the values of the average stopping powers into agreement with the ICRU stopping power values.

CONCLUSION

We conclude that no simple and physically meaningful manipulation of the water vapor code for protons in the energy range between 1 and 10 MeV could bring the stopping powers into agreement with currently accepted values. Thus, it is better to use liquid oscillator strengths and utilize codes specifically developed for the liquid phase to generate track structure analyses of charged particles traversing media of biological interest.

ACKNOWLEDGEMENTS

This work was supported by the Office of Health and Environmental Research (OHER) of the U. S. Department of Energy under contracts DE-AC03-76SF00098 (S.B.C. and J.B.S.) and DE-AC06-76 RLO 1830 (W.E.W.).

REFERENCES

Berkowitz, J. (1979), Photoabsorption, Photoionization, and Photoelectron Spectroscopy, Academic Press, New York, ISBN 0-12-091650-9.

Heller Jr., J.M., Hamm, R.D., Birkhoff, R.D., Painter, L.R., (1974), J. Chem. Phys. 60, 3483–3486.

ICRU (1984), Stopping Powers for Electrons and Positrons, ICRU Report 37 (International Commission on Radiation Units and Measurements, Bethesda, Maryland).

ICRU (1993), Stopping Powers for Protons and Alpha Particles, ICRU Report 49 (International Commission on Radiation Units and Measurements, Bethesda, Maryland).

LaVerne, J.A., Mozumder, A. (1986), Effect of phase on the stopping and range distribution of low-energy electrons in water, J. Phys. Chem. 90, 3242–3247.

Paretzke, H.G. (1987), Radiation track structure theory, in: "Kinetics of Nonhomogeneous Processes," G.R. Freeman, ed., John Wiley & Sons, New York, ISBN 0-471-81324-9.

Paretzke, H.G., Turner, J.E., Hamm, R.N., Ritchie, R.H., and Wright, H.A. (1991), Comparative study of electron energy deposition and yields in water in the liquid and vapor phases, Radiat. Res. 127, 121–129.

Turner, J.E., Paretzke, H.G., Hamm, R.N., Wright, H.A., and Ritchie, R.H. (1982), Radiat. Res. 92, 47–60.

Zeiss, G.D., Meath, W.J., MacDonald, J.C.F., and Dawson, D.J. (1975), Radiat. Res. 63, 64–82.

DISCUSSION

Inokuti: You showed a few cluster cross-sections for protons and for helium and they refer to the same velocity.

Curtis: Yes, they are at the same velocity.

Inokuti: Then why is it that there is so much difference?

Paretzke: It's because there is four times more ionization.

Inokuti: I have one more question. Looking at these things, I would expect if you compared distributions, the difference would be quite large.

Ritchie: It has to be.

Miller: Yes.

Hamm: Between the two codes?

Inokuti: Yes, between the two codes, and the distribution is something that is often measurable.

Miller: Yes.

Inokuti: It would be very instructive. If cluster cross-section depends so much, then distribution should depend even more.

Ritchie: Well, Bob you've done that already, haven't you?

Hamm: Yes, for electrons.

Curtis: This has not been done for the heavy ions.

THREE STATISTICAL TECHNOLOGIES WITH HIGH POTENTIAL IN BIOLOGICAL IMAGING AND MODELING

Moshe Fridman* and J. Michael Steele**

Department of Statistics
University of Pennsylvania
Philadelphia PA 19104-6302

ABSTRACT

The three technologies that are surveyed here are (1) wavelet approximations, (2) hidden Markov models, and (3) the *Markov chain Renaissance.* The intention of the article is to provide an introduction to the benefits these technologies offer and to explain as far as possible the sources of their effectiveness. We also hope to suggest some useful relationships between these technologies and issues of importance on the agenda of biological and medical research.

INTRODUCTION

The purpose of this article is to review some of the most significant recent progress in statistical theory and to focus attention to the extent that is possible on the assistance these advances offer to biological and medical technology.

The first topic we engage is the theory and application of *wavelets*, which is possibly the most far-reaching development in all of applied mathematics over the last ten years. The emerging technology has important implications for all domains of signal processing, or wherever one works to reconstruct a sound, an image, or a more elaborate object such as a three-dimensional representation of a human organ. The roots of the theory of wavelets can be traced to sophisticated questions of harmonic analysis, but the explosive development would never have taken place if the basic ideas were not simple, easy to implement on computers, and demonstrably superior to earlier technologies in some important instances. The features of simplicity and broad impact are common to all of material of this review.

The second topic we take up is the technology of *hidden Markov models.* These models offer a natural tool for dealing with one of the fundamental problems in stochastic modeling: many naturally generated stochastic processes exhibit temporal heterogeneity that is driven by an underlying (but unobservable) change in the signal generating system. Because of substantial changes in computational technology, we now find that the range of uses for the methods of the hidden Markov model (HMM) are much more substantial than had previously been imagined.

*Research partially supported by NSF DMS92-11634
**Research partially supported by ARO Grant DAAL03-91-G-0110 and NSF DMS92-11634

The source of the strength of the HMM seems to be due to its ability to acknowledge the relationships between changing regimes where on a short term basis one could adequately model the observed data by a homogeneous process. A second source of the strength of HMM's are their exceptional ability to incorporate structural features of the phenomena under study into the structural features of the model. Often the topology of the HMM (the number of states, the transition matrix structure, and the observed sequence distribution) is designed to incorporate as many features of the observed sequence as the underlying science can justify. Although such modeling is not *a priori* effective, history has done much to support the practice. The HMM has been applied with telling success in a variety of scientific contexts.

The third topic we explore is actually a broader development that nowadays often goes under the banner of the *Markov Chain Renaissance.* There are two particular subjects in this domain that help focus our review: *simulated annealing* and the *Gibbs sampler.* Each of these topics has been subjected to intensive study over the last ten years. Together with some closely related developments – like the uses of Markov chain simulation methods in the theory of algorithms and the theory of random walks on finite structure like groups – simulated annealing and Gibbs sampling have lead to a rebirth of research interest in discrete time, finite state space Markov chains, whence the notion of a *Renaissance.*

The method of simulated annealing offers an approach to optimization problems that are particularly common in computationally intensive statistical problems such as those provided by image analysis and related inverse problems. Still, the method is exceptionally general and also gives insights into many problems of combinatorial optimization. Similarly, Gibbs sampling is a broadly applicable tool for the analysis and understanding of multivariate distributions that had been viewed as computationally intractable. The most notable successes in the application of the Gibbs sampler have been in making possible a numbers of natural Bayesian procedures, including some that are of importance in imaging.

All of the work review here has the capacity for further theoretical development and more extensive application. The collection of applications of these tools to issues in biological and medical technology is already extensive, but the natural expectation is that we have seen only a small fraction of the important possibilities that lie ahead.

WAVELETS

The purpose of this section is to introduce the central ideas and the main benefits that come from wavelet analysis. Although these notions are *per force* mathematical, our intention is always to keep one eye trained on suggestions for the empirical sciences, especially for those that are dependent on effective imaging and image storage. The development recalled here of wavelet theory substantially follows the notation and conceptualization of Daubechies (1992) and Chui (1992a) both of which offer a down to earth initiation to the practical study of wavelets. The much anticipated volume of Meyer (1993) also can be expected to offer an excellent starting point. The recent survey given in Strang (1993) offers many useful insights on the relationship of wavelets to Fourier and fast Fourier transforms.

The mathematical high road to the study of wavelets if offered by the three volume treatise of Meyer (1990) and Coifman and Meyer (1991). There are also two recent edited volumes by Chui (1992b) and Ruskai, M.B. *et al* (1992) that focus substantially on applications. In fact, any electronic bibliographic search of the recent scientific literature tends to yield an almost oppressively large number of current contributions. Although for the moment wavelet analysis may seem exotic, the likely expectation is that wavelet application will evolve to become part of the tool kit one is expected to bring to any problem in signal analysis.

Scale and Location

The scientific benefits from wavelet analysis come principally from their ability to help us focus on an object at many different scales of resolution. The implementation of this idea is illustrated most clearly by consideration of the continuous wavelet transform, though eventually one finds that the exciting applications almost all evolve from the discrete cousins.

The continuous wavelet transform begins with a "suitable" function ψ and then creates the scale location family

$$\psi_{ab}(x) = |a|^{-\frac{1}{2}}\psi(\frac{x-b}{a}),$$

where the scaling normalization $|a|^{-1/2}$ has been chosen so that all of ψ_{ab} have the same L^2 norm. The features of ψ that qualify it for suitability are not cut in stone; but the *mother wavelets* ψ that have proved to be most useful are typically smooth, have compact support, and have the property that m of its moments vanish for some $m \geq 1$. A big part of the art of wavelet analysis remains the choice (or construction) of the mother wavelet that is most suitable for the task at hand.

The wavelet transform is given by the mapping $f \mapsto \langle f, \psi_{ab}\rangle$, or more explicitly by

$$T^{WAV}(a,b) = |a|^{-\frac{1}{2}} \int_{-\infty}^{\infty} f(x)\psi(\frac{x-b}{a})dx.$$

Just as one has an inversion formula for the Fourier transform, there is a formula that lets us invert the wavelet transform:

$$f(x) = \frac{1}{C_\psi}\int_{-\infty}^{\infty}\int_{0}^{\infty} T^{WAV}(a,b)\psi_{ab}(x)\frac{dadb}{a^2},$$

and where the normalizing constant C_ψ is given by

$$C_\psi = \int_{-\infty}^{\infty} \frac{1}{|u|}|\hat{\psi}(u)|^2 du < \infty$$

where $\hat{\psi}$ is the Fourier transform of ψ.

Source of Power

Even if one agrees that the source of technological power in the wavelet transform is the fact that the wavelet coefficients $\langle \psi_{ab}, f\rangle$ weigh f over different scales "a" and locations "b", there are still mysteries to be resolved concerning the sources of efficiency of wavelet transformations and representations. Still, three key points emerge:

(1) Wavelets are responsive to the empirical fact that some natural phenomena are related to scale. This fact is observable in the natural occurrence of scaled similarities (as in snowflakes) and also in more diverse phenomena, such as ranges of validity of approximations, averaging methods of differential equations, and renormalization methods of statistical physics.

(2) Wavelets can take advantage of sharper localization and smoothness properties than traditional Fourier (or windowed Fourier) methods. The availability of both scale and location parameters provides an extra degree of freedom that provides some relief from constrictive phenomena of Fourier analysis that say in essence that f and $\hat{f}$ can not both be too concentrated; for example, f and $\hat{f}$ cannot both have compact support.

(3) Wavelets also seem to have aspects of redundancy that provide useful robustness properties. This redundancy turns out to be useful in several respects, and, in particular, it permits substantial data compression for images by permitting the wavelet transform to be stored with limited precision. This fact is one that suggest there may be an important role for HDTV image compression and transmission.

In the next few subsections, we will take a look at some of the mathematics that underlies the technological effectiveness of wavelet representations. Each of these subsections calls of the lectures of Daubechies (1992) for notation, organization, and insight. The first of these details some simple and calculations that show rather generally how approximation properties of some discrete series are usefully abstracted through the language of *frames* that goes back to Duffin and Schaeffer (1952).

Frames and Practical Inversion

Any set $\{\psi_j : j \in J\}$ of elements of a Hilbert space $\mathcal{H}$ is called a *frame* provided there are constants $0 < A, B < \infty$ such that for all $f \in \mathcal{H}$ we have

$$A||f||^2 \leq \sum_{j \in J} |\langle f, \psi_j \rangle|^2 \leq B||f||^2.$$

Certainly, if $\{\psi_j : j \in J\}$ is a complete orthonormal basis for $\mathcal{H}$, then $\{\psi_j : j \in J\}$ is a frame with $A = B = 1$, but the reason for bothering with this notion is that there are frames that exhibit a much different character than bases. One example to keep in mind consists of three vectors in $I\!R^2$ with 120 degree angles; for such a set $\{\psi_1, \psi_2, \psi_3\}$ one can check that in the required inequality we have $A = B = \frac{3}{2}$. This frame is certainly not a basis, and it also helps us see that when $A = B > 1$ the frame carries a measure of redundancy.

Given any frame, we define a frame operator $F : \mathcal{H} \mapsto \ell^2$ by taking

$$(Ff)_j = c_j \equiv \langle f, \psi_j \rangle,$$

and we note that the adjoint operator $F^* : \ell^2 \mapsto \mathcal{H}$ defined by the relationship $\langle Ff, c \rangle_{\ell^2} = \langle f, F^*c \rangle_{\mathcal{H}}$ can be given explicitly by the formula

$$f^*c = \sum_{j \in J} c_j \psi_j.$$

One can check using the definitions that F^*F is invertible and that we have the bounds $A\,Id \leq F^*F \leq B\,Id$ and $B^{-1}Id \leq (F^*F)^{-1} \leq A^{-1}Id$.

Now, if $\{\psi_j : j \in J\}$ is any frame and F is the associated frame operator, we can define a new frame $\{\tilde{\psi}_j : j \in J\}$ by taking

$$\tilde{\psi}_j = (F^*F)^{-1}\psi_j.$$

This *dual frame* turns out to be the key to a practical inversion formula, since one can check just from the definitions that the frame operator $\tilde{F}$ associated with $\{\tilde{\psi}_j : j \in J\}$ satisfies

$$\tilde{F}F = Id = F\tilde{F},$$

or, in long-hand, we have the Frame Resolution of the Identity:

$$\sum_{j\in J}\langle f, \psi_j\rangle\tilde{\psi}_j = f = \sum_{j\in J}\langle f, \tilde{\psi}_j\rangle\psi_j.$$

This last formula offers many suggestions of how to approximate f, and despite the abstract simplicity of the result, the technological implications are substantial–even in a world where closely related identities have been known for more than a hundred years.

Reconstruction Calculations

The bound $A\,Id \leq F^*F \leq B\,Id$ suggests that one may think of F^*F as a crude approximation of $\frac{1}{2}(A+B)Id$. When we replace $\tilde{\psi}_j = (F^*F)^{-1}\psi_j$ by the approximation $2(A+B)^{-1}\psi_j$ in the first equality in our resolution of the identity equation and work out the remainder term we find the *First Frame Approximation*:

$$f = \frac{2}{A+B}\sum_{j\in J}\langle f, \psi_j\rangle\psi_j + Rf,$$

where we have introduced a remainder term Rf that is explicitly defined by

$$R = Id - \frac{2}{A+B}F^*F.$$

Now, the norm of R is less than 1, so we do have an approximation of some sort, but one needs to judge how good the approximation may be and to explore the ways in which it might be improved. The first step is to consider the operator inequalities

$$-\frac{B-A}{B+A}\,Id \leq R \leq \frac{B-A}{B+A}\,Id,$$

from which a traditional argument gives the norm bounds $||R|| \leq (B-A)/(A+B) = r/(2+r)$ where $r = -1 + B/A$.

The fact that R is linear with norm less than one carries the seed of self-improvement. By the definition of R we have

$$(F^*F)^{-1} = \frac{2}{A+B}(Id - R)^{-1}$$

so we get a series representation for $\tilde{\psi}_j$ by

$$\tilde{\psi}_j = (F^*F)^{-1}\psi_j = \frac{2}{A+B}\sum_{k=0}^{\infty} R^k\psi_j.$$

This suggest how we can do better than just approximate $\tilde{\psi}_j$ by $2\psi_j/(A+B)$; we can take as many terms of the geometric sum as we like. Specifically, we can create an $N+1$ term approximation by taking

$$\tilde{\psi}_j^N = \frac{2}{A+B}\sum_{k=0}^{N} R^k \psi_j = \tilde{\psi}_j - \frac{2}{A+B}\sum_{k=N+1}^{\infty} R^k \psi_j = [Id - R^{N+1}]\tilde{\psi}_j.$$

There is a simple, instructive computation of the error that is made by using $\tilde{\psi}_j^N$ in place of $\tilde{\psi}_j$ in the resolution of the identity. First we note

$$\Delta_N = f - \sum_{j\in J}\langle f, \psi_j\rangle \tilde{\psi}_j^N = \sum_{j\in J}\langle f, \psi_j\rangle(\tilde{\psi}_j - \tilde{\psi}_j^N),$$

so expressing $\tilde{\psi}_j - \tilde{\psi}_j^N$ in terms of R gives

$$\Delta_N = \sum_{j\in J}\langle f, \psi_j\rangle R^{N+1}\tilde{\psi}_j$$

$$= R^{N+1}\sum_{j\in J}\langle f, \psi_j\rangle\tilde{\psi}_j = R^{N+1}f,$$

where the last step invokes the resolution of the identity. Finally, taking norms we see $||\Delta_n|| \leq ||R||^{N+1}||f||$, so using the approximation $\tilde{\psi}_j^N$ in lieu of $\tilde{\psi}_j$ gives us an approximation that conveys geometrically fast.

An important feature of the $\tilde{\psi}_j^N$ is that when we write out $\tilde{\psi}_j^N$ in an expansion in terms of ψ_ℓ we can get a simple recursion for the coefficients. In particular, if we write

$$\tilde{\psi}_j^N = \sum_{\ell\in J}\alpha_{j\ell}^N\psi_\ell$$

we find

$$\alpha_{j\ell}^N = \frac{2}{A+B}\delta_{\ell j} + \alpha_{j\ell}^{N-1} - \frac{2}{A+B}\sum_{m\in J}\alpha_{jm}^{N-1}\langle\psi_m, \psi_\ell\rangle.$$

As Daubechies (1992, page 63) points out, this expression may look "daunting" but in practice many of the $\langle\psi_m, \psi_\ell\rangle$ are negligibly small.

Closing the Loop

Our excursion into the theory of frames serves to give a sense of the organizing role that Hilbert space methods give to the theory of wavelets, but eventually one has to leave the soft analysis of Hilbert space to engage the hard analysis that provides us with mother wavelets ψ for which the doubly indexed set of functions given by

$$\psi_{m,n}(x) = a^{-m/2}\psi(a^{-m}x - nb)$$

provide us with honest frames with useful scientific approximation properties.

The easiest example to write down is the *Sombrero Function* given by the second derivative of the Gaussian hump $e^{x^2/2}$:

$$\psi(x) = \frac{2}{\sqrt{3}}\pi^{-1/4}(1-x^2)e^{-x^2/2}.$$

For $a = 2$ and $b = 1$ this function provides a frame that has $A = 3.223$ and $B = 3.596$.

This particular frame is not as nice as one would hope; specifically, ψ does not have compact support. Still, even this function yields a frame that is quite useful, since the tails of the Gaussian decay so rapidly. The world's catalog of good mother wavelets and associated frames is increasing rapidly, and some useful general principals are starting to emerge about how one "designs" a good wavelet.

Designer Wavelets

Mallat (1989) and Meyer (1990) used the idea of incrementing the information needed to represent a picture at one level of resolution to a more refined level of resolution to articulate a set of concrete mathematical ideas that are evolving as instrumental in the design of wavelet approximations. The emerging theory goes under the name of multiresolution analysis. The abstract set up requires a set of approximation closed subspaces V_j of $L^2(\mathbb{R})$ that satisfy the four nesting and self-similarity conditions:

$$\cdots V_2 \subset V_1 \subset V_0 \subset V_{-1} \subset V_{-2} \subset \cdots,$$

$$\overline{\cup_{j \in \mathbb{Z}} V_j} = L^2(\mathbb{R}),$$

$$\overline{\cap_{j \in \mathbb{Z}} V_j} = \{0\},$$

and finally the condition that adds teeth to this abstract structure by tying the V_j's all together

$$f \in V_j \Longleftrightarrow f(2^j \cdot) \in V_0.$$

The moral principal of multiresolution analysis is that whenever one finds a collection of subspaces satisfying the four preceding conditions, then there is an orthonormal wavelet basis $\psi_{j,k} = 2^{-j/2}\psi(2^{-j}x - k)$ such that the projection operators P_j onto V_j satisfy the key identity

$$P_{j-1} = P_j + \sum_{k \in \mathbb{Z}} \langle f, \psi_{j,k} \rangle \psi_{j,k}.$$

Beyond mere morality, there is a process that is quite often successful in finding a mother wavelet that accommodates the multiresolution. This process is not simple enough to recall here but it is simple enough for one to take seriously in the context of any scientific problem for which there are logical subspaces V_j satisfying the conditions detailed above. This is also the path by which the first basis of compactly supported wavelets was developed in Daubechies (1988), the work that perhaps most pointedly initiated the current flurry of wavelet activity.

Last Wavelet Steps

The preceding review has been reasonably complete in detailing the most basic definitions and properties of wavelet analysis. With luck there is also some suggestion of where one is likely to find effective applications, though the most compelling process is to review those collections of applications that have been edited and the current applications as they appear. Still, there is one more conceptual step that even an introduction like this should address, and that is wavelet constructions for higher dimensions. The bottom line is that there are at least two trustworthy constructions, one based on tensor products and one based on the use of lattices in $I\!R^d$ for $d > 1$. For further discussion of these points one should consult Gröchenig (1991), Gröchenig and Madych (1992), and Kovačević, J. and Vetterli, M. (1991),

HIDDEN MARKOV MODELS

We have already described the key qualitative feature of HMM's — they accommodate the relationships between differing regimes of homogeneity in processes that are only locally homogeneous. This feature is evident in almost all of the important HMM applications including (1) speech recognition (cf. the survey and extensive bibliography of Juang and Rabiner (1991)), (2) the study of DNA composition where one uses sequences of bases in a DNA molecule to identify types of molecule segments (Churchill (1989)), (3) hypothesis testing in the study of different haematopoiesis theories, where data on counts of specific bone marrow cells are used to determine the number of unobserved active stem cells from which all blood cells develop, and information on the number of active stem cells is essential for the determination of a correct haematopoiesis model (Guttorp *et al.* (1990)), (4) modeling of ion channels (or large proteins that span cell membranes), where data on electrical single channel currents in neuronal membranes are obtained to study the multiple conductance levels of the ion channel (Chung *et al.* (1990), Fredkin and Rice (1992)), and (5) electrocardiology, where sequences of electrocardiac wave patterns that represent different states of the heart are recognized from EKG data (Coast *et al.* (1991)). The variety of these applications surely suggests that HMM's are a flexible tool, but to understand how HMM's actually serve in these contexts we naturally need to engage the mathematical details of the model. As a first step, we need to lay out some basic definitions.

Model Definition

Let $A = (a_{ij})$ be an $N \times N$ Markov transition matrix, and let $\Pi = (\pi_i), 1 \leq i \leq N$ denote an arbitrary probability distribution. If we view Π as an initial distribution, we can define a discrete-time finite-state Markov chain $\{Q_t\}_{t=1}^{\infty}$ by (Π, A). We further denote the N possible states of the Markov chain by $\{S_1, S_2, \ldots, S_N\}$, and for each $1 \leq i \leq N$, we associate a probability density function $b_i(\cdot)$.

For the triple $(A, B = \{b_i\}_{i=1}^{N}, \Pi)$ we observe a stochastic process $\{O_t\}_{t=1}^{\infty}$. Given that $Q_t = S_i$, the random variable O_t has density $b_i(\cdot)$ and are assumed independent of $\{O_s : s \neq t\}$.

The sequence $\{O_t\}$ is referred to as the *observed sequence or observation sequence*, while the sequence $\{Q_t\}$ is unobserved and referred to as the *hidden, or unobserved sequence*. The probability density function for the sequence $O = (O_1, \ldots, O_T)$ is given by

$$P_{(A,B,\Pi)}(O_1 = o_1, \ldots, O_T = o_T) = \sum_{1 \leq q_1, \ldots, q_T \leq N} P_{(A,B,\Pi)}(O, Q) =$$

$$\sum_{1 \leq q_1, \ldots, q_T \leq N} \pi_{q_1} b_{q_1}(o_1) a_{q_1 q_2} b_{q_2}(o_2) \cdots a_{q_{T-1} q_T} b_{q_T}(o_T).$$

One often has to distinguish the situation where the $b_j(\cdot)$ are probability mass functions, and in such cases we refer to the *discrete value model*. Typically, one assumes that there is a set of parameters of interest that determine the distributions assumed for the hidden and observable processes. Here we will assume that there is an $n \geq 1$ and an open subset Λ of the Euclidean n space such that for each $\lambda \in \Lambda$ we have a one-one correspondence $\lambda \leftrightarrow (A(\lambda), B(\lambda), \Pi(\lambda))$. The set Λ is defined to be the parameter space of the model.

After having defined and parameterized the model, we turn to the three fundamental questions that arise when applying HMM's.

1. Probability **Evaluation** Problem. Given the observation sequence $O = (O_1, O_2, \ldots, O_T)$ and a model $\lambda = (A, B, \Pi)$, how can we compute $P_\lambda(O)$ in the discrete value model? Correspondingly, how do we compute the likelihood function in the continuous value model?
   We have obtained an expression for the likelihood $P_\lambda(O)$, but the expression is only of theoretical value for it contains far too many summands to be evaluated numerically.

2. Parameter **Estimation** Problem. Given the sequence O, how can we estimate the model parameters $\lambda = (A, B, \Pi)$? Moreover, if we proceed by maximum likelihood, how do we calculate λ to maximize $P_\lambda(O)$?

3. State sequence **Identification** Problem. Given the observation sequence O and the parameter set λ, how can we compute a state sequence $q = (q_1, q_2, \ldots, q_T)$ that has maximal conditional probability $P_\lambda(Q \mid O)$?

Maximum likelihood is the main estimation principle that has been used in the solutions to the parameter estimation and the best sequence of states determination problems, but there are other estimation principals that have been successfully used for these problems, such as the *state optimized likelihood criterion* in Juang and Rabiner (1990). Here, the usual likelihood objective function is replaced by the state optimized likelihood function

$$\max_q \sum_q \pi_{q_1} b_1(o_1) \prod_{t=2}^{T} a_{q_{t-1}q_t} b_{q_t}(o_t).$$

However, we shall not elaborate on alternative methods for maximum likelihood estimation.

Except in some trivial cases, one cannot provide a closed-form solution for the maximum likelihood estimates of the parameters or hidden sequence associated with the hidden Markov model. Hence, one has to call on algorithms that were developed to address the three problems mentioned above. Each of these algorithms offers computational challenges that must be met for the method to be effective.

In addition to the computational difficulties in applying HMM's, there exist a number of theoretical and implementational HMM problems. We next list a number of HMM limitations and some suggested solutions.

Identifiability Identifiability has been studied extensively in connection with mixture distributions. A closely related problem is the identifiability of the HMM parameter. Identifiability makes the estimation of λ an unambiguous problem, and it is a necessary condition for existence of a consistent estimate for λ. One of the subtle features of the HMM is that the parameterization is unidentifiable. The most obvious way to see this is to consider a permutations of the Markov chain state indices, i.e., a relabeling of the states. As a consequence, all the results on estimation of the HMM parameters and their properties, apply only up to equivalence classes of parameters that define the same distribution for the observation sequence. For the discrete value model, Petrie (1969) discusses the identifiability problem, and in the continuous value model there is a detailed discussion in Leroux (1992).

For both models, if A is irreducible and aperiodic, so that it has a unique stationary distribution, and the observed sequence densities are distinct, then the only ambiguity remaining in parameter values originates from the symmetry of the likelihood function with respect to permutations in state labels, but not from any other changes in parameter values.

Local Optimality All the methods for HMM likelihood maximization can only offer a local maximum point of the likelihood. The particular local maximum point obtained will frequently depend upon the location of the initialization used in these iterative methods. For an interesting approach to the maximum likelihood estimation of HMM parameters using the simulated annealing method, see Paul (1985).

Model Dimensionality The number of states N and the number of distinct symbols M for the discrete observation distribution case have to be chosen *a priori* to appropriately specify an HMM. Misspecification of these parameters implies incorrect dimensions of the model's parameter space, λ. This could potentially lead to incorrect solutions of the estimation and state identification problems stated above.

Duration Models One of the inherent limitations of HMM's is the constraint that the Markov model imposes on the amount of time that the unobserved process can stay in a given state. The probability of staying in state S_i for d observations is $p_i(d) = (a_{ii})^{d-1}(1 - a_{ii})$, that is the time to exit state S_i is geometrically distributed with success probability $(1 - a_{ii})$. This geometric state duration distribution is inappropriate for many applications. Several alternatives for implementing different state duration models leading to *hidden semi-Markov models* (HSMM) have been proposed, and a review of the alternative models can be found in Rabiner (1989).

Another model that has been considered by Levinson (1986) in an attempt to alleviate some of these problems, is to use a parametric state duration density instead of the nonparametric $p_i(d)$ used in the HSMM. In particular, the Normal family and the Gamma family have been considered. For these models, reestimation formulas have been derived and successfully used in applications.

A variety of other implementational issues have been discussed in the literature. Some of the most important ones can be found in Rabiner (1989).

HMM Development and Computational Complexity

Early contributions to identifiability and statistical inference problems for functions of finite Markov chains were given in papers by Blackwell and Koopmans (1957), Gilbert

(1959), and Baum and Petrie (1966), but the landmark papers by Baum and Eagon (1967), Petrie (1969), and Baum, Petrie, Soules and Weiss (1970) seem to be the first to introduce solutions to the first two fundamental problems in HMM's mentioned above. These landmark papers contain a set of theorems that lead to iterative reestimation procedure for the maximum likelihood estimation of the HMM parameters.

Although the early work in this field tended to restrict attention to the case where the observed sequence of signals is a sequence of univariate random variables with log-concave densities $b_i(\cdot)$, $1 \leq i \leq N$, the work by Liporace (1982), Juang (1985), and Juang, Levinson and Sondhi (1986) offers extensions to multivariate stochastic observations of Markov chains with densities $b_i(\cdot)$ that are mixtures of log-concave and elliptically symmetric densities. As for the estimation of the hidden state sequence, an algorithm based on dynamic programming methods called the *Viterbi algorithm* was introduced by Viterbi (1967) and has been further studied in Forney (1973).

The computational and analytical complexity inherent in HMM's originates from the complicated nature of its likelihood function. In this section we briefly present three algorithms that offer computationally feasible solutions to the three fundamental questions posed above. We shall only focus on the main ideas of the algorithms. The interested reader can find an excellent exposition of the procedures in Rabiner (1989).

The Evaluation Problem

The evaluation problem is a question in computational efficiency. A naive evaluation of $P_\lambda(O)$ is an infeasible computation that involves N^T possible state sequences. Instead we can invoke a simple but powerful procedure called the *Forward-Backward* procedure, that evaluates $P_\lambda(O)$ in an order of $N^2 \times T$ operations.

The key idea is that instead of proceeding T steps at a time separately with each possible state sequence realization as in the naive evaluation, the Forward-Backward evaluation proceeds one step at a time simultaneously with all possible state sequence realizations. The latter process allows for a recursive calculation of partial sequence probabilities that make this procedure so useful.

Formally, given a model λ and an observed sequence $(o_1, \ldots, o_T)$, we define *forward* and *backward* probabilities as follows:

$$\alpha_t(i) \stackrel{def}{=} P_\lambda(O_1 = o_1, O_2 = o_2, \ldots O_t = o_t, Q_t = S_i),$$

$$\beta_t(i) \stackrel{def}{=} P_\lambda(O_{t+1} = o_{t+1}, O_{t+2} = o_{t+2}, \ldots O_T = o_T | Q_t = S_i).$$

Recursive formulas for the forward and backward probabilities are readily calculated using Bayes' Rule, the Markov property of the state sequence Q, and the conditional independence of $\{O_t, 1 \leq t \leq T\}$ given Q. Namely, we have

$$\alpha_t(i) = \left[\sum_{j=1}^{N} \alpha_{t-1}(j) a_{ji}\right] b_i(o_t), \quad \text{with} \quad \alpha_1(i) = \pi_i b_i(o_1),$$

$$\beta_t(i) = \sum_{j=1}^{N} a_{ij}\, b_j(o_{t+1})\, \beta_{t+1}(j), \quad \text{with} \quad \beta_T(i) = 1,$$

where $2 \le t \le T$, $1 \le i \le N$.

Finally, we obtain the following formulae,

$$P_\lambda(O) = \sum_{i=1}^{N} \alpha_T(i),$$

$$P_\lambda(O, Q_t = S_i) = \alpha_t(i)\beta_t(i), \qquad \text{for any } 1 \le t \le T,$$

$$P_\lambda(O, Q_t = S_i, Q_{t+1} = S_j) = \alpha_t(i) a_{ij} b_j(o_{t+1}) \beta_{t+1}(j).$$

The first formula can serve as a tool for likelihood based model comparisons. Under a given model λ, the last two formulae provide a method to obtain *a posteriori* estimates

$$\hat{\pi}_i = \frac{\alpha_1(i)\beta_1(i)}{\sum_{j=1}^{N} \alpha_T(j)},$$

$$\hat{a}_{ij} = \frac{\sum_{t=1}^{T-1} \alpha_t(i) a_{ij} b_j(o_{t+1}) \beta_{t+1}(j)}{\sum_{t=1}^{T-1} \alpha_t(i)\beta_t(i)}.$$

for the initial and transition probabilities. Also, note that we can find the most likely state at time t by $\hat{q}_t = \arg\max_i P_\lambda(Q_t = S_i|O)$. The sequence $(\hat{q}_1, \hat{q}_2, \ldots, \hat{q}_T)$ is often called the *Maximal Aposteriori Probability*, or MAP, estimate of $(q_1, q_2, \ldots, q_T)$

The Estimation Problem

The most widely used optimization technique for the maximum likelihood estimation of the HMM parameters is known as the *Baum-Welch* algorithm. The key observation, noted originally by Baum and Eagon (1967) in the case of a discrete value model, is that there is a transformation $\tau : \Lambda \to \Lambda$ of the parameter space such that the transformed parameter $\tau(\lambda)$ is guaranteed to increase the likelihood $L(\cdot)$. There exist several ways to arrive at the form of the transformation τ, such as standard constrained optimization techniques or the *a posteriori* approach illustrated above, but the one that proved to be the most useful is based on an auxiliary function that is closely related to the Kullback-Leibler number introduced in Kullback and Leibler (1951). Baum *et al.* (1970) derive the explicit form of the transformation τ for the continuous value model with log-concave density functions $b_i(\cdot)$, $1 \le i \le N$, and prove that fixed point solutions of τ are locally optimal points of the likelihood function.

Formally, define the auxiliary function

$$Q(\lambda, \lambda') = \sum_q P_\lambda(O, Q = q) \ln P_{\lambda'}(O, Q = q),$$

where the summation is over all feasible paths q through the state product space. Let

$$\tau : \lambda \to \hat{\lambda} \stackrel{def}{=} \arg\max_{\lambda'} Q(\lambda, \lambda').$$

The Baum-Welch algorithm begins with a feasible initial estimate of the parameter values $\lambda = \lambda_0$, to which the transformation τ is applied to obtain a new estimator $\hat{\lambda}$. The process is iterated by replacing the old values in λ by the newly obtained values $\hat{\lambda}$, until a fixed point of τ is approximated. For this procedure we have the following important result established in Baum *et al.* (1970).

Theorem *Under the above assumption on the densities $b_j(\cdot)$, we have for all $\lambda \in \Lambda$ that $L(\tau(\lambda)) \geq L(\lambda)$. Moreover, equality can hold if and only if λ is a critical point of L, or equivalently, λ is a fixed point of τ.*

The significance of this result can be brought out in several ways:

1. The explicit form of the transformation τ is given by a set of so called *reestimation formulas* that express the new value for each parameter as a function of its old value. These formulas are obtained by differentiating $Q(\lambda, \lambda')$ with respect to each one of the primed parameters and equating the derivatives to zero. Part of the usefulness of the Baum-Welch approach is the form of the auxiliary function that greatly facilitates the manipulation of the primed parameters. also, note that only first derivatives are required by the Baum-Welch algorithm.

2. The reestimation formulas involve probability expressions of the sort handled by the Forward-Backward procedure. Hence, to reduce the computations to a feasible order, one usually invokes the Forward-Backward procedure within each iteration of the Baum-Welch algorithm.

The Identification Problem

We are often interested in uncovering the true state sequence that led to a given observation sequence O. Although the probability measure $P_\lambda(O)$ does not explicitly involve a specific state sequence realization it can often provide useful insight into the structure of the mechanism that generates the observations.

When the HMM has state transitions with zero probabilities and we choose to maximize separate state subsequences, the optimal state sequence resulting from this process may not even be a valid state sequence. The Viterbi algorithm overcomes this problem by maximizing over the entire state sequence. Prior to discussing the algorithm, we describe a trellis structure that captures the features of the problem.

Consider an $N \times T$ vertex trellis structure whose vertices are arranged in T columns corresponding to time slots 1 through T and N rows representing the N states of the Markov chain. Directed edges connect between all possible pairs of edges with positive transition probabilities. Clearly, this construction has the property that for every possible state sequence there is a unique corresponding path through the trellis and vice versa. For a given model λ, we attach weights to the edges and initial vertices in a way that our problem becomes to find the longest path through the trellis. Denote the jth vertex in the tth time slot by S_j^t. The edge of the trellis connecting vertex i at time slot $t-1$ with vertex j at time slot t have weight $a_{ij}b_j(o_t)$, $1 \leq i, j \leq N$, $2 \leq t \leq T$ and the initial weight for vertex S_i^1 is $\pi_i b_i(o_1)$, $1 \leq i \leq N$. We will calculate the weights of partial state sequences, as we move along the time slots, by multiplying the weights along the edges of the path. For a particular complete path q, the weights product along the path's edges results in

$$\pi_{q_1} b_{q_1}(o_1) \prod_{t=2}^{T} a_{q_{t-1}q_t} b_{q_t}(o_t) = P_\lambda(Q = q, O).$$

Let $q(S_j^t)$ denote any path segment from time 1 to t, ending at state S_j, $1 \leq j \leq N$. Let $\hat{q}(S_j^t)$ denote the longest such path segments, also called the *survivors*. Assume for simplicity that $\hat{q}(S_j^t)$ is uniquely defined for any (j, t), or else choose one such path arbitrarily. Then for any time $t > 1$ there are N survivors in all, one for each possible terminating state (vertex) of the partial path.

The main observation is that the longest complete path must begin with one of these survivors. If it did not, we could have found a longer path segment from time 1 to t which would be a contradiction. Thus, at any time t, we need to remember only the N survivors $\hat{q}(S_j^t)$, $1 \leq j \leq N$ and their lengths. To get from time t to $t+1$, we need only extend all time-t survivors by one time unit. This is done by selecting for each time-$(t+1)$ state S_k^{t+1}, $1 \leq k \leq N$, the time-t survivor that is longest when extended . The length of the new survivors is recalculated by multiplying the length of its last edge times the total length of the corresponding old survivors. The algorithm proceeds indefinitely, advancing one time unit at a time, without the number of survivors ever exceeding N. As was the case for the Forward-Backward procedure, computations are on the order of $N^2 \times T$ operations.

Recent Advances in HMM's

The popularity of HMM's continues to grow rapidly both in applied and theoretical work. On the theoretical side an important part of the research focuses on inferential properties of likelihood methods. Leroux (1992) established under mild conditions the consistency of the maximum likelihood estimate for the continuous value model, and thus complemented the pioneering work by Baum and Petrie (1966) and Petrie (1969) where the consistency and asymptotic normality of the maximum likelihood estimates had been established under the discrete value model. Another important step was taken in Bickel and Ritov (1993) where the log-likelihood for continuous value HMM's is shown to obey the local asymptotic normality conditions of LeCam as a consequence of which asymptotically efficient analogs of the maximum likelihood estimates can be constructed and the information bound that gives their asymptotic variance can be estimated.

Aggoun and Elliot (1992) consider the case of a continuous time Markov chain observed in Gaussian noise. Finite dimensional normalized and unnormalized predictors are obtained for the state of the chain, for the number of jumps from one state to another, and for the occupation time in any state.

In the hidden sequence estimation area, some simplified estimation procedures for both the filtration and interpolation problems in a two-state HMM are proposed and analyzed in Khasminskii, Lazareva and Stapleton (1993). These estimates have been designed to perform well for the case where $a_{01} = \epsilon\lambda, a_{10} = \epsilon\mu; \epsilon \to 0$ and $0 \leq \lambda, \mu \leq 1$, thus creating a similarity to the change point detection problem. Kogan (1991) gives conditions under which the estimated chain, as given by the Viterbi algorithm, has a lower recognition error rate than the alternative MAP estimated chain, for the two-state HMM case with symmetric transition matrix.

An extension of the basic HMM paradigm to the regression setting is proposed by Fridman (1993). The *hidden Markov model regression* offers a way to extend the benefits of HMM's to problems that are naturally studied through regression analysis. In general terms, it is assumed in HMM regression that the regression parameter values depend on the Hidden Markov chain state. As a result, given that the state at time t is S_i, we have that $Y_t = X_t'\beta_i + \sigma_i\epsilon_t$, where the error terms ϵ_t are i.i.d. $N(0,1)$. There is a connection to the *switching regression model* introduced in Quandt (1972) and Quandt and Ramsey (1978) is a special case of HMM regression for a Markov transition matrix with the property $a_{ij} = a_j$.

MARKOV CHAIN RENAISSANCE

The subject of discrete time discrete space Markov chains have receive greatly increased attention over the last five years because of several developments in the theory of algorithms. One of these developments concerns simulated annealing, which we introduce in the next section. Two other potentiating developments were the Gibbs sampler and the invention of Markov chain methods for making uniform selections from large discrete sets, like the set of all matchings in a graph. Only the first of these is engaged in this survey.

An Algorithm for All Problems

A bewildering variety of substantial computational problems can be cast in to the framework of determining the minimum of a real-valued function on a finite set, $f : S \rightarrow \mathbb{R}$. For example, if $\{x_1, x_2, ..., x_n\}$ denotes a set of n points in the plane and S denotes the set of permutations of the n-set $\{1, 2, ..., n\}$, then by taking

$$f(\sigma) = \sum |x_{\sigma(k)} - x_{\sigma(k+1)}|$$

we see that the determination of an element of the set of minimizers of f,

$$S^* = \{s^* \in S : f(s^*) = \min\{f(s) : s \in S\},$$

is the same as solving the famous *traveling salesman problem*, or *TSP* . Naturally, it is no surprise that the *TSP* is only one of a hord of problems that can be put into the specified form —the form is so general as to seem to impose virtually no restriction. Rather, the surprise comes from the fact that we can still say something useful, even at so great a level of generality.

Naturally, one has to cut some slack. The *TSP*, like many of the interesting problems of combinatorial optimization, can be regarded as computationally intractable. It is an element of the class of *NP-Complete* problems, and, as a consequence, it is extremely unlikely that one can ever obtain a good algorithm for the *TSP*, if one regards a good algorithm as one that can provide an optimal solution in an amount of time that grows only polynomially in the size of the input. Because of this natural barrier of intractability, many problems like the *TSP* have been studied in the context of approximate solutions.

It is remarkable that one can give anything like a general recipe for making progress toward the general problem of determining an element of S^*. We can in fact provide a sequence of such recipes. We begin with one which is not quite practical, but it still offers some genuine insight, and —even in its naive form –it is cleverer and more practical than the often useless idea of exhaustive search.

A skeletal version of the recipe is simple. First, we introduce a probability measure μ_t on S by defining

$$\mu_t(s) = exp(-tf(s))/Z(t)$$

where $Z(t)$ is chosen just in order to normalize everything properly, i.e.

$$Z(t) = \sum_{s \in S} exp(-tf(s)).$$

Second, we select a "large" t. Third, and finally, we just choose an element of S at random according to the measure μ_t.

This is a phony recipe in a couple of ways; but it has something to it, and its faults can be substantially ameliorated. But even before seeing how it can be made honest, we should get an idea why it might work. The essential point is that as $t \to \infty$ the measure μ_t becomes increasingly concentrated on the states for which s is small. Formally, it is not hard to prove that

$$\lim_{t \to \infty} \mu_t(S^*) = 1$$

so ultimately $\mu(s)$ is concentrated on the best choices for s. This limit result would have to be supplemented by more precise continuity statements if we were to try use it to justify our recipe along the lines presented thus far, but a wiser course —and one the development of the theory has actually taken— is to improve our recipe at least a couple more times before aiming at the convergence theorems that offer an honest justification.

An Honest Version

So, how do we honestly pick an element of S according to the measure μ_t? The denominator $Z(t)$ in the definition of μ_t can have many billions of summands in even an "easy" problem, so naive methods of drawing a sample according to μ_t are meaningless. Luckily, there is a brilliant trick due to Metropolis, Rosenbluth, Rosenbluth, Teller, and Teller (1953) that provides the key. As an incomplete historical note, it is interesting to record that the fifth of these authors is the Edward Teller who is an acknowledged father of the hydrogen bomb.

The essence of the trick is that one can make progress by viewing μ_t as the stationary distribution of a Markov chain with state space S. For this to buy us anything, we need to be able to simulate the steps of the chain, but we have lots of room to maneuver in this aim, because, after all, we get to invent the chain.

To restrict our search for a good Markov chain to particularly tractable situations, we will impose a graph structure on S, and we will only consider Markov chains were one always moves to a neighbor in the graph (or else stays still, as is sometimes useful in a Markov chain to guarantee aperiodicity). In many applications, the set S has a graph structure is naturally at hand, but in any event the restriction is modest since all it really means is that for each $s \in S$ we have a set $N(s) \subset S - \{s\}$ that we call the neighbors of s.

As we start our hunt for chains that have μ_t as their stationary measure, we might also pay particular attention to those that are reversible, since the stationarity equations then can be replaced by the much simpler *total balance* equation. As a first step, we may as well also restrict our attention to chains for which the graph on S is *regular*, i.e. $|N(s)| = N$, some constant, for all $s \in S$. By making sure that the cardinality N is feasibly small, we will be able to simulate the behavior of any Markov chain that only makes transitions to neighboring states.

To make matters precise, we need transition matrix p_{ij}, where $p_{ij} = 0$ except for $j \in N(i)$ and for which μ_t is the stationary measure. Hunting, as we are, in the context of reversible chains, we want our transition probabilities to satisfy the *total balance* condition:

$$\mu_t(i)p_{ij} = \mu_t(j)p_{ji}$$

Determination of a suitable p_{ij} is now pretty easy, but instead of continuing with more heuristic groping, it seems better to look at one good answer and its associated story.

Before writing down the transition probabilities, it is useful to indicate first how to simulate the behavior of the required Markov chain $\{X_n, n \geq 0\}$. The story goes like this: (1) if the current state X_n is equal to i, first choose a neighbor j of i at random, (2) if the value of f at j improves on the value of f at i, then move to the chosen state i, but (3) if the value of f would not be improved by the move (so that $f(j) \geq f(i)$), then make the move with probability $\exp(-t(f(j) - f(i)))$ while otherwise staying at the state i. Formally, for $i \neq j$ we have

$$P(X_{n+1} = j \mid X_n = i) = \begin{cases} 1/N & \text{if } f(j) < f(i) \\ (1/N)\exp(-t(f(j) - f(i)) & \text{otherwise} \end{cases}$$

The transition probabilities in case $i = j$ are just those needed to pick up the leftovers:

$$P(X_{n+1} = i \mid X_n = i) = 1 - \sum_{j:j\neq i} P(X_{n+1} = j \mid X_n = i).$$

A useful point to check at this juncture, is that a chain with the transition function given above does satisfy the total balance condition, so to make a choice from S according to the measure μ_t all we need to do is start at an arbitrary point $X_0 = s$ and run the Markov chain for a "long time" after which X_n will be a realization from a distribution that is approximately equal to μ_t.

To see how simulation of X_n can be practical even when direct simulation of μ_t is not, just reconsider the *TSP*. In an n-city problem the cardinality of S is $(n-1)!$, the number of cyclic permutations, but we can introduce a natural graph on these permutations where the degree of each vertex is bounded by n. The graph is defined by saying that two permutations are considered adjacent if we can go from one to the other by a an "interchange operation" given by picking two non-intersecting edges of the cycle, breaking the cycle at each of these edges, and building a new cycle by reconnecting the two disconnected components of the cycle in the opposite way from their initial connection.

At this point we have in hand a method for "solving" all problems in combinatorial optimization—but, of course, it solves some problems better than others. After the next section, we will review the performance of the method, but even as it sits, it has some victories. One of these that is entertaining to code is the famous "Eight Queens" problem: How can one place eight queens on a chessboard in such a way that no pair of queens are mutually attacking? This problem is one that is often assigned in programming courses where backtracking algorithms are studied, but it is also a nice one to study with our naive sampling method. Once one chooses an appropriate reduction of the problem (say, to "rook good" configurations of queens) it is not hard to find an appropriate graph on the set of configurations (pick a pair of queens and switch the two in the only way that preserves "rook-goodness").

A More Honest Version

The only scurrilous part of our recipe that remains is that of our choice of t and our choice of how long to run the Markov chain. To many people it is unsatisfying to say that the choices are simply up to them, and luckily the search for a more adaptive procedure turns out to be a source of insight, and there is a fortuitous charm in combining the two problems into one.

Specifically, we consider a sequence of values t_n such that $t_n \to \infty$ as $n \to \infty$, and now let $\{X_n : n \geq 1\}$ evolve as before except on the nth step we use t_n in place of t. Letting the t_n grow to infinity, provides us a way to combine the issues of picking large t and a suitably large n. The issue now is to obtain the conditions on the sequence $\{t_n : n \geq 1\}$ that suffice to provide us some proper assurance that the method is effective.

A central result in this direction is due to Hajek (1988). In addition to answering the basic question as asked, it also provides us with some special insight. A key notion in Hajek's theorem is that of a *height*. Specifically, we say a state s *communicates with* S^* *at height* h if $h(s)$ is the smallest number such that there is a path from s to some state of $t \in S^*$ for which each state u on the path satisfies

$$f(u) \leq f(s) + h(s).$$

Theorem If h^* is the largest of the set of heights $\{h(s) : s \in S\}$, then we have

$$\lim_{n \to \infty} P(X_n \in S^*) = 1,$$

if and only if

$$\sum_{n=1}^{\infty} \exp(-t_n h^*) = \infty.$$

One consequence of this result is that the choice $t_n = log(n)/h$ is sufficient to provide convergence if and only if h satisfies $h \geq h^*$. As a practical matter one seldom knows h^*, but still taking $t_n = log(n)/h$ for a speculatively chosen h is a common occurrence, or at least it it is common among those bother with changing t as n evolves. Many people have found that in their application the change provided by $t = log(n)/h$ is too slow to be worth the trouble and hence they fall back on the naive process of just picking a "big" t. This approach makes one miss out on a lot of engaging theory; but, as a matter of practice, it can work out well, and picking t directly is not substantially more *ad hoc* than picking h.

Origins and Aspects of Metaphor

There are three ideas that were stirred together to make the theory just described. The first idea is that we might get a small value of $f(s)$ if we pick a point at random according to μ_t. The second idea— an old but wonderful one—is that a simple Markov chain can be use to help us make the selection process practical. The third idea is that we can link the processes of letting X_n "run for a long time" and of "picking a big t" by letting X_n evolve as a time inhomogeneous.

The final algorithm that combines all of these ideas goes under the engaging name of *Simulated Annealing*. It was introduce independently by Kirkpatrick, Gelett and Vecchi (1983) and Cerny (1985), and it was first described in the context of a physical analogy that has continued to exert considerable influence. It is traditional nowadays in the discussion of simulated annealing to a call on the language of statistical mechanics: (1) μ_t is called the Gibbs distribution, (2) $Z(t)$, the partition function, (3) $T = K/t$, temperature, and (4) the sequence $T_n = K/t_n$, the cooling schedule. There is a certain beauty in this way of describing the optimization process, and the metaphor brings with it a rich collection of intuitions and experience from statistical mechanics. Still, there is some merit in taking a bare-bones look without introducing language which - though apt and properly evocative - can make simple ideas look more mystical than they otherwise might. A second benefit of the less fortified description is that it offers us a different set of opportunities future for development than those offered by the physical insights. Finally, the usual presentation does not pick apart the three steps as thoroughly as we have done here, so the bare-bones route also suggests opportunities for innovation that precede even the proto-simulated annealing algorithm.

How Well Does One Do?

The jury is out on many aspects of the simulated annealing algorithm, and it is unlikely that a definitive understanding of its merits will come about anytime soon. One difficult issue is that there is no *single* simulated annealing algorithm. One has a whole family of algorithms for every problem one might study because of the need to choose (1) the graph structure to be used (2) one candidate from among many f that would yield the given S^* of interest, and (3) some specific values for $\{t_n\}$, or the equivalent.

It seems almost impossible to make all of these selections in a way that would lead to a definitive evaluation, but some reliable experience has evolved. In particular, the papers of Johnson, Aragon, McGeoch, and Schevon (1990,1991,1992) provide many useful comparisons.

Metropolis-Hastings Algorithm

The method sketched above for generation of an observation from the Gibbs distribution has an extension that has the benefits of being less *ad hoc* , more general, and providing a connection to the final topic of this review: the Gibbs sampler. The extension is due to Hastings (1970) and it tells us how an arbitrary Markov chain can be modified to provide one with a specified stationary distribution. Quite simply, to provide a Markov chain $\{X_n : 1 \leq n < \infty\}$ that has stationary distribution π on the state space $\mathcal{S}$, the Metropolis-Hasting Algorithm begins with an arbitrary transition function $q(x, y)$ and makes modifications. In particular, a new transition function defined by

$$p(x,y) = \begin{cases} q(x,y)\alpha(x,y) & \text{if } x \neq y \\ 1 - \sum_z q(x,z)\alpha(x,z) & \text{if } x = y \end{cases}$$

where α is defined by

$$\alpha(x,y) = \begin{cases} \min\{\frac{\pi(y)q(y,x)}{\pi(x)q(x,y)}\} & \text{if } \pi(x)q(x,y) > 0 \\ 1 & \text{if } \pi(x)q(x,y) = 0. \end{cases}$$

The Gibbs Sampler

The Gibbs sampler shares with simulated annealing the feature of being a Monte Carlo integration method that proceeds by a Markovian updating scheme. The product that is delivered by the Gibbs sampler is an observation from a multivariate distribution, and the raw material that is required for the algorithms is a collection of conditional distributions from which one can easily draw observations.

Suppose that we have a collection of k real, possibly vector-valued, random variables $U_1, \ldots, U_k$ whose full conditional densities, denoted by $f_{U_i}(\cdot|U_j; j \neq i), 1 \leq i \leq k$, have a simple known form. By that we mean that the full conditional densities are available for sampling, given values of the appropriate conditioning random variable.

Our interest is to simulate an observation from the joint density $f_{(U_1,\ldots,U_k)}$ with the eventual intention of gaining insight into the joint density or some other quantity that can be estimated using such observations, such as an estimate of the marginals f_{U_i}.

The idea is to generate a sample of k-tuples from the joint density using only the available full conditional densities. By simulating a large enough sample, any population characteristic can be approximated to a desired degree of accuracy. Before we formally describe the method, we introduce some simplifying notation from Gelfand and Smith (1990). Densities are denoted generally by brackets, so joint, conditional, and marginal forms appear as $[X,Y], [X|Y]$, and $[X]$ correspondingly. Multiplication of densities is denoted by $*$ and marginalization by forms such as $[X|Y] = \int_{Z,W}[X|Y,Z,W]*[Z|W,Y]*[W|Y]$. We assume that the joint density exists and is strictly positive over the product sample space. Besag (1974) shows that this condition ensures that knowledge of the full conditional densities uniquely defines the full joint density.

The Gibbs sampler algorithm generates an approximation to an observation from $f_{(U_1,\ldots,U_k)}$ by iterations of a k-step process, each complete pass of which corresponds to a step of a Markov chain. Specifically, given an arbitrary set of values $(U_1^{(0)}, U_2^{(0)}, \ldots, U_k^{(0)})$ we draw

$$\begin{aligned} U_1{}^{(1)} &\sim [U_1|U_2^{(0)}, U_3^{(0)}, \ldots, U_k^{(0)}] \\ U_2^{(1)} &\sim [U_2|U_1^{(1)}, U_3^{(0)}, \ldots, U_k^{(0)}] \\ &\vdots \\ U_k^{(1)} &\sim [U_k|U_1^{(1)}, U_2^{(1)}, \ldots, U_{k-1}^{(1)}]. \end{aligned}$$

Upon completion of this first iteration of the algorithm, we obtain the vector $U^{(1)} = (U_1^{(1)}, U_2^{(1)}, \ldots, U_k^{(1)})$. Next, we generate $U^{(2)}$ using conditioning values taken from $U^{(1)}$. After i such iterations we arrive at $(U_1^{(i)}, U_2^{(i)}, \ldots, U_k^{(i)})$.

The great usefulness of the algorithm comes from the following theoretical results, established by Geman and Geman (1984). Under mild conditions,

$$(U_1^{(i)}, U_2^{(i)}, \ldots, U_k^{(i)}) \xrightarrow{d} [U_1, \ldots, U_k]$$

as $i \to \infty$, and hence for each $1 \leq j \leq k$, $U_j^{(i)} \xrightarrow{d} [U_j]$. The rate of convergence here (in the sup norm) is geometric in i. Furthermore, for any measurable function G of $U_1, \ldots, U_k$ whose expectation exists, an ergodic theorem holds, namely

$$\lim_i i^{-1} \sum_{l=1}^{i} G(U_1^{(l)}, U_2^{(l)}, \ldots, U_k^{(l)}) \xrightarrow{a.s.} E(G(U_1, \ldots, U_k)).$$

As a result, Gibbs sampling through m replications of the aformentioned i iterations produces m i.i.d. k-tuples of the form $(U_{1r}^{(i)}, \ldots, U_{kr}^{(i)})$, $1 \leq r \leq m$, with the desired joint density. Gelfand and Smith (1990) recommend a density estimate for $[U_j]$, $1 \leq j \leq k$ having the form,

$$[\hat{U}_j]_i = m^{-1} \sum_{r=1}^{m} [U_j | U_t = U_{tr}^{(i)}; t \neq j].$$

The the Gibbs sample can be viewed from many directions, but one fruitful perspective is to consider it as an adaptation of the Metropolis-Hastings algorithm (Metropolis *et al.* (1953), and Hastings (1970)). The Gibbs sample seems to have been first formally developed by Geman and Geman (1984) in the context of image reconstruction, though as with most good ideas the roots of the Gibbs sampler can be traced to many suggestive sources.

In the statistical framework, Tanner and Wong (1987) used a similar technique in their substitution sampling approach to missing data problems. Gelfand and Smith (1990) showed the applicability of the Gibbs sampler to general parametric Bayesian computations and a variety of other conventional statistical problems. The Gibbs sampler approach, along with several other computer-intensive statistical methods, are reshaping many traditional methods in statistics.

Further reviews of uses of Monte Carlo Markov Chain (MCMC) methods for Bayesian computations and inference can be found in Besag and Green (1993), and in Smith and Roberts (1993). Tierney (1991) gives an outline of MCMC methods for exploring posterior distributions. Geyer (1991) explores the use of MCMC in likelihood based inference. A Gibbs sampler approach to generalized linear models with random effects is given in Zeger and Karim (1991). Hierarchical Bayesian analysis of changepoint problems are approached using the Gibbs sampler in Carlin, Gelfand and Smith (1992). The simulation tool of Gibbs sampling is employed in Albert and Chib (1993) to generate marginal posterior distributions for all parameters of interest in an AR model subject to Markov mean and variance shifts that is closely related to HMM's. This method is expedient since the conditional posterior distributions of the states given the parameters, and the parameters given the states, all have form amenable to Monte Carlo sampling. A variety of applications of the Gibbs sampler in Medicine are reviewed in Gilks *et al.* (1993).

An important question that has been recently given a considerable amount of attention in the literature is what is the "best" way to extract information from a Gibbs sampler sequence? More specifically, the two issues at steak are,

Convergence What is a "long enough" run of a Gibbs sampler?

Sampling How can we design an efficient Gibbs sequence, or sequences, sampling strategy?

As was the case in choosing a cooling scheme for the simulated annealing procedure, different ways of extracting information from a Gibbs sequence have been suggested, and it seem unlikely that there exist one optimal solution to this problem. An excellent bibliography on the Gibbs sampler, and the convergence rate and output sampling problem can be found in the special discussion paper of Smith and Roberts (1993).

CONCLUDING REMARKS

We have reviewed the theory and applications of wavelets, of hidden Markov models, of simulated annealing, and of the Gibbs sampler. This is almost a litany of the major steps in statistical science over the last ten years, so perhaps one has done about as well as possible just to get a taste of the possibilities for applications. Still, because the ideas behind these developments are fundamentally simple, perhaps also just enough detail has been given so that the central mathematical facts might be honestly understood.

The work reviewed here is far from done, and the best is surely years ahead of us. Pointers have been given throughout the review to many articles and books that develop our topics with much greater detail, though the most compelling work is not to be found in the books but rather in the marriage of the most basic parts of these technologies to problems of importance in science.

REFERENCES

Wavelets

Chui, C.K. (1992a), *An Introduction to Wavelets.* Academic Press, New York.

Chui, C.K.(1992b), *ed.*, *Wavelets: A Tutorial in Theory and Applications.* Academic Press, New York.

Coifman, R.R. and Meyer, Y. (1991), "Remarques sur l'analyse de Fourier á fenêtre," *C.R. Acad. Sci. Paris Sér. I*, 259-261.

Daubechies, I. (1988), "Orthonormal bases of compactly supported wavelets," *Comm. Pure Appl. Math.*, **41**, 909-996.

Daubechies, I., (1992) *Ten Lectures on Wavelets*, SIAM Publications, Philadelphia PA.

Duffin, R.J. and Schaeffer, A.C. (1952), "A class of nonharmonic Fourier series," *Trans. Amer. Math. Soc.*, **72**, 341-366.

Gröchenig, G.K. (1991), "Describing functions: atomic decompositions versus frames," *Monatsh. Math.*, **112**, 1-42.

Gröchenig, K. and Madych, W.R. (1992), "Multiresolution analysis, Haar bases and self-similar tilings of $\mathbb{R}^n$," *IEEE Trans. Inform. Theory*, **38**, 556-568.

Kovačević, J. and Vetterli, M. (1991), "Nonseparable multidimensional perfect reconstruction filter banks and wavelet bases for $\mathbb{R}^n$," *IEEE Trans. Inform. Theory*, **38**, 533-535.

Mallat, S. (1989), "Multiresolution approximation and wavelets," *Trans. Amer. Math. Soc.*, **315**, 69-88.

Meyer, Y. (1990), *Ondelettes ét opéateurs, I: Ondelettes, II: Opéateurs de Calderón-Zygmund, III: Opéateurs multilinéaires.* Hermann, Paris.

Meyer, Y. (1993), *Wavelets: Algorithms and Applications.* SIAM Publications, Philadelphia, PA.

Strang, G. (1993), " Wavelet Transforms versus Fourier Transforms", *Bulletin of the American Mathematical Society* **28**,2, 288-305.

Ruskai, M.B., Beylkin, G., Coifman, R.R., Daubechies, I., Mallat, S. Meyer, Y. and Raphael, L. (1992), *eds., Wavelets and their Applications.* Jones and Bartlett, Boston.

HMM

Aggoun, L. and Elliot, R.J. (1992), "Finite dimensional predictors for hidden Markov chains," *System Control Lett.*, **19**, 335-340.

Baum, L.E. and Eagon, J.A. (1967), "An inequality with applications to statistical estimation for probabilistic functions of Markov processes and to a model of Ecology," *Bull. Amer. Math. Soc.*, **73**, 360-363.

Baum, L.E. and Petrie, T. (1966), "Statistical inference for probabilistic functions of finite state Markov chains," *Ann. Math. Statist.*, **37**, 1554-1563.

Baum, L.E., Petrie, T., Soules, G. and Weiss, N. (1970), "A maximization technique occurring in the statistical analysis of probabilistic functions of Markov chains," *Ann. Math. Statist.*, **41**, 164-171.

Bickel, P.J. and Ritov, Y. (1993), "Inference in hidden Markov models I Local asymptotic normality in the stationary case," Technical Report, Department of Statistics, University of California, Berkeley.

Blackwell, D. and Koopmans, L. (1957), "On the identifiability problem for functions of finite Markov chains," *Ann. Math. Statist.*, **28**, 1011-1015.

Chung, S.H., Moore, J.B., Xia, L., Premkumar, L.S. and Gages, P.W. (1990), "Characterization of single channel currents using digital signal processing techniques based on hidden Markov models," *Phil. Trans. Roy. Soc. Lond. Ser. B*, **329**, 265-285.

Churchill, G.A. (1989), "Stochastic models for heterogeneous DNA sequences," *Bull. Math. Biol.*, **51**, 79-94.

Coast, D.A., Cano, G.G. and Briller, S.A. (1991), "Use of hidden Markov models for Electrocardiographic signal analysis," *J. of Electrocardiology*, **23**, 184-191.

Forney, G.D. (1973), "The Viterbi algorithm," *Proc. IEEE*, **61**, 268-278.

Fredkin, D.R. and Rice, J.A. (1992), "Bayesian restoration of single channel patch clamp recordings," *Biometrics*, **48**, 427-448.

Fridman, M. (1993), "Hidden Markov model regression," *Unpublished Thesis*, University of Pennsylvania.

Gilbert, E.J. (1959), "On the identifiability problem for functions of finite Markov chains," *Ann. Math. Statist.*, **30**, 688-697.

Guttorp, P., Newton, M.A. and Abkowitz, J.L. (1990), "A stochastic model for haematopoiesis in cats," *IMA J. Math. Appl. Med. Biol.*, **7**, 125-143.

Juang, B.H. (1985), "Maximum Likelihood estimation for mixture multivariate stochastic observations of Markov chains," *AT & T Tech. J.*, **64**, 1235-1249.

Juang, B.H., Levinson, S.E. and Sondhi, M.M. (1986), "Maximum likelihood estimation for multivariate mixture observations of Markov chains," *IEEE Trans. Inform. Theory*, **IT-32**, 307-309.

Juang, B.H. and Rabiner, L.R. (1990), "The Segmental K-Means algorithm for estimating parameters of hidden Markov models," *IEEE Trans. Acoust., Speech, Signal Process*, **ASSP-38**, 1639-1641.

Juang, B.H. and Rabiner, L.R. (1991), "Hidden Markov models for speech recognition," *Technometrics*, **33**, 251-272.

Khasminskii, R.Z., Lazareva, B.V. and Stapleton, J. (1993), "Some procedures for state estimation of a hidden Markov chain with two states," Technical Report, Department of Statistics and Probability, Michigan State University.

Kogan, J.A. (1991), "Optional reconstruction of Markov sequences through indirect observations," *6th USSR-Japan Symp. on Probab. and Math. Statist.*, Kiev.

Kullback, S. and Leibler, R.A. (1951), "On information and sufficiency," *Ann. Math. Statist.*, **22**, 79-86.

Leroux, B.G. (1992), "Maximum-likelihood estimation for hidden Markov models," *Stochastic Process. Appl.*, **40**, 127-143.

Levinson, S.E. (1986), "Continuously variable duration hidden Markov models for automatic speech recognition," *Computer, Speech and Language*, **1**, 29-45.

Liporace, L.A. (1982), "Maximum Likelihood estimation for multivariate observations of Markov sources," *IEEE Trans. Inform. Theory*, **IT-28**, 729-734.

Paul, D.B. (1985), "Training of HMM recognizers by simulated annealing," *Proc. ICASSP*, New York: IEEE, 13-16.

Petrie, T. (1969), "Probabilistic functions of finite state Markov chains," *Ann. Math. Statist.*, **40**, 97-115.

Quandt, R.E. (1972), "A new approach to estimating switching regressions," *J. Amer. Statist. Assoc.*, **67**, 306-310.

Quandt, R.E. and Ramsey, J.B. (1978), "Estimating mixtures of Normal distributions and switching regressions," *J. Amer. Statist. Assoc.*, **73**, 730-738.

Rabiner, L.R. (1989), "A tutorial on hidden Markov models and selected applications in speech recognition," *Proc. IEEE*, **77**, 257-285.

Viterbi, A.D. (1967), "Error bounds for convolutional codes and an asymptotically optimal decoding algorithm," *IEEE Trans. Inform. Theory*, **IT-13**, 260-269.

Markov Renaissance

Albert, J.H. and Chib, S. (1993), "Bayes inference via Gibbs sampling of autoregressive time series subject to Markov mean and variance shifts," *J. Bus. & Econ. Statist.*, **11**, 1-16.

Besag, J. (1974), "Spatial interaction and the statistical analysis of lattice systems (with discussion)," *J. Roy. Statist. Soc. Ser. B*, **36**, 192-236.

Besag, J. and Green, P.J. (1993), "Spatial statistics and Bayesian computations," *J. Roy. Statist. Soc. Ser. B*, **55**, 25-37.

Cerny, V. (1985), "A thermodynamic approach to the traveling salesman problem: An efficient simulation", *J. Optim. Theory Appl.,* **45**, 41-51.

Gelfand, A.E. and Smith, A.F.M. (1990), "Sampling-based approaches to calculating marginal densities," *J. Amer. Statist. Assoc.,* **85**, 398-409.

Geman, S. and Geman, D. (1984), "Stochastic relaxation, Gibbs distributions, and the Bayesian restoration of images," *IEEE Trans. Pattn Anal. Mach. Intell.*, **6**, 721-741.

Geyer, C.J. (1991), "Markov chain Monte Carlo maximum likelihood," *Computer Science and Statistics: Proc. 23rd Symp. Interface*, Fairfax Station: Interface Foundation, 156-163.

Gilks, W.R., Clayton, D.G., Spiegelhalter, D.J., Best, N.G., McNeil, A.J., Sharples, L.D. and Kirby, A.J. (1993), "Modelling complexity: Applications of Gibbs sampling in medicine," *J. Roy. Statist. Soc. Ser. B*, **55**, 39-52.

Johnson, D.S., Aragon, C., McGeoch, L. and Schevon, C. (1990), "Optimization by simulated annealing: An experimental evaluation, Part I: Graph partitioning," *Oper. Res.*, **37**, 865-892.

Johnson, D.S., Aragon, C., McGeoch, L. and Schevon, C. (1991), "Optimization by simulated annealing: An experimental evaluation, Part II: Graph coloring and number partitioning," *Oper. Res.*, **39**, 378-406.

Johnson, D.S., Aragon, C., McGeoch, L. and Schevon, C. (1992), "Optimization by simulated annealing: An experimental evaluation, Part III: The traveling salesman problem," in preparation.

Hajek, B. (1988), "Cooling schedules for optimal annealing," *Math. Oper. Res.*, **13**, 311-329.

Hastings, W.K. (1970), "Monte Carlo sampling methods using Markov chains and their applications," *Biometrika*, **57**, 97-109.

Kirkpatrick, S., Gelett, C.D. and Vecchi, M.P (1983), "Optimization by simulated annealing," *Science*, **220**, 621-630.

Metropolis, N., Rosenbluth, A.W., Rosenbluth, M.N., Teller, A.H. and Teller, E. (1953), "Equation of state calculation by fast computing machines," *J. Chem. Phys.*, **21**, 1087-1092.

Smith, A.F.M. and Roberts, G.O. (1993), "Bayesian computation via the Gibbs sampler and related Markov chain Monte Carlo methods (with discussion)," *J. Roy. Statist. Soc. Ser. B*, **55**, 3-23.

Tanner, M.A and Wong, W. (1987), "The calculation of posterior distributions by data augmentation (with discussion)," *J. Amer. Statist. Assoc.*, **82**, 528-550.

Tierney, L. (1991), "Exploring posterior distributions using Markov chains," *Computer Science and Statistics: Proc. 23rd Symp. Interface*, Fairfax Station: Interface Foundation, 563-570.

Zeger, S. and Rizaul Karim, M. (1991), Generalized linear models with random effects: A Gibbs sampling approach," *J. Amer. Statist. Assoc.*, **86**, 79-86.

MONTE CARLO APPROACH IN ASSESSING DAMAGE IN HIGHER ORDER STRUCTURES OF DNA

Aloke Chatterjee, James B. Schmidt, and William R. Holley

Division of Life Sciences, Lawrence Berkeley Laboratory
University of California, Berkeley
1 Cyclotron Road, Berkeley, California 94720

ABSTRACT

We have developed a computer model of nuclear DNA in the form of chromatin fiber. The fibers are modeled as an ideal solenoid consisting of twenty helical turns with six nucleosomes per turn. The chromatin model, in combination with our Monte Carlo theory of radiation damage induced by charged particles, based on general features of track structure and stopping power theory, has been used to evaluate the influence of DNA structure on initial damage. An interesting feature has emerged from our calculations. Our calculated results predict the existence of strong spatial correlations in damage sites associated with the symmetries in the solenoidal model. We have calculated spectra of short fragments of double stranded DNA produced by multiple double strand breaks induced by both high and low LET radiation. The spectra exhibit peaks at multiples of ~85 base pairs (the nucleosome periodicity), and ~1000 base pairs (solenoid periodicity). Preliminary experiments to investigate the fragment distributions from irradiated DNA, made by B. Rydberg at Lawrence Berkeley Laboratory, confirm the existence of short DNA fragments and are in substantial agreement with the predictions of our theory.

INTRODUCTION

We are currently in the process of developing a general theory of radiation-induced DNA damage. The theoretical model uses Monte Carlo techniques to calculate DNA strand break production by both direct and indirect effects of ionizing radiation, and is based on general features of track structure, stopping power theory and the detailed geometry of DNA structures (1–3). We have recently incorporated into our calculations a computer model of 30nm chromatin fiber (4). The chromatin model takes the form of an idealized solenoid consisting of twenty turns of a right handed helix with 6 nucleosomes per turn. Each nucleosome-linker unit contains 171 base pairs and the whole fiber contains more than 20,000 base pairs. The computer codes implementing the computational scheme outlined above provide detailed pictures of the extent of damage induced by each randomly generated charged particle which passes through or near the DNA fiber. Figures 1 and 2 are examples of typical events generated by the Monte Carlo program STBRSOL.

In this paper we call attention to a general prediction based on the results of these theoretical calculations which arises from spatial correlations in radiation damage related to DNA structure. Our results predict that substantial numbers of short fragments of DNA (20 bp to 20 kbp), which are normally not detectable in standard double strand break assays, are produced by irradiation of eucaryotic DNA with ionizing radiation.

SOLENOID MODEL

DNA within the living cell is generally accepted as the principal target upon which ionizing radiation acts to produce biologically significant effects. Many of the consequences of exposure to ionizing radiation, cell survival, mutation, transformation, to name a few, correlate well with DNA damage induction. DNA double strand breaks correlate particularly well with a number of radio-biological effects. Our previous theoretical model for the interaction of charged particles with DNA in solution utilized a simple particle track model and a detailed model of B-form DNA in a "cell-like" environment (1–3). Damage due to both the direct deposition of energy on the DNA by the primary and secondary electrons and the indirect damage due to chemical attack by diffusing radical species was included. Since it is likely that the cell's response to radiation damage is dependent on the spatial distribution of damage along the DNA molecule, it is important to understand the spatial correlation between particle track and DNA goemetry. This necessitates a thorough understanding of the way in which DNA is packaged into chromatin and distributed throughout the nucleus. In this paper we extend our previous work to particulate irradiations

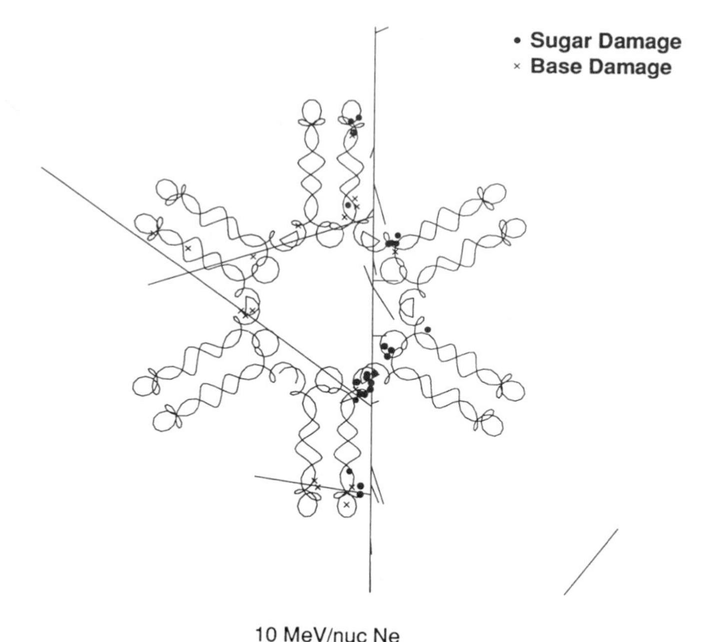

Figure 1. Typical interaction pattern of 10 MeV/u neon track with 30 nm chromatin fiber as calculated by Monte-Carlo program STBRSOL. Solid circles indicate ionization (direct) or radical attack (indirect) on sugar moiety and x indicates base damage.

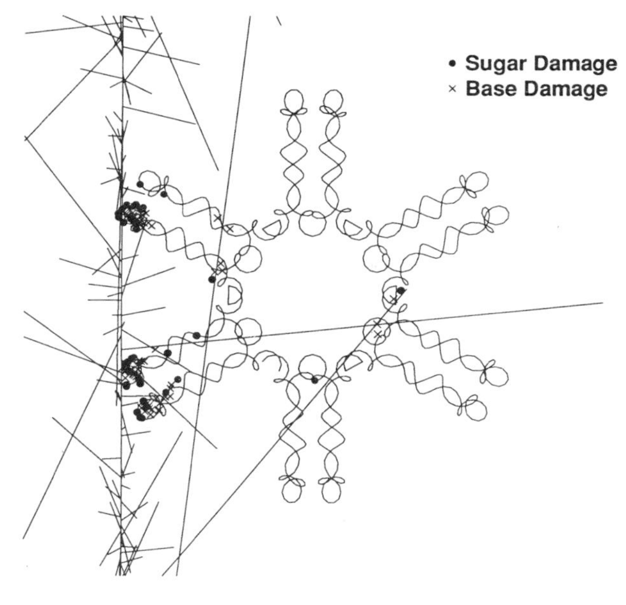

Figure 2. Typical interaction pattern of 10 MeV/u iron track. Note the presence of at least three distinct regions of dense multiple sugar and base damage sites on the DNA solenoid. This event probably produced two short fragments.

of DNA packaged in a 30 nm diameter solenoidal chromatin fiber. Both direct and indirect effects due to the primary and low energy secondary electrons (i.e., electrons with energies up to 2 keV) are treated in a single Monte Carlo calculation. Damage yields due to high energy delta rays (i.e., >2 ke V) are folded into the Monte Carlo results by finding numerical solutions to a system of integro-differential equations.

Damage to nuclear chromatin by single charged particle irradiation has been treated by combining a detailed model for the 30 nm chromatin fiber with a model for the interaction of charged particles in aqueous solution. The model was encoded in a FORTRAN program, which we named STBRSOL for "strand breakage in solenoidal DNA." Methods used by Chatterjee and Holley (5) to model the interaction of charged particles with linear DNA and nucleosomal DNA have been extended to the irradiation of the 30 nm chromatin fiber (4). In addition, improvements have been made to the charged particle track model. Previously the interactions of the primary particle and the secondary electrons were treated separately. The track model presented here includes the interactions of the primary along with all secondary electrons with energies up to 2000 eV. The model has also been improved so that the density of OH radicals in a delta ray closely approximates that of the ellipsoidal delta ray model proposed by Magee and Chatterjee (6). Lastly the model has been modified to account for the changing *dE/dx* along the length of the low energy delta ray in both the direct effect and indirect effect calculations.

A short segment of a 30 nm diameter chromatin fiber has been modeled as a right handed "super-solenoid-like" structure with 20 turns, 11 nm per turn for a total length of 220 nm. One turn of the super-solenoid contains six approximately cylindrical nucleosomes each with a diameter of 10.5 nm and a length of 7.073 nm. A gap of 0.5 nm exists between adjacent turns of the solenoid. The nucleosomes are oriented so that their central axes are parallel to the helical vector of the super-solenoid giving approximately a 22 degree tilt to the nucleosomes. The centers of the nucleosomes are located 9.79 nm from the axis of the super-solenoid to give the required 30 nm outer diameter. The 9 nm diameter interior of the solenoid is devoid of DNA. Nucleosomes consist of a 171 base pair sequence of B-form DNA wound in a left handed helix to give 2 turns per nucleosome. The nucleosome is often defined to be comprised of a core particle consisting of the histone octomere and the DNA which is most tightly bound to it, along with linker DNA flanking both sides of the core particle. The average number of base pairs in the core particle is typically 146. Hence in our model the linker DNA consists of the 25 base pairs situated midway between two adjacent nucleosome core particles, and is located almost entirely in the interior of the chromatin fiber. For simplicity the same randomly derived DNA sequence having a GC to AT ratio of 0.86 has been used for all of the nucleosomes. Spatial coordinates based on X-ray diffraction data (7) were obtained for linear DNA then bent to conform to the desired nucleosomal helix. In this model DNA is represented by a collection of points with Cartesian coordinates which correspond to the position of each sugar and base moiety. In total just under 20.7 kilobase pairs have been represented in the chromatin fiber model.

TRACK STRUCTURE

Given the model for the 30 nm super-solenoidal chromatin fiber the problem becomes one of simulating both damage due to the direct deposition of energy by charged particles on the fiber and damage due to attack by radicals formed in the surrounding solution upon irradiation. In describing the energy deposition events of accelerated charged particles it is often useful to make a distinction between glancing collisions, events in which the projectile particle interacts weakly with the electromagnetic field of the target molecule's electrons, and knock-on collisions, events in which the projectile interacts strongly with a particular bound electron in the target molecule. The use of Bohr's classical adiabatic criterion leads to a radius about the primary trajectory within which all glancing collisions occur with a random distribution (1). This radius, the glancing radius or core radius, is given by,

$$r_g = \beta c \left(\frac{4\pi n_e e^2}{m_e} \right)^{-1/2} \tag{1}$$

where b is the velocity of the incident particle relative to the speed of light, n_e is the number density of valence electrons, m_e is the mass of the electron, and e is the electronic charge. The core depends only on the velocity of the incident particle and not on the particle's charge. Since nearly all of the oscillator strength for glancing collisions in water is included in energy deposition events involving 100 eV or less (8), 100 eV has been selected as the partition between glancing and total collisions. It has previously been shown that each type of process accounts for approximately half of the energy lost by the incident charged particle to the medium. This approximation is good for heavy charged particles with incident energy in the range of 1 MeV/u to 1 GeV/u (4). For electrons in the energy range 100 eV to 10 MeV the ratio of glancing to knock-on energy loss is approximately 0.6 (6). We have modeled particulate irradiations by dividing the particle track into three parts: primary glancing interactions, low energy delta rays ($\leq$ 2 keV), and high energy delta rays (> 2 keV). The glancing collision calculation takes into account both the direct deposition of energy onto the chromatin by the primary particle and indirect damage resulting from attack by radicals produced in the surrounding solution by glancing collisions between the primary particle and water molecules. Glancing collisions are assumed to occur in a cylindrical volume with radius equal to the glancing radius (Equation 1). Low energy delta rays are defined to be those secondary electrons with initial energies of 2000 eV or less. This energy limit corresponds to a delta ray with a root mean square (RMS) penetration depth approximately equal to one half the length of the chromatin fiber model and was used in the model to distinguish between delta rays which had some significant probability of depositing energy close (i.e., on the order of the length of the chromatin fiber) to primary glancing collisions.

Together the primary particle glancing collisions and the low energy delta ray interactions comprise the portion of the modeled irradiation which is simulated explicitly using Monte Carlo techniques. The contribution to DNA damage yields due to the high energy delta rays is included using a combination of Monte Carlo simulation and numerical integration techniques (6,9).

INDIRECT EFFECT

In assessing the damage to the chromatin fiber by the indirect effect one must take into consideration the reactive chemical species which are formed in the aqueous environment of the nucleus following irradiation, their movement through diffusive processes, and their interaction with each other, the chromatin, and other molecules which are present in the nucleus. The four major reactive species formed from the excitation and ionization of water molecules are H•, OH•, e_{aq}^-, and H_3O^+. Most sugar moiety indirect effect damage in DNA results from interactions with OH radicals, primarily through H-abstraction, and leads to a scission in the DNA sugar-phosphate backbone (10). The focus of this research is on DNA strand breakage hence the calculation of the indirect effect on the DNA was restricted to OH radical attack. However all of the major water radiolysis products were taken into consideration in radical-radical recombination reactions.

The first step in modeling the attack of radicals involves the creation and positioning of the radicals within the glancing radius and the delta rays. Once this has been accomplished the diffusion of the radicals and their interaction with the DNA can be simulated. The stochastic nature of radical production has been modeled by randomly placing spurs of radicals within the cylinder defined by the glancing collision radius. The mean amount of energy required to produce an OH radical is 17 eV (11); therefore, up to six OH radicals can result from a single glancing collision. Given the frequency distribution for the number of OH radicals in a spur one can calculate the expected number of radicals per collision. The mean energy required to form a spur is just the product of the mean number of OH radicals in a spur and the mean energy per OH radical formed:

$$\langle \mathbf{E}_{sp} \rangle = \langle n_{OH} \rangle \cdot \langle \mathbf{E}_{OH} \rangle \approx \langle n_{OH} \rangle \cdot 17eV \, . \tag{2}$$

The projected mean free path along the primary trajectory between spur formation sites is given by the ratio of the mean energy per spur to the rate of energy loss by the primary through glancing collisions:

$$\langle d_{sp} \rangle = \frac{2\langle \mathbf{E}_{sp} \rangle}{dE/dx} \, . \tag{3}$$

A spur "center" is assumed to occur along the particle trajectory by a Poisson process, hence the projected distance between spurs is a random variable with an exponential distribution of mean $\langle d_{sp} \rangle$. This distribution is utilized to randomly position spurs along the particle trajectory. Finally the positions of each OH radical relative to the spur centers are determined by generating three dimensional Gaussian random deviates. For a spur of average size (i.e., approximately 3 OH radicals) the standard deviation is 2 nm, representing one half the expected diameter. OH radicals are prohibited from being created within the space "occupied" by the DNA or the histones.

The placement of OH radicals within the delta rays is accomplished through an algorithm analogous to that used for the primary glancing collisions.

The second phase in modeling the indirect effect due to OH radical attack involves the simulation of OH• migration and interaction with the DNA. Diffusion of an OH radical has been modeled by producing randomly oriented discrete steps. The mean jump size depends on the distance between the radical and the chromatin. This allows one to expend a minimum amount of computational time when the radical is far from the chromatin without leading to scoring errors by making big steps and jumping through DNA targets when close to the chromatin. The diffusion simulation has been described in detail previously (12). Each OH radical which has been generated is allowed to diffuse until it meets one of four possible fates: (a) it reacts with another radical (e.g., H•, e^-_{aq}, or another OH•), (b) it is scavenged by a molecule which is present in the nucleus (e.g., proteins, RNA, DNA other than the target DNA in the calculation, and other molecules present in the nucleus), (c) it diffuses into a region which is defined to be occupied by chromatin protein, or (d) it reacts with a sugar or base site on the DNA.

Once a new position for a given OH radical has been calculated a determination is made as to whether or not the OH• reacts with another radical or a scavenger molecule during the time required to make the step. Either type of reaction results in its removal from the simulation. The determination is based on a composite OH survival curve which gives the probability of the radical surviving such encounters from it's creation to some later time. The method of calculating the survival curve for OH radicals reacting with other radicals has been described previously (2,13). The recombination reactions between water radicals formed by the primary glancing collisions and in the delta rays are treated independently in the calculations. The reaction of OH radicals with scavenger molecules is assumed to be a random process corresponding to a time constant for OH loss of 6.67×10^{-10} s which is consistent with experiment (14).

Once a determination has been made that an OH radical has indeed survived a particular time step the coordinates of the radical are updated to it's new position. The new position of the radical is used to determine whether or not the radical reacts with either the proteins of the nucleosome or solenoid interiors, or any of the sugar or base sites on the DNA. If the radical jumps into a region occupied by protein a reaction is assumed to occur which removes the radical from consideration. A surviving radical is then tested to determine whether or not it reacts with the DNA. The occurrence of a reaction between a radical and a specific DNA target is based on a scheme developed by Smoluchowski (15).

If an OH radical steps within a critical sphere defined by the Smoluchowski radius for each site on the DNA, a reaction is said to have occurred and a damage event is registered for the indicated site. The Smoluchowski radii for the DNA moieties have been calculated to be 0.36, 0.36, 0.51, 0.36, and 0.1 nm for the bases A, T, G, and C, and ribose respectively. The simulated diffusion process is iterated repeatedly until the terminal fate of each OH radical is determined.

DIRECT EFFECT

Given the positions of the primary trajectory and it's delta rays the direct effect calculation proceeds by identifying those DNA sites on the chromatin fiber which are close enough to be targets of direct energy deposition events. Only those sugar and base moieties whose positions are located either within the primary glancing radius or the delta ray volumes are considered to be potential targets. We utilize Bragg's rule for the partitioning of energy loss between individual atoms or molecules to obtain the mean energy deposited on sugar and base groups located within the track volume. The mean energy deposited through glancing collisions between a projectile particle and a collection of atoms, A, is given by

$$\langle E_{g,A} \rangle = \frac{2e^4 z^2}{m_e v^2 r_g^{\ 2}} \cdot \sum_{i \in A} n_i Z_i \ln\left[\frac{2mv^2}{I_i}\right] \tag{4}$$

where n_i is the number of the ith atomic species in moiety A, Z_i is the atomic number of the ith species, and I_i is the mean ionization potential of the ith atomic species. Clearly the mean energy deposited on a moiety is dependent on the projectile's charge and velocity. The mean energy deposited per glancing collision, $<E_g>$, is obtained by integrating the product of the energy and the energy loss function for charged particles over the entire range of possible energy transfers in a glancing collision. This integration calculated numerically yields a value of 29.9 eV for $<E_g>$ (1). The mean number of glancing collisions with a specified collection of atoms is just

$$\mu_{g,A} = \frac{\langle \mathbf{E}_{g,A} \rangle}{\langle E_g \rangle} \tag{5}$$

where $\langle \mathbf{E}_{g,A} \rangle$ is the mean energy deposited on moiety A. In order to model the stochastic nature of the direct energy depositions, the assumption is made that the number of energy depositions occurring on any given group of atoms within the track volume is a random variable possessing a Poisson distribution with mean $m_{g,A}$. In the model we assume that one energy deposition event on a moiety is sufficient to induce damage. Therefore the probability that a moiety is damaged is simply

$$P_{d,A} = 1 - e^{-\mu_{g,A}}. \tag{6}$$

Random numbers uniformly distributed between 0 and 1 are generated for each moiety within the track volume and if the number is less than $P_{d,A}$, a damage event is registered; otherwise, the moiety remains undamaged. For the sugar moiety one damage event is assumed to be both necessary and sufficient to produce a single strand break in the DNA molecule. Finally, double strand breaks are scored when single strand breaks occur on opposite strands within 10 basepairs of each other.

RESULTS

STBRSOL was developed and implemented on Sun Microsystems Sparcstation computers. Results were obtained for a range of atomic numbers from protons to iron ions. The energies of the incident particles ranged from 1 to 1000 MeV/u. The number of simulated tracks which were generated for a particular particle energy and z varied between 1.4×10^3 and 2.5×10^5.

Figures 1 and 2 depict typical interactions of charged particles with the short chromatin fiber model. In Figure 1 a 10 MeV/u Ne particle passes through the chromatin near the solenoid axis and in Figure 2 a 10 Me/u Fe particle passes through the chromatin approximately 15 nm from the axis. In each case the primary's trajectory intersects the chromatin at more than one location. Secondary electrons have also been produced, some of which intersect the chromatin. Direct and indirect damage has been incurred from both the primary particle and the secondary electrons. In addition, the high energy density of the Fe track is seen to have resulted in the production of multiple damage sites in close proximity.

As suggested by the individual track interaction diagrams, a charged particle, especially at high LET, can produce damage at more than one site. Multiple double strand breaks, in particular, will generate short fragments of DNA. The predicted distributions of short fragments for a selection of low energy beams are shown in Figure 3. The x-axis covers a range up to 1500 bp. Two features are of interest. The first is the general prediction of production of significant numbers of fragments. Second is the existence of structure in the distributions. The shapes of the fragment spectra are of course related to symmetries in the assumed structure in the model of DNA. The peak at about 85 bp represents the revolution period about the nucleosome. The broader peak at about 1000 bp corresponds to one turn about the solenoid. One can even see fine structure in the 1 kbp peak which corresponds to nucleosomal structure convoluted with the solenoidal structure. Fragments are also produced by low LET beams such as x-rays and high energy low z particles but our results currently have poor statistics.

Figure 4 is a plot of the fragment spectrum for 200 MeV/u iron which shows that longer fragments are also produced. In addition to the peaks at multiples of 85 bp and at 1000 bp, there is perhaps a suggestion of another small peak at about 2 kbp, corresponding to 2 full turns about the solenoid.

We have integrated these distributions to obtain normalized yields of short fragments. Figure 5 is the fraction of the total genomic DNA produced in short fragments per Gray of radiation. The solid lines are our calculated results for He, Ne and Fe for a range of LETs, and our predicted x-ray point is the open circle. Bjorn Rydberg (16), at Lawrece Berkeley Laboratory, using a combination of pulse field and ordinary gel electrophoresis, has succeeded in detecting and measuring these yields. His preliminary results for three iron energies are shown here as the solid diamond points, his Ne result is the filled square, and the solid circle represents his experimental x-ray point. Agreement with the experimental

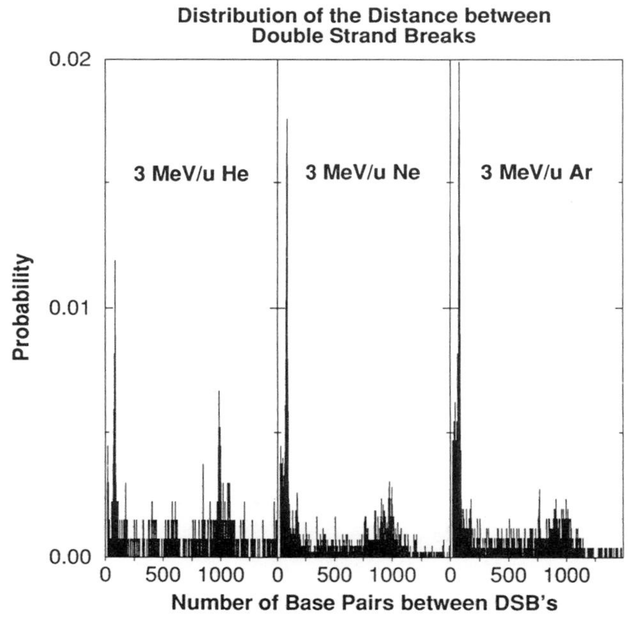

Figure 3. Probability distributions for production of short DNA fragments in the range 20–1500 bp for three different low energy beams. Note peaks at multiples of ~85 bp (nucleosome period) and ~1000 bp (solenoid period).

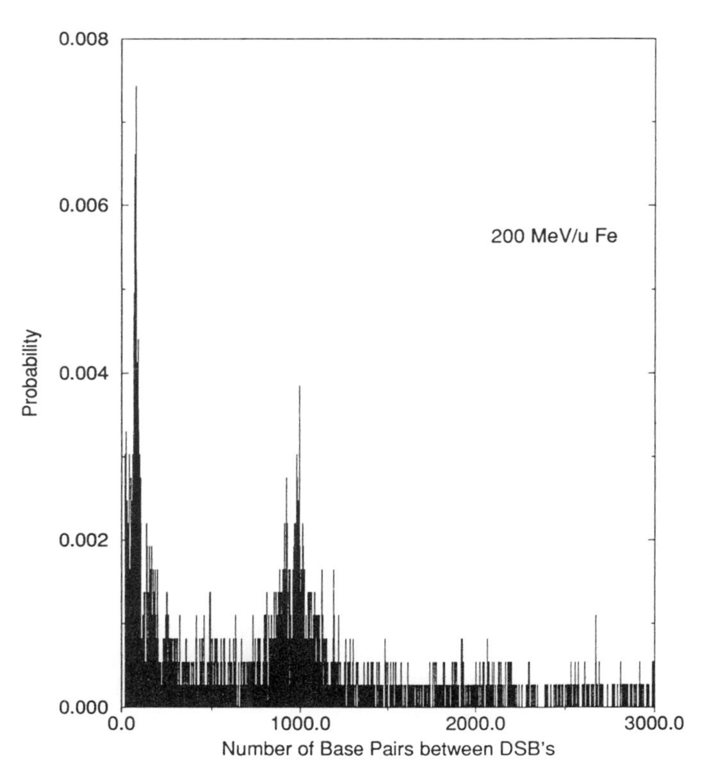

Figure 4. Probability distribution for production of short DNA fragments (in range 20–3000 bp) for 200 MeV/u iron beam. In addition to 85 and 1000 bp peaks there is a possible enhancement in the probability at ~2000 bp corresponding to 2 solenoid turns.

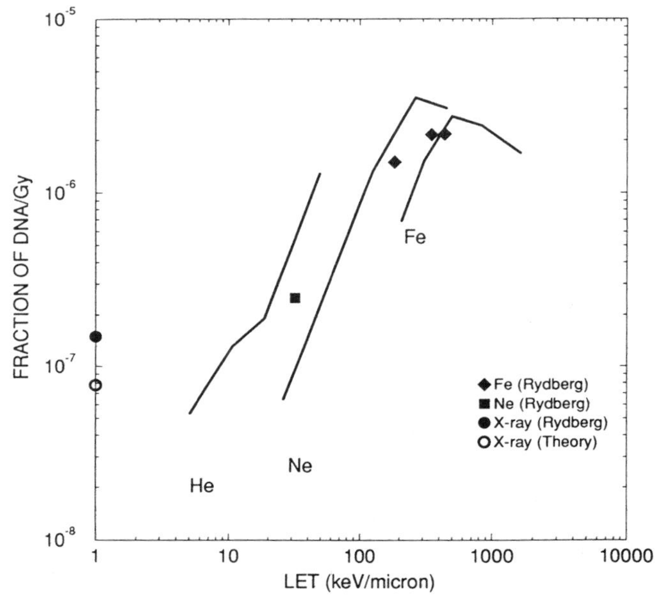

Figure 5. Fraction of total DNA in a cell produced in small fragments (≤4kbp) by one Gray of various qualities of radiation plotted as a function of LET. Solid curves are theoretical predictions for helium, neon and iron beams. The open circle is the theoretical prediction for low LET (x-ray) radiation. Solid points are preliminary experimental results of B. Rydberg.

data to within a factor of two is very encouraging. To give a feeling for the frequency of occurrence of short fragments, our results indicate that at relatively low LET (x-rays and high energy Neon) a few percent of the conventionally measured double strand breaks have an accompanying short fragment while at the higher LET's, on average, there are comparable numbers of large and small fragments.

DISCUSSION

What are possible implications of the existence of substantial numbers of short fragments in the range 20 bp to a few kbp induced by radiation? It is possible that the presence of short fragments will complicate or interfere with DNA repair processes. Short fragments may likely be excluded or left out when DSB's are rejoined leading to enhanced rates of mutations and short deletions, and may also affect the production of large deletions and translocations as well.

The effects described here (i.e., production of short fragments of DNA) result from symmetries or near symmetries in DNA structure. Our current calculations have been made for ideal, perfectly repeated, structures. Some details in the calculated spectra, such as the 85 bp fine structure superimposed on the 1000 bp peak evident in Fig. 3 will probably be smoothed out by local variations from fiber to fiber when averaged over all the DNA in a typical eucaryotic cell. Careful comparison of theory and experiment may eventually provide information on the extent to which interphase DNA exists in solenoid fiber form and the range of variations in structure.

The existence of short DNA fragments implies frequent production of relatively closely spaced double strand breaks by ionizing radiation, especially high LET radiation. Preliminary experimental results suggest that short fragments (kbp's) are not rejoined to the main body of DNA as efficiently as large fragments (Mbp's) leading to enchanced production of deletions and translocations.

At the very least it is clear that the presence of radiation induced short DNA fragments not detectable in standard assays of double strand break formation, as well as the possible existence of intermediate length fragments (10's of kbp – 100's of kbp) associated with the very poorly understood higher order organization of interphase DNA must be taken into account in interpreting measurements of the relative yield of double strand breaks at high and low LET.

ACKNOWLEDGMENT

Supported by the Office of Health and Environmental Research (OHER), U.S. Department of Energy, under Contract No. DE-AC03-76SF00098 and NASA Specialized Center of Research and Training (NSCORT) in radiation health under Contract Order W-18265.

REFERENCES

1. W. R. Holley, A. Chatterjee, and J. L. Magee, Production of DNA strand breaks by direct effects of heavy charged particles, *Radiation Research*, **121**, 161–168 (1990).
2. A. Chatterjee and W. R. Holley, A general theory of DNA strand break production by direct and indirect effects, *Radiation Protection Dosimetry*, **31**, 241–247 (1990).
3. A. Chatterjee and W.R. Holley, Biochemical Mechanisms and Clusters of Damage for High-LET Radiation,*Adv. Space. Res.* 16, (2) 33-(2) 33-(2)43 (1992).

4. J. B. Schmidt, *Heavy Ion Induced Lesions in DNA: A Theoretical Model for the Initial Induction of DNA Strand Breaks and Chromatin Breaks*, Ph. D. Thesis, University of California, Berkeley (1993).
5. A. Chatterjee and W. R. Holley, Early chemical events and initial DNA damage, in *Physical and Chemical Mechanisms in Molecular Radiation Biology*, eds. W. A. Glass and M. N. Varma, Plenum Press, New York, 1991.
6. J. L. Magee and A. Chatterjee, Theory of the chemical effects of high-energy electrons, *J. Phys. Chem.*, **82**, 2219–2226 (1978).
7. S. Arnott and D. W. L. Hukins, Optimized parameters for A-DNA and B-DNA, *Biochem. Biophys. Res. Commun.*, **47**, 1504–1509 (1972).
8. A. Mozumder, Charged particle tracks and their structure, in *Advances in Radiation Chemistry*, eds. M. Burton and J. L. Magee, Vol. I, pp. 1-102, Wiley-Interscience, New York, 1969.
9. A. Chatterjee and W.R. Holley, Energetic electron tracks and DNA strand breaks, *Nucl. Tracks Radiat. Meas.* 16, #2/3, 127–133 (1989).
10. C. Von Sonntag, U. Hagen, A. Schon-Bopp, and D. Schulte-Frohlinde, Radiation-induced strand breaks in DNA: chemical and enzymatic analysis of end groups and mechanistic aspects, *Adv. Radiat. Biol.* **9**, 109–142 (1981).
11. A. Chatterjee and J. L. Magee, Radiation chemistry of heavy-particle tracks. 2. Fricke dosimeter system, *J. Phys. Chem.*, **84**, 3537–3543 (1980).
12. A. Chatterjee and W.R. Holley, Energy deposition mechanisms and biochemical aspects of DNA strand breaks by ionizing radiation, *Int. J. of Quant. Chem.,* Vol. XXXIX, 709–727 (1991)
13. A. Chatterjee, J. L. Magee, and S. K. Dey, The role of homogeneous reactions in the radiolysis of water, *Radiation Research*, **96**, 1–19 (1983).
14. F. Hutchinson, Chemical changes induced in DNA by ionizing radiation, *Prog. Nucleic Acid Res. Mol. Biol.*, **32**, 115–154 (1985).
15. M. V. Smoluchowski, Drei Vortrage uber diffusion, brownsche molekular-bewegung und koagulation von kolloidteilchen, *Physik Zeitschr.*, **17**, 557 (1916).
16. B. Rydberg, Clusters of DNA damage induced by ionizing radiation: formation of kb-sized DNA fragments, submitted to *Int. J. of Radiat. Biol.*

DISCUSSION

Wilson: In your solenoidal model between the nucleosomes, there is linker DNA. Do you assume, as you go around the solenoid, that successive nucleosomes are adjacent beads or are they opposite?

Chatterjee: Adjacent.

Miller: This is not the cross linker model. This is a normal solenoid.

Holley: This is a normal solenoid, and this particular model that we've used has minimal linkers. Basically two full turns around the histone core and then it connects onto the next. We're thinking about ways of incorporating additional linker.

Miller: The H1 has long tails on it. Where do they go in your structure?

Chatterjee: They get all folded in the center.

Miller: Each one is completely in the center.

Holley: We don't model those explicitly.

Chatterjee: There's no room. If you use 30 nanometers, you know, you keep that distance constant. You don't have any other room except for the center.

Zaider: Is there any evidence that histones scavenge hydroxyl radicals?

Chatterjee: Yes.

Zaider: Why should they?

Chatterjee: Because OH radicals can abstract ·H from constituent molecules of the histones and becomes water just like from the sugar sites. Histones also have sites where H can be extracted from.

Holley: John Ward has some information on that. I don't remember the details but at the Dallas radiation research meeting he discussed measurements which indicated that histones have something like 60 percent of the scavenging capacity of the DNA itself.

Chatterjee: If you just compare the beads on a string data (i.e., with histones) and the linear DNA data, you will find that half the protection comes because of the presence of histones.

Holley: If the histones just block the OH radicals then those OH radicals are still available. If it scavenges them, then they're not.

Chatterjee: Radicals don't live too long.

Turner: Do you keep track of the individual atomic coordinates?

Chatterjee: Yes.

Terrisol: What is the difference between step 2 and step 4?

Chatterjee: Step 2 involves reactions between water radicals only i.e., the decay of ·OH is due to the reactions with ·H, another ·OH, and so on. In step 4, the surviving ·OH reacts with DNA sites.

Terrisol: Normally they are together.

Chatterjee: We do that for two reasons. Initially when the radicals are formed, they interact with themselves first before they see the DNA. Of course that depends on the DNA concentration.

Paretzke: Absolutely not true.

Chatterjee: Why not?

Paretzke: Because if you have the overall interaction probability of OH radical with the DNA is essentially confined to 20 angstroms or so. The probability to have another OH radical within 20 angstroms is not too high.

Chatterjee: Depends upon the density of ·OH radicals. If the density is very high then the radical reaction will dominate.

Paretzke: Sure. It can be, but there is a competition and you cannot treat them in two different stages.

Chatterjee: For Low LET, what you are saying could be true. The effectiveness of OH with high LET goes down simply because of the sibling radical reactions and that cannot happen if what you are saying is true.

Paretzke: You have both effects and you cannot split them apart in a computation.

Chatterjee: For a 40 MeV electron calculation, the initial (~10^{-12} sec.) G value of OH is 5.88. And then you have put this 500 millimolar scavenger in the system with or without scavenger. The solid line is pure water and the dashed line is with the tris buffer. This is the kicker which matches what Jonas thinks. When you put 500 millimolar of tris as a scavenger the scavenger doesn't find ·OH till about a nanosecond according to the calculations. Experimentally it is difficult to ascertain.

Zaider: Depends how much scavenger you have.

Chatterjee: A scavenger concentration of 500 millimolar represents 40-angstrom or 30-angstrom distance for ·OH migration. You can increase the scavenger concentration, you will see something earlier in time scale for scavenger concentration which we think is equivalent for cells, you don't see till late unless you got to very, very high scavenger concentration. Then you will see earlier effects of scavenger.

Terrisol: What is the concentration of tris?

Chatterjee: This one I can't remember, but generally we use 500 millimolar for cells. 500 millimolar gives us a characteristic distance of about 30 to 40 angstroms. In the beginning, the water radical decay curves are controlled mostly by the sibling water reactions unless the track is very close to the DNA.

Zaider: I know that this is the party line, but I'm uncomfortable with this number of 2 nanometers because there's no firm evidence, as far as I know, for this number. I know that everybody accepts this number.

Paretzke: There is experimental evidence.

Zaider: But it all amounts to using the deterministic approach. When you do it stochastically, you come up with a rather larger number.

Paretzke: How much?

Zaider: We obtain about 6 or 7 nanometers at the scavenger concentration John Ward gave as representing the cell. This is essential in the sense that if the hydroxyl radical is very close to the DNA, you get a tight distribution of the spacing of DNA damage while if it comes from further apart it will be more diffuse.

Holley: One interesting comment relevant to your comment, Marco, is that we have recently been doing calculations with different scavenger concentrations. With the solenoidal model, since there's so much scavenging introduced by the DNA and the histones themselves, the results seem to be rather insensitive to the external scavenging concentration we used. We changed it from .5 to .1.

Zaider: According to the data you have shown the RBE goes down.

Chatterjee: Yes! That's a new thing we learned.

Miller: Were these experiments done in cells?

Chatterjee: Yes and followed by gel electrophoresis measurements.

Miller: Wouldn't it be a better test of our model to do this in some sort of reconstituted chromatin system like with John Ward? That's lots of ways you can get small fragments themselves, degradation, for example, nucleases.

Chatterjee: I'm sure they measured those backgrounds.

Holley: Bjorn Rydberg worried about many background sources, such as Okazaki fragments.

Chatterjee: Measurements in a cell is all right as long as you can stop the repair at very low temperature.

Miller: Inside the cell, the radiation breakage could be followed by lots of nuclease degradation because you have all these free ends floating around, and that would change the size distribution.

Chatterjee: It can. You are right.

Zaider: I would also be very careful at the claim that they can measure a hundred base pair fragment.

Chatterjee: Well, Bjorn is not here. I don't know. I think Bjorn, who has done the measurements, is a very good experimentalist and I do trust his data.

Holley: Ordinary gel electrophoresis will handle hundred base pairs. They do it all the time. They make experiments with SV 40 DNA which has been broken into smaller pieces. They are planning to go to acrylamide gels, which will, I think, have even more sensitivity.

Sachs: There are two things — I didn't understand the point about the scavenger concentration being necessarily different in the calculation for single and double stranded DNA and the 30 nanometer fiber calculation.

Chatterjee: Suppose you have a linear DNA molecule. Now, the OH can come to this from all directions. There is a rate constant for that. Now we adjust the scavenger concentration in that system, 500 millimolar, thereby giving us a characteristic diffusion length for OH about 30 angstroms or so. Now you start folding the DNA molecule itself. There will be regions of DNA which will be protected from the radical attack.

Sachs: You put scavenger where there's histones as well?

Chatterjee: Yes. And there's overscavenging. That's why we feel that the experimental data is at a higher yield than the theoretical calculations.

Sachs: The other point is that it may well be that the quality of double strand breaks changes and would possibly account for some of the LET curves. But in addition, they should be closer together at high LET, and that could also account for a difference of RBE in survival versus induction.

Zaider: Would you explain that? I don't understand how.

Sachs: If you make a lot of them close together, then, of course, you will have more pair wise interactions and those are what are doing the killing.

Zaider: Pair wise intersection of what?

Sachs: Of double strand breaks.

Zaider: That's not what he's talking about.

Sachs: He said we have a lower RBE for double strand breaks than for killing.

Zaider: Single or double strand break production has nothing to do with other double strand breaks.

Sachs: On the same slide he said we have an increasing RBE for cell killing.

Zaider: That's well known.

Sachs: And then he said that indicates a difference in the quality of the double strand breaks. I believe there is a quality difference. But I'm saying an additional explanation could be that for high LET, the double strand breaks are on average closer together. The discrepancy between the second to last and last column could be, apart from any difference in the quality of the double strand breaks, due to a proximity effect — is that right?

Zaider: It may very well be.

Miller: Isn't the spur radius for OH comparable to this 20 to 40 angstroms?

Chatterjee: Yes.

Miller: If the spur size is 20 angstroms and the average diffusion length in the cell is 20 angstroms, then it looks like the spurs have got to overlap the DNA in order to be effective. I don't see how you can decouple these two problems.

Chatterjee: The spur can be at any distance from the DNA. But the fact is that they're diffusing. Some of the OH's are interacting within the spur while they're diffusing out. They may escape the sibling recombination process and can attack the DNA. So I don't see the link between those two.

Miller: It seems to me that you could basically forget about all the spurs that didn't fall in DNA, that they really have a low probability to have any effect.

Chatterjee: It could very well be this 20, 30 angstroms is a folding distance.

Inokuti: To what extent do you have the effects of molecular thermal motion? Also, can you say something about the temperature effect?

Chatterjee: Temperature no, I cannot say anything about temperature effect.

Holley: You mean in the diffusion of the OH radicals or in the DNA itself?

Inokuti: OH, everything.

Zaider: It's totally unimportant.

Inokuti: How do you know?

Zaider: Because of the energy involved.

Inokuti: I would imagine that scavenging action has a lot to do with whether the critical site is open or closed.

Zaider: I mean relative to the kind of approximations we are using.

Terrisol: Rates have been determined at a certain temperature: 300K and they use the same value. They use value at room temperature for the rates between OH and DNA.

Zaider: That's correct.

Varma: When you say double strand break, you didn't really define what double strand breaks are. And then you talk about the qualitative difference between double strand breaks. If you counted all the double strand breaks, maybe you wouldn't get that.

Chatterjee: First of all, you are right. In my talk I did not define what I call double strand breaks. In our case, two breaks on opposite strands, within less than 10 base pairs, is what we call double strand breaks. That's our definition.

Zaider: Your calculations show the RBE is going down.

Paretzke: The experiments don't contradict this. If you select your parameters, your energy properly in this field of parameters, I can show you an increased RBE by selecting the proper energies of the proper ions.

Miller: Since your model contains both direct and indirect, I mean you could conceivably score like in the production double strand breaks, how often is the combination direct and indirect.

Chatterjee: Yes, you are right.

Miller: Have you done any of that type of scoring?

Chatterjee: We want to do that. You are absolutely right. I should have done it two years ago.

Holley: We have done that, but we don't have the numbers here right now.

Chatterjee: The question is how many double strands are caused by direct effect and how many by indirect effect and so on.

Miller: And combination.

Chatterjee: Let's say one strand is broken by OH and the other strand by direct ionization.

Holley: We have all that information in detail.

Chatterjee: We have the information, but we have never looked at the numbers yet.

Varma: You didn't answer my question. Ten base pairs apart, biologists measure the double strand breaks, what is their definition? I'm not sure if it's the same definition as yours. How do you compare it?

Curtis: Something is measured and they call it double strand breaks, but there may be a whole bunch of double strand breaks that are being missed.

Varma: But the point is that the definition in your theoretical calculation, you can change it from 15 base pairs apart, 10 base pairs apart, or 20 base pairs apart, and you get a completely different result. And the point is how do you reconcile that with the experimental measurements that are being made?

Zaider: That's the question. Do you get a different result?

Chatterjee: First of all, when OH radical attacks on two strands within the 10 base pairs, then the definition can matter. But when direct ionization happens, it's always within a few base pairs anyway. So it doesn't matter for the direct part.

Varma: I understand the theoretical part. All I'm trying to understand is that from theoretical point of view or a computational point of view, you can generate a set of numbers that gives you the RBE for double strand breaks depending upon your different definitions, whether it is 10, 20, 30, or 40, and you will get completely different results. Now, when you compare that with the measurements that you talk about, where they don't know whether this is 10 base pair or 20 base pair apart or 50 base pairs apart.

Nikjoo: Experimental data by various workers on dsb indicates value of 8–10 bps apart on opposite strand. We chose 10 bp in our calculations as has been used also by Chatterjee and Ito.

Varma: I'm not questioning that. I'm perfectly willing to accept that it's between 10 and 15. What I'm saying is that the biologist, when he measures it, doesn't measure only those DNA double strand breaks.

Zaider: It's not a matter of definition. It's a matter of what happens. But I don't think your result is so sensitive. Going from 10 to 20, will it change?

Varma: All your experimental data was higher by about a factor of two, if I'm not mistaken. All the data looked to me like we're off by a factor of two.

Zaider: The direct effect will go up a lot.

Chatterjee: Direct effect is not seen by scavenger concentration. That will not change. But the iron, where you saw the very dense, where the radical mechanism will be absent, is coming very close. So I'm not going to adjust any parameters. I'm going to try first to see what is happening by this overscavenging. How much are we overscavenging?

Zaider: Direct effect.

Chatterjee: Both.

Zaider: Direct effect must go up.

Chatterjee: This is normalized.

Zaider: That's okay. It should go up.

Chatterjee: There's no question about it.

Zaider: I want to find an explanation, a logical explanation, why does it increase the LET?

Paretzke: Energy you select determines whether it goes up or down.

Holley: One of the reasons that the iron points are low is that you do have additional double strand breaks. These are double strand breaks that are calculated as far apart and the experimental numbers only include those double strand breaks which lead to large fragments and those results don't include the short fragments.

Varma: Shouldn't you include those then if you want to compare?

Holley: The theoretical curves and the experiments here don't include the short fragments.

Chatterjee: I did show you a curve of short fragments as well.

A NUCLEOSOME MODEL FOR THE SIMULATION OF DNA STRAND BREAK EXPERIMENTS

Michel Terrissol

Centre de Physique Atomique
Université Paul Sabatier
118 Route de Narbonne
31062 Toulouse, France

Ekkehard Pomplun

Abteilung Sicherheit und Strahlenschutz
Forschungszentrum Jülich GmbH
Postfach 1913
52425 Jülich, Germany

ABSTRACT

Using a set of Monte Carlo simulation models, track structures of ^{125}I Auger electrons generated in liquid water are superimposed on a nucleosome DNA model able to precisely localize energy deposition events on sub-molecular units of the DNA strands. After scoring direct hits taking place during the physical phase (at about 10^{-15} s) the radiation chemistry of the whole system is simulated between 10^{-12} and 10^{-8} s, taking into account all reactions between water radio-chemical species, radicals, sub-molecular units of DNA (Ribose, Adenine, Thymine, Guanine and Cytosine), and scavengers like Tris or Formate ions.

The model's possibility to distinguish between direct and indirect hits has been utilized to introduce different assumptions for strand break induction by both hit modes. The number of SSB and DSB as well as their local distribution will be given and compared with experimental and theoretical results from the literature.

INTRODUCTION

A complex computer model has been developed consisting of major components for the generation of a radiation spectrum, of biomolecular structures, and of electron track structures in liquid water. As radiation source ^{125}I is assumed here due to the possibility to exactly localise its position in the DNA structure and due to its large biological toxicity as consequence of the emission of short-ranging Auger electrons. A special emphasis has been

Computational Approaches in Molecular Radiation Biology
Edited by M.N. Varma and A. Chatterjee, Plenum Press, New York, 1994

put to the extension from a linear DNA plasmid model[1,2] with length of 14 nm and about 2500 atoms to a nucleosome model representing the double helix of DNA with 146 basepairs and more than 9000 atoms[3] surrounding the histons (Fig. 1). By introducing few assumptions for the induction of strand breakage (see Methods) this model yields a series of results confirming the plausibility of this nucleosome model as well as characterising the applied radiation in terms of energy deposition and strand break pattern. In a first approach also the lethality of cells with ^{125}I-labelled DNA will be related to those DSB induced by direct radiation hits.

METHODS

Monte Carlo simulated ^{125}I Auger cascades[4] are used as input to a 4-dimensional (x,y,z,t) electron transport in liquid water.[5] Iodine atoms are considered individually, but uniformly distributed among all the thymine bases of the nucleosome by replacing its methyl group (71 positions in the model). Complete transport of Auger electrons is then simulated decay by decay,[6] in a working sphere of radius 8 nm (same centre as nucleosome), surrounding the nucleosome and filled with liquid water (Fig. 2): the physical step (up to 10^{-15} s),[6] the physico-chemical step (10^{-15} s to 10^{-12} s)[7] and the chemical step (10^{-12} s to seconds)[7] lead to the creation of species e^-_{aq}, $OH\cdot$, $H\cdot$, H_3O^+, H_2O_2, OH^-, H_2, HO_2 diffusing and reacting: well known reactions of water species among themselves (26 reactions), as well as reactions with DNA sub-units[8] (Table I).

During the physical step, (up to 10^{-15} s) : an ionisation located inside the van der Waals radius of a phosphate-group or sugar atom is stored (time, co-ordinates and energy deposited on this atom) as a direct single-strand break (SSB), and this event is removed from the initial track. From 10^{-15} s to 10^{-8} s when a deoxyribose-monophosphate reacts it is transformed in a sub-product and cannot react twice : an indirect SSB is scored (time, co-ordinates and total energy deposited in the working sphere by the Auger electrons belonging to the decay).

Double-strand breaks (DSB) are defined per ^{125}I decay as two SSB on opposite strands within a distance of up to 10 basepairs.

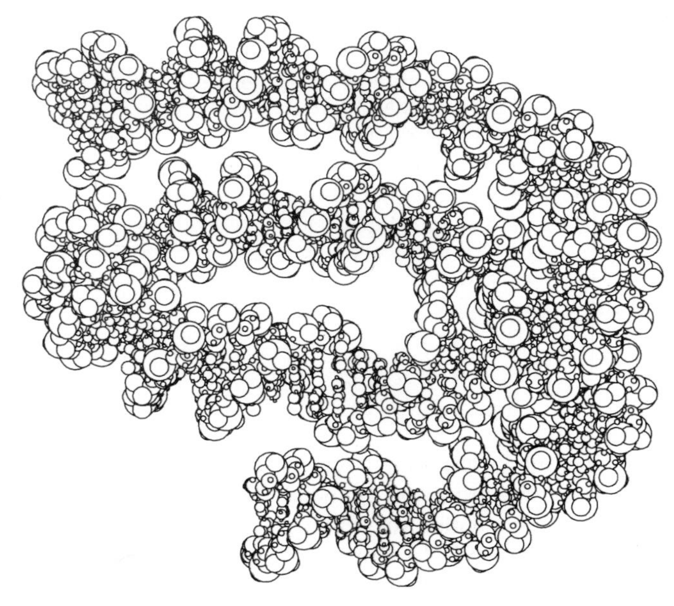

Figure 1. DNA nucleosome target model (146 basepairs, diameter: 11 nm, pitch: 2.7 nm).

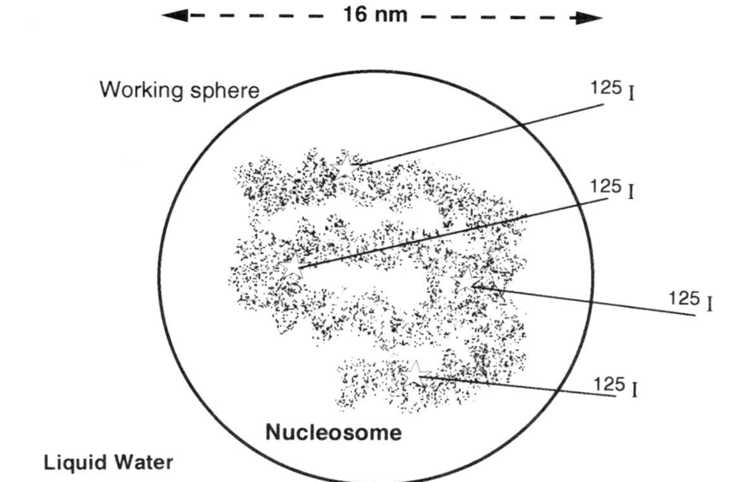

Figure 2. Schematic representation of the simulation model.

RESULTS

A characteristic of the ^{125}I radiation as well as the nucleosome model can be derived from Fig. 3. In most of all two DSB inducing decays, the distance between these two DSB is less than 20 basepairs reflecting the short range action of ^{125}I. The second maximum at about 80 basepairs is expected due to the fact the DNA helix has completed one turn around the histons so that these breaks, although 80 basepairs apart, are very near the decay site. Quantitatively, the same results are found by considering the DSB yield in dependence of the distance between the contributing SSB.

Figure 4 shows that during the physical phase (at about 10^{-15} s after the decay) only very small amounts of energy will be deposited on the target atoms whereas the large fraction of absorbed energy is transferred to the secondary electrons interacting and depositing this energy at later times via radicals.

Table I : Reactions radii in nanometers of main water species with DNA sub-units calculated with values of reaction rate constants from Buxton et al.[8]

	OH	e-aq	H
Deoxyribose-monophosphate	0.085	0.0003	0.0006
Adenine	0.288	0.265	0.0019
Cytosine	0.288	0.382	0.017
Guanine	0.425	0.413	0.000
Thymine	0.302	0.53	0.011

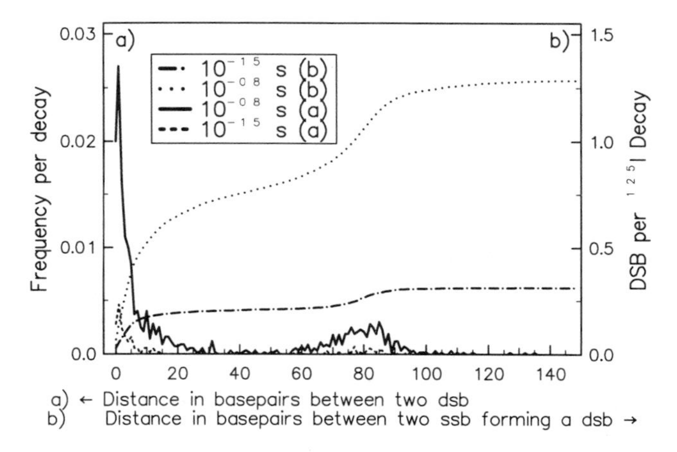

Figure 3. Spatial distribution of SSB and DSB yield as function of SSB distance.

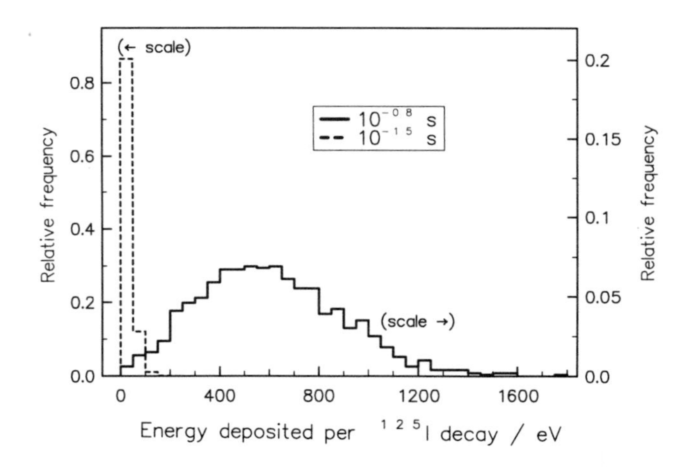

Figure 4. Frequency distribution of the size of energy deposition events during the physical phase (up to 10^{-15} s) and after 10^{-8} s.

By applying the assumptions for strand breakage, the energy deposition is converted into a break pattern. Figure 5a gives the absolute number of decays inducing different types or combinations of breaks and the thereby deposited energy. The relative number of these breaks per decay are plotted in Fig. 5b. Low energy values correspond with high fractions of no break situations or simple SSB whereas with increasing amounts of energy the former decreases very steeply and the more complex breaks determine the pattern. A continuous increase can be seen with two or more DSB.

Figure 6a shows a flat and broad distribution for the number of SSB per decay with the highest probability in the range between 2 and 4 and a maximum number of 14 breaks per ^{125}I decay. The DSB number distribution results in about 50% of all breaks with no DSB at all. Comparison of mean values from these distributions with some experimental data[9,10,11,12] is shown Table II.

The mean value of about 1 DSB per decay reported from many experiments is realised by the fact that in about 20% of all decays more than 1 DSB is produced. Differentiating directly and indirectly induced DSB (Fig. 6b) leads to quite interesting results which could perhaps give hints for the assumed correlation between DSB and cell death. Survival experiments have shown that about 40 to 60 ^{125}I decays in a cell are necessary to inhibit proliferation. On the other hand, it has been suggested recently by Barendsen[13,14] that the interaction of two 'paired' DSB results ultimately in cell lethality. On this basis, it is interesting to find for decays producing 2 DSB by direct interaction a frequency of about 0.02; i.e., each 50th decay would be a lethal one, a number fitting well in the above mentioned range.

CONCLUSION

An atomic scale nucleosome target structure has been introduced here into track structure analysis in order to simulate the geometric DNA arrangement in cells during irradiation to study molecular damage. The modular set-up of the whole computer model allows the extension to other radiation sources (e.g., Tritium) or to add scavenging species in the working sphere (e.g., formate ions or Tris) to reflect more exactly the cells conditions. We intend to complete our nucleosome model to take into account linkage with neighbouring ones to extend to chromatin fibre and to include new direct interaction cross-sections for electron on DNA sub-units.

Table II : Mean number of SSB and DSB per ^{125}I decay.

Break	This work	Experiment
SSB	4.2	3 – 6
DSB	0.8	0.9 – 1.0

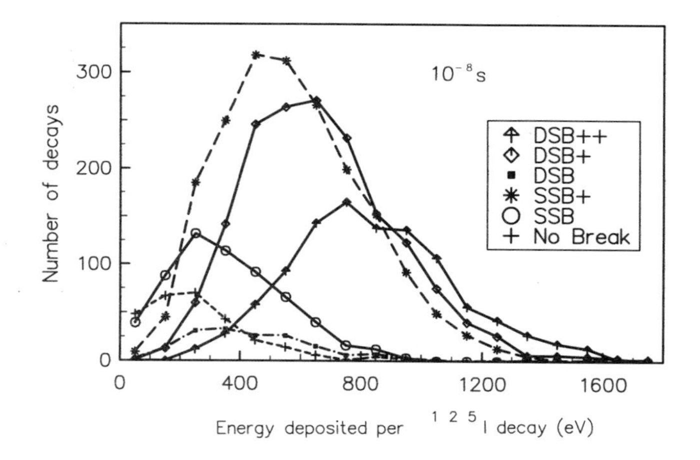

Figure 5a. Break pattern induced by a total number of 5680 decays. 'SSB': one single SSB; 'SSB+': two or more SSB on one strand; 'DSB': one DSB; 'DSB+': one DSB plus additional SSB; 'DSB++': two or more DSB.

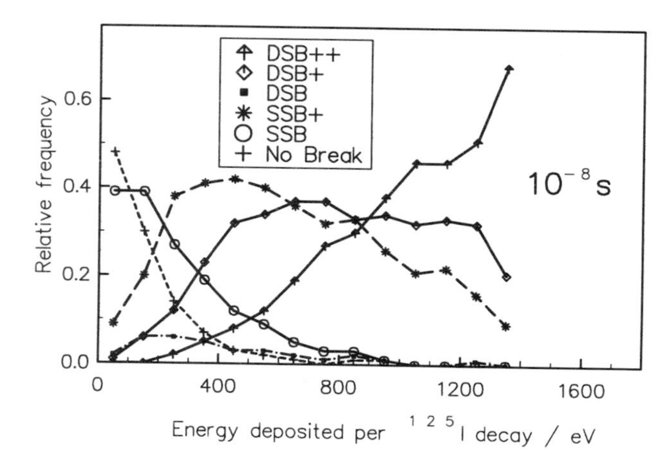

Figure 5b. Relative frequency of break induction in dependence of energy deposited.

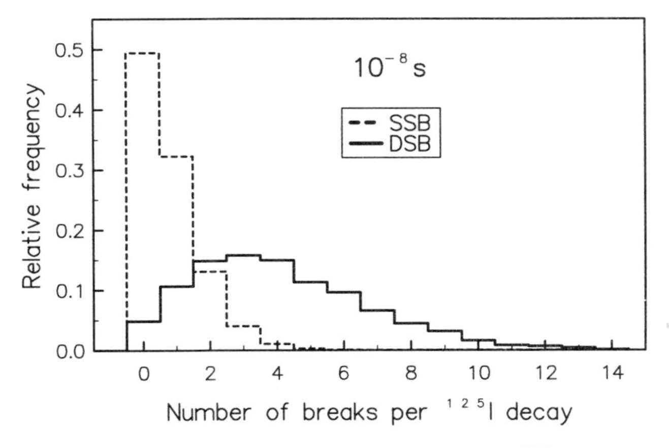

Figure 6a. Number distribution of SSB and DSB per ^{125}I decay.

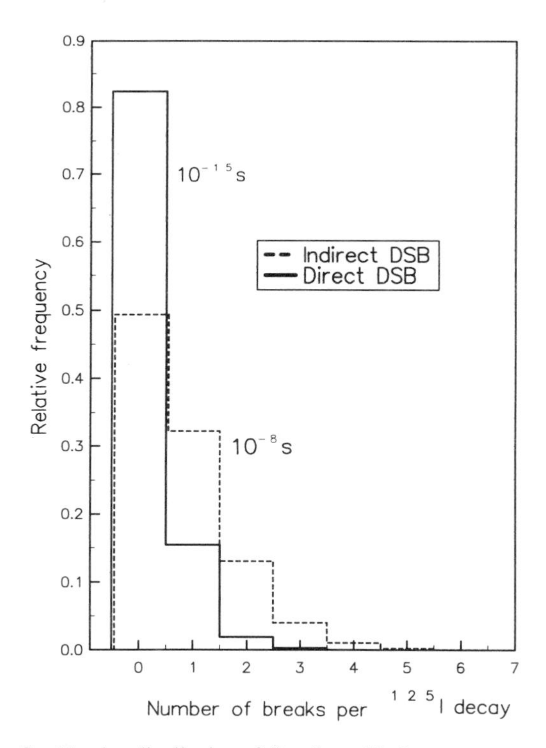

Figure 6b. Number distribution of directly and indirectly induced DSB.

REFERENCES

1. E. Pomplun. A New DNA Target Model for Track Structure Calculations and its First Application to ^{125}I Auger Electrons. *Int. J. Radiat. Biol.* 59:625–642 (1991).
2. M. Terrissol and E. Pomplun. Computer Simulation of DNA-incorporated ^{125}I Auger Cascades and of the Associated Radiation Chemistry in Aqueous Solution. *Radiation Protection Dosimetry* 51:171–181 (1994).
3. E. Pomplun. DNA Helix and Nucleosome Models in the Track Structure Analysis of Beta Particles and Auger Electrons from Incorporated H-3 and I-125, Radiation Research, Volume 1, J.D. Chapman, W.C. Dewey, G.F. Whitmore (eds.) Ninth International Congress of Radiation Research, 1991, Toronto.
4. E. Pomplun, J. Booz and D.E. Charlton. A Monte Carlo Simulation of Auger Cascades. *Radiat. Res.* 111:533–552 (1987).
5. E. Pomplun, M. Roch and M. Terrissol. Simulation of Strand Break Induction by DNA-incorporated ^{125}I, *Biophysical Aspects of Auger Processes,* AAPM Symposium Series No. 8 (1992) 137–152, Eds. RW Howell, VR Narra, KSR Sastry, DV Rao.
6. M. Terrissol. Méthode de Simulation du Transport d'Electrons d'Energies Comprises entre 10 eV et 30 keV. *Thesis,* Université Paul Sabatier, n° 839, Toulouse (1978).
7. M. Terrissol and A. Beaudré. Simulation of Space and Time Evolution of Radiolytic Species Induced by Electrons in Water. *Radiat. Prot. Dosim.* 31:171–175 (1990).
8. G.V. Buxton, C.L. Greenstock, W. P. Helman and A. B. Ross. Critical Review of Rate Constants for Reactions of Hydrated Electrons, Hydrogen Atoms and Hydroxyl Radicals in Aqueous Solution. *J. Phys. Chem. Ref. Data*, **17**, No. 2, (1988).
9. S. Sundell-Bergman and K.J. Johanson. Repairable and Unrepairable DNA Strand Breaks Induced by Decay of ^{3}H and ^{125}I incorporated into DNA of Mammalian Cells, *Radiat. Environ. Biophys.* 18:239–248 (1980).
10. L.E. Feinendegen, P. Henneberg and G. Tisljar-Lentulis. DNA Strand Breakage and Repair in Human Kidney Cells after Exposure to Incorporated ^{125}I and ^{60}Co γ-rays. *Current Topics in Radiation Research Quarterly*, 12:436–452 (1977).
11. R.E. Krisch, F. Krasin and C.J. Sauri. DNA Breakage, Repair, and Lethality Accompanying ^{125}I Decay in Microorganisms. *Current Topics in Radiation Research Quarterly*, 12:355–368 (1977).
12. P.K. Lemotte and J.B. Little. DNA Damage Induced in Human Diploid Cells by Decay of Incorporated Radionuclides. *Cancer Res.* 44:1337–1342 (1984).
13. G.W. Barendsen. Mechanisms of Cell Reproductive Death and Shapes of Radiation Dose Survival Curves of Mammalian Cells. *Int. J. Radiat. Biol.* 57:885-896 (1990).
14. G.W. Barendsen. Sub lethal Damage and DNA Double Strand Breaks have Similar RBE-LET Relationships: Evidence and Implications. *Int. J. Radiat. Biol.* 63:325–330 (1993).

A COMPUTATIONAL APPROACH TO THE RELATIONSHIP BETWEEN RADIATION INDUCED DOUBLE STRAND BREAKS AND TRANSLOCATIONS

William R. Holley and Aloke Chatterjee

Division of Life Sciences, Lawrence Berkeley Laboratory
University of California, Berkeley
1 Cyclotron Road, Berkeley, California 94720

ABSTRACT

A theoretical framework is presented which provides a quantitative analysis of radiation induced translocations between the abl oncogene on CH9q34 and a breakpoint cluster region, bcr, on CH 22q11. Such translocations are associated frequently with chronic myelogenous leukemia. The theory is based on the assumption that incorrect or unfaithful rejoining of initial double strand breaks produced concurrently within the 200 kbp intron region upstream of the second abl exon, and the 16.5 kbp region between bcr exon 2 and exon 6 interact with each other, resulting in a fusion gene. For an x-ray dose of 100 Gy, there is good agreement between the theoretical estimate and the one available experimental result. The theory has been extended to provide dose response curves for these types of translocations. These curves are quadratic at low doses and become linear at high doses.

INTRODUCTION

Carcinogenesis is a multi-step process. It is generally believed that from 5 to as many as 8 genetic changes are required to generate a human tumor cell [Peto et al. (77), Vogelstein et al. (88)]. Much experimental information is available on radiation induction of cell transformation and mutagenesis. To properly organize this data a general and comprehensive theoretical framework is needed. Over the years we have made considerable progress in understanding and describing theoretically many of the mechanistic steps involved in the initial damage of biological material (DNA) by essentially all qualities of ionizing radiation [Holley et al. (90), Chatterjee and Holley (91a), Schmidt et al.]. The next step, we think, is to try to understand the mechanisms involved in some of the intermediate processes along the pathways to mutagenesis, cell transformation and ultimately carcinogenesis. By understanding and mathematically modeling these steps we hope to be able to help bridge the large gap between the initial damage stage (mostly physical and

Computational Approaches in Molecular Radiation Biology
Edited by M.N. Varma and A. Chatterjee, Plenum Press, New York, 1994

physico-chemical in nature) and the risk assessment stage, and hopefully place any necessary extrapolation procedures on a firm theoretical basis. We have, therefore, recently initiated a program of theoretical studies in which we try to understand, in terms of mathematical models, some of the fundamental mechanisms underlying the complex of processes involved in mutagenesis and carcinogenesis. Some of the biological end points we are attempting to model are listed as follows:

Double Strand Breaks
Chromatin Breaks
Point Mutations
Small Deletions
Large Deletions
Translocations
Cell Transformation

A very general model has been suggested previously which attempts to organize and correlate the results of experiments designed to induce, by radiation, the neoplastic transformation of already partially transformed cell lines such as C3H 10T1/2 [Chatterjee & Holley (91b)]. The mechanisms considered in this earlier study included radiation induced point mutations and large deletions. The basic assumption underlying these calculations, and in fact underlying the theory to be described in this paper, is that double strand breaks (dsb) are the critical DNA lesions which act as the essential precursor to most radiation induced mutations [Little (93)]. The impetus for this particular study is based on the existence of the so called Philadelphia chromosome which is found associated very frequently with chronic myelogenous leukemia. The Philadelphia chromosome is a chromosome aberration which consists of a reciprocal translocation between the abl oncogene on CH9q34 and a break point cluster region or bcr on CH22q11.

Takashi Ito and collaborators [Ito et al. (93)] recently announced results of an experiment which measured x-ray induced abl-bcr translocations similar to the Philadelphia chromosome. They did this by detecting the presence of fusion gene mRNA using reverse transcription and PCR.

The experiment detected translocations which place the bcr exon 2 upstream of the abl exon 2. The mRNA was used because distances on the translocated genomic DNA would in general be too large for PCR amplification. One reason this is a feasible experiment is that the abl gene offers a very large target region for break points for translocations. The intron upstream of abl exon 2 has been measured to be about 200 kbp [Bernards et al. (87)]. The mRNAs from fusion genes were detected in which the break point at bcr apparently could come anywhere within the intron following exon 2, to the intron between exon 5 and exon 6. The total bcr target region adds up to about 16.5 kbp [Heisterkamp et al. (85)] (See Fig. 1).

For 10^8 cells irradiated with 100 Gy of x-rays, Ito et el. (93), found 5 fusion gene events with mRNA ranging from the classical Philadelphia chromosome rearrangement where bcr exon 2 is directly upstream of abl exon 2 to the case where the bcr exon 1 through exon 5 are juxtaposed ahead of the abl exon 2 (Fig. 2).

We have used the following plausible assumptions as the basis of a simple model which seems to be consistent with Ito's results and which has some rather interesting properties. We assume

1. random production of double strand breaks by low LET radiation leads to concurrent induction of a dsb somewhere within the 200 kbp intron region upstream of the second abl exon and somewhere within the 16.5 kbp region between bcr exon 2 and exon 6.

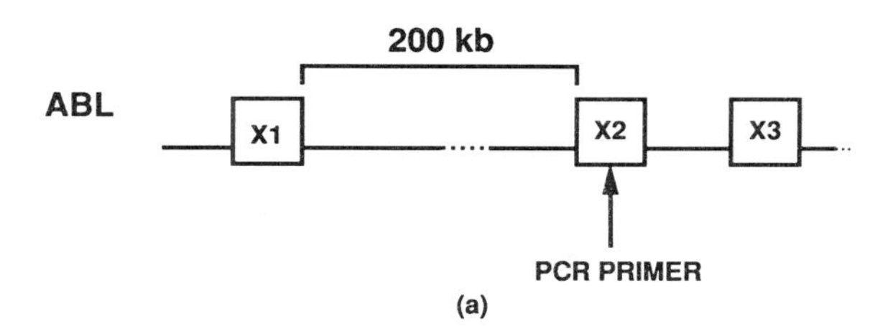

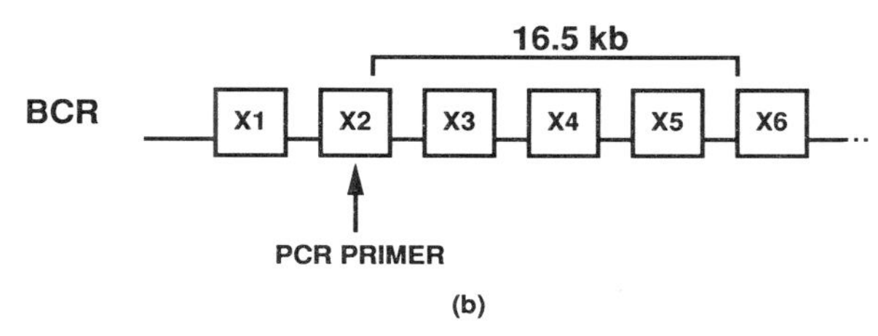

Figure 1. Simplified exon-intron genomic maps showing the location of the reverse transcription PCR primers, as well as the cumulative intron target sizes for (a) abl and (b) bcr.

2. some fraction R (~ 0.9) of these dsb's rejoin more or less faithfully, allowing for the possibility of point mutations or very short deletions.

3. some fraction, which for purposes of these calculations we have taken as ~1.0, of the remaining broken ends from unrejoined double strand breaks eventually join at random to other broken ends.

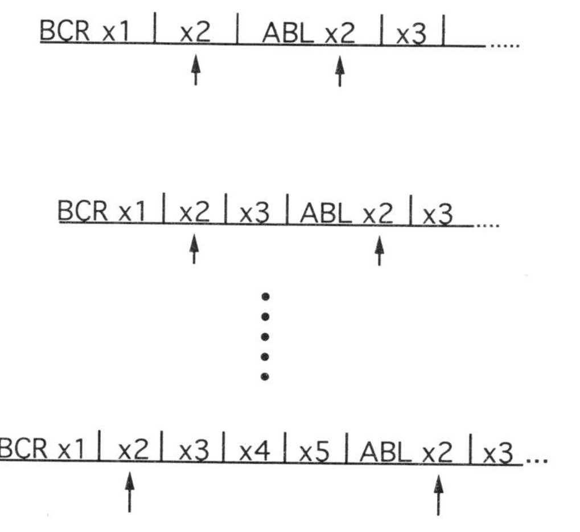

Figure 2. Typical fusion gene mRNA seen in experiment of Ito et al. (93). Arrows indicate approximate locations of PCR primers used in detection of the corresponding DNA afer reverse transcription.

Based on these assumptions we can write down the probability of translocation as the joint product of the probability of an "unrejoined" dsb occurring in the abl target region times the probability of an "unrejoined" dsb in the bcr region times the probability that the abl end will join up with the bcr end.

In the high dose limit, that is, when there are many dsb's produced in each cell, these factors can be written as shown here in Eq. (1)

$$P_{tr} = \frac{[2 \bullet Y \bullet D \bullet N_{abl} \bullet (1 - R)] \bullet [2 \bullet Y \bullet D \bullet N_{bcr} \bullet (1 - R)]}{[2 \bullet Y \bullet D \bullet N_{dna} \bullet (1 - R)]} \tag{1}$$

The factors of 2 in the numerator are simply for the 2 copies each of the abl and bcr genes, Y is the Yield of dsb per gray per BP, D is dose, N_{abl} and N_{bcr} are the number of base pairs in the abl and bcr target region respectively, N_{dna} is the total number of base pairs in the cell, and (1-R) represents the fraction of dsb's which do not rejoin "correctly". The term in the denominator is just the total number of unrejoined DNA ends (2 times the total number of "unrejoined" double strand breaks) and respresents the probability of the abl end matching up at random with the right bcr end.

Surprisingly the resultant expression is linear in dose as shown in the following equation, although one naively expects quadratic dose behavior from a process which is obviously second order in dsb production.

$$P_{TR} = \frac{2 \bullet Y \bullet D \bullet N_{abl} \bullet N_{bcr} \bullet (1 - R)}{N_{dna}} \tag{2}$$

If we evaluate this probability for the following values of parameters appropriate for the Ito experiment, we end up with 9.9 calculated translocations in 10^8 irradiated cells to be compared with the 5 translocations measured experimentally:

$$\begin{aligned} D &= 100 \text{ Gy} \\ Y &= 6 \times 10^{-9} \text{ dsb/(Gy} \bullet \text{bp)} \\ N_{abl} &= 200 \text{ kbp} \\ N_{bcr} &= 16.5 \text{ kbp} \\ N_{dna} &= 6 \times 10^{9} \text{ bp} \\ R &= 0.9 \end{aligned}$$

Equation 2 has been derived under high dose conditions where the total number of unrejoined double strand breaks is much larger than 1 and the use of average values is justified. For low LET radiation, the probability of inducing a double strand break by any one incident particle or photon is small and the total number of dsb's per cell for dose D should follow the Poisson distribution appropriate for the average total number of dsb's. The number of unrejoined dsb's in a cell should also be Poisson distributed, with a mean value given by $\bar{n} = Y\, D\, N_{dna}\, (1\text{-}R)$. Using this assumption we can derive a more general expression for the translocation probability, valid for all doses. We use the same prescription as before with a somewhat more general notation to calculate the contribution to the translocation probability from exactly n unrejoined dsb's. The total translocation probability is then obtained by summing over n from 2, the minimum needed in our model, to infinity.

Assuming n unrejoined dsb's are randomly positioned on the nuclear DNA, the mean number of dsb's within a total target region N_a is nN_a/N_{dna}. Given a dsb in "a", the probability that one of the remaining n-1 breaks lies within the second target region N_b is $(n-1)N_b/N_{dna}$. Assuming random joining of broken ends, the fraction of the time that the end at a finds the correct end at b to form the desired translocation is 1/2(n-1). The product of these three terms multiplied by the Poisson probability, $\bar{n}^n e^{-\bar{n}}/n!$, of having exactly n unrejoined breaks where $\bar{n}$ is the average gives the contribution, $P_{tr}(n)$, to the translocation probability:

$$P_{tr}(n) = \frac{nN_a}{N_{dna}} \frac{(n-1)N_b}{N_{dna}} \frac{1}{2(n-1)} \frac{\bar{n}^n e^{-\bar{n}}}{n!} \tag{3}$$

After some simplification of terms, the total translocation probability, P_T, can be written

$$P_T = \frac{N_a N_b}{2N_{dna}^2} \sum_{n=2}^{\infty} n \frac{\bar{n}^n e^{-\bar{n}}}{n!} = \frac{N_a N_b}{2N_{dna}^2} \bar{n}(1 - e^{-\bar{n}}) \tag{4}$$

In the high dose limit this expression reduces to

$$P_T \rightarrow \frac{N_a N_b}{2N_{dna}^2} \bar{n} = \frac{N_a N_b}{2N_{dna}} Y\,D\,(1-R) \tag{5}$$

which is identical to Eq. 3 when N_a and N_b are replaced by $2N_{abl}$ and $2N_{bcr}$ respectively. In the low dose limit Eq. 4 reduces to

$$P_T \rightarrow \frac{N_a N_b}{N_{dna}^2} \bar{n}^2 = N_a N_b\, Y^2 D^2 (1 - R)^2/2 \tag{6}$$

DISCUSSION

The general expression for the translocation probability, Eq. 4, clearly has a complex dose dependence. The formula exhibits the quadratic behavior (Eq. 6) expected from a two hit process at low dose, but undergoes a transition to linear behavior at high dose (Eq. 5). This transition is due to a linearly increasing multiplicity of possible joining sites in competition with any particular (i.e., a-b) translocation. This behavior is illustrated in Fig. 3. The solid curve is a plot of the logarithm of P_T from Eq. 4 as ordinate versus the log of the dose, evaluated for the parameters listed earlier which are appropriate for bcr-abl translocations. The data point represents the experimental result of Ito et al., (5 events), and the error bars correspond to a standard deviation of $\pm \sqrt{5}$ events expected from Poisson statistics. The dashed curve is the linear extrapolation of the high dose behavior to low doses. The quadratic nature of the low dose region is shown more clearly in Fig. 4 where the same curves are plotted on a linear scale over the dose range 0.0 to 0.50 Gy.

To show the sensitivity of the calculation to values of the parameters, some of which are not well-known at present, Fig. 5 is a log-log plot of the theoretical probability of abl-bcr translocations evaluated for two sets of parameters representing a reasonable range of values. The upper curve corresponds to increasing the fraction of unrejoined DSB's by a factor of three to 30% ($R = 0.7$), and the lower curve corresponds to reducing this fraction to 5% ($R = 0.95$) and also assuming a lower value for the low LET (x-ray) DSB yield ($Y = 3 \times 10^{-9}$). As can be seen from Fig. 5, these sets of parameters easily span a range of calculated values consistent with the experimental data point available. More stringent tests of the model will have to await additional experimental data, either higher precision translocation measurements, preferably at lower doses, and/or precise independent measurements of such

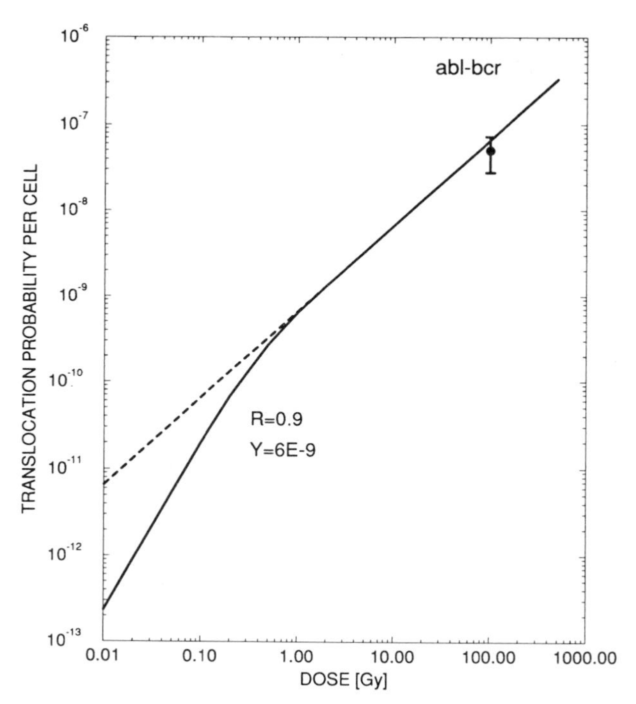

Figure 3. Probability of radiation induction of translocations per cell versus radiation dose in Gy plotted on a log-log scale. Solid curve: Theoretical abl-bcr translocation probability. Dashed curve: Linear extrapola-tion of abl-bcr translocation probability to low doses. Data Point: Experimental result of Ito et al. (93).

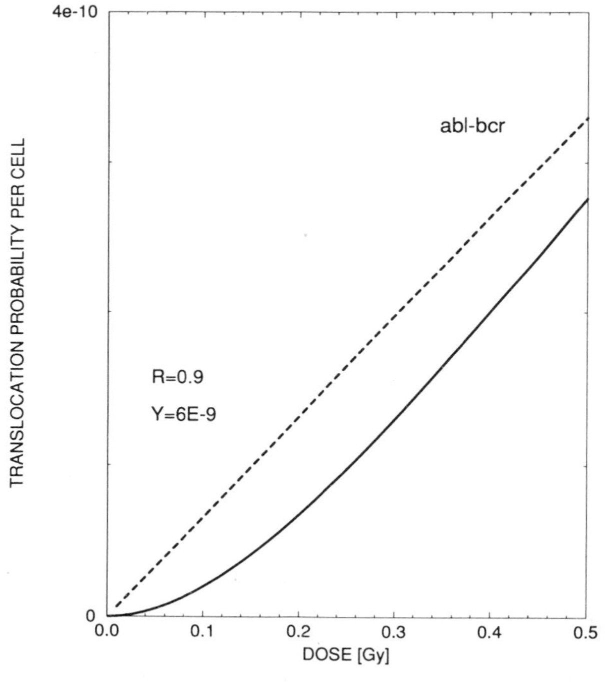

Figure 4. Theoretical abl-bcr translocation probability per cell versus radiation dose plotted on a linear scale over the low dose region 0.0 to 0.50 Gy. The solid curve shows clearly the quadratic behavior at low dose. The dashed curve is the linear extrapolation of the high dose behavior.

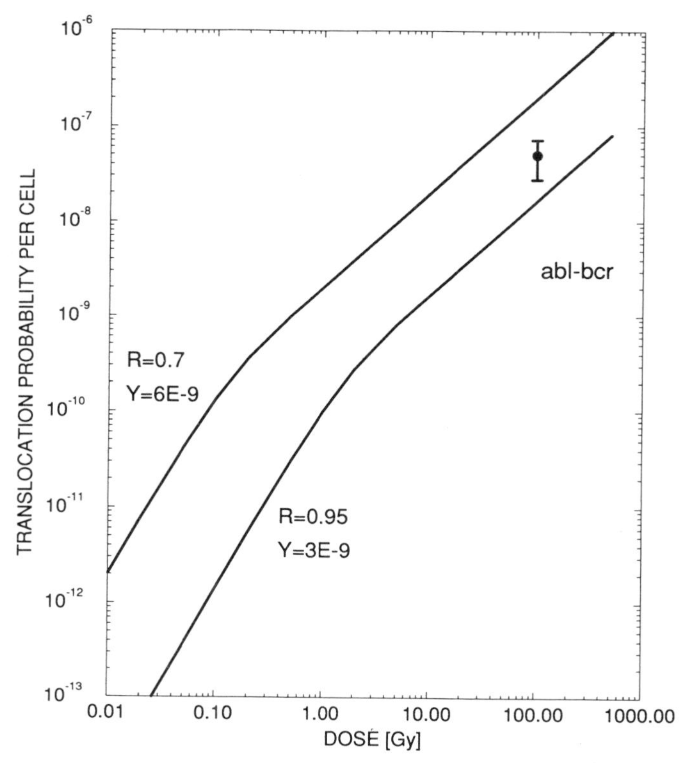

Figure 5. Theoretical abl-bcr translocation probabilities per cell versus radiation dose for two sets of parameters representing a reasonable range of values.

quantities as DSB yields, Y, and rejoining fractions, R, in the same cell system and under the same conditions as translocation experiments.

We have cast this calculation in terms of (interchromosomal) translocations, but it is clear that other aberrations can be tested with similar methods. For example, the yield of large deletions (i.e., intrachromosomal exchanges), can be calculated using the same prescription, resulting in an identical expression to Eq. 4, where the target areas (N_a and N_b) for double strand breaks are now on the same chromosome.

As we have indicated, the current model is applicable to low LET radiation only. Under this condition the assumption of random spatial distributions of double strand breaks is reasonable, as is the assumption of Poisson statistics for the number of DSB's in a cell. Conversely, in the high LET case, where many double strand breaks can be produced by passage of a single particle through the nucleus, the spatial distribution of breaks will, in general, not be random, nor will the number of breaks in a cell follow Poisson statistics. A more detailed model of the chromosomal organization within the cell nucleus, including chromosomal localization, is necessary to properly take into account the non-random, spatially correlated nature of the double strand break distributions from high LET irradiation. Such studies are in progress.

ACKNOWLEDGEMENT

Supported by the Office of Health and Environmental Research (OHER), U.S. Department of Energy, under Contract No. DE-AC03-76SF00098 and NASA Center of Research and Training (NSCORT) in radiation health under Contract Order W-18265.

REFERENCES

Bernards, A., Rubin, C.M., Westbrook, C.A., Paskind, M., and Baltimore, D., 1987, The first intron in the human c-abl gene is at least 200 kilobases long and is a target for translocations in chronic myelogenous leukemia, *Molec. and Cell. Biol.* 7:3231–3236.

Chatterjee, A., and Holley, W.R., 1991a, Energy deposition mechanisms and biochemical aspects of DNA strand breaks by ionizing radiation, *Int. J. Quant. Chem.*, 391:709–727.

Chatterjee, A., and Holley, W.R., 1991b, Problems in mechanistic theoretical models for cell transformation by ionizing radiation, *in:* "Biophysical Modeling of Radiation Effects," K.N. Chadwick, G. Moschini and M.N. Varma, eds., Adam Hilger, New York, pp. 193–200.

Heisterkamp, N., Stam, K., Groffen, J., Klein, A., and Grosveld, G., 1985, Structural organization of the bcr gene and its role in the Ph′ translocation, *Nature* 315:758–761.

Holley, W.R., Chatterjee, A., and Magee, J.L., 1990, Production of DNA strand breaks by direct effects of heavy charged particles, *Radiat. Res.* 121:161–168.

Ito, T., Seyama, T., Mizuno, T., Hayashi, T., Iwamoto, K.S., Dohi, K., Nakamura, N., and Akiyama, M., 1993, Induction of BCR - ABL fusion genes by *in vitro* x-irradiation, *Jpn. J. Cancer Res.* 84:105-109.

Little, J.B., 1993, Cellular, molecular, and carcinogenic effects of radiation, Hematology/Oncology Clinics of North America, 7 (2), 337–352.

Peto, R., 1977, Epidemiology, multistage models and short term mutagenicity tests, *in:* "Origins of Human Cancer, Book C," J.D. Wilson and J.A. Winstin, eds., Cold Spring Harbor Laboratory, Cold Spring Harbor, NY, pp. 1403–1428.

Schmidt, J.B., Holley, W.R., and Chatterjee, A., DNA single and double strand break yields in a 30 nm diameter chromatin fiber - a theoretical model. (submitted to *Radiat. Res.*).

Vogelstein, B., Fearon, B.A., Hamilton, S.R., Kern, S.E., Preisinger, A.C., Lappert, M., Nakamura, Y., White, R., Smits, A.M.M., and Bos, J.L., 1988, Genetic alterations during colorectal tumor development, *New Eng. J. of Med.* 319:525–532.

DISCUSSION

Varma: In your model, the target size along with segments of exons and introns have been considered in order to calculate the translocation frequencies. No consideration has been given with respect to sequence specificity. However, biologists seem to emphasize such specificity with respect to translocation events. Can you explain this discrepancy?

Holley: Radiation has the property that the initiation points of the double strand breaks, at least with low-LET radiation are random. This is in contrast with the formation of spontaneous breaks followed by homologous recombination.

Chatterjee: I agree with Holley's answer in general. However, I can think of a situation where hydroxyl mediated damage, especially strand breaks, may not be quite random. It is known that Guanine has the highest reactivity with hydroxyl radical and hence it is very likely that the sugar adjacent to Guanine may be shielded from •OH attack. In that case strand breaks may preferentially occur at the A-T sites. This possibility can be tested with specially designed oligonucleotides, either rich in A-T or rich in G-C. If it turns out that the initial damage by •OH has sequence specificity, it can be taken into account in theoretical modeling.

Miller: I have just two points to make. The first one may be relative to what Varma is saying is that in our laboratory we have performed a few experiments in deletion mutations and have identified the end points of deletion mutations as they fall within the HPRT gene and those deletion break points are not random and they are induced by radiation. So definitely there are sequence effects. In your specific case, I'm not saying one way or the other. The other point is that I just want to alert you to another mechanism that is in the literature primarily put forth by Roger Cox at MRC. According to Roger, the ends of the breaks in this process are not stable by any means. In fact, when you make double strand breaks in the chromosome, unless you protect them in some way, enzymes are present and the ends. And in fact, the magnitude of this chewing up can be comparable to the sizes of the genes that you are considering. It seems like you have to consider the possibility in the process that you broke the gene, but the nucleations will essentially chew all the coding sequences up before those ends have had time to find one another; just another phenomenon that occurs there, and you should think about it.

Sachs: It really should be emphasized that the assumed kinetics here are the following: First you have a linear process, restitution (faithful rejoining), competing with a linear process, breaking. Thereafter, you have a pair-wise rejoining process competing with nothing. That is what gives you the linear dose behavior. That may be quite good at these very high doses and it is a standard model, but of course at much lower doses, when you observe quadratic yields you must believe in a different kind of kinetics where, at the last step, a pair-wise process competes with a linear process (restitution) and the linear process does the majority of the work.

INDEX